D1523506

The Bounds of Reason

The Bounds of Reason
Game Theory and the Unification of the Behavioral Sciences

Herbert Gintis

Princeton University Press
Princeton and Oxford

Published by Princeton University Press, 41 William Street, Princeton, New Jersey 08540
In the United Kingdom: Princeton University Press, 6 Oxford Street, Woodstock, Oxfordshire OX20 1TW

Library of Congress Cataloging-in-Publication Data
Gintis, Herbert
The Bounds of Reason: Game Theory and the Unification of the Behavioral Sciences / Herbert Gintis
p. cm.
Includes bibliographical references and index.
ISBN 978-0-691-14052-0 (hardcover: alk. paper) 1. Game theory. 2. Practical reason. 3. Human behavior. 4. Social sciences–Methodology. 5. Psychology. I. Title.
HB144.G552009
519.3–dc22
2008036522
British Library Cataloging-in-Publication Data is available
The publisher would like to acknowledge the author of this volume for providing the camera-ready copy from which this book was printed
This book has been composed in Times and Mathtime by the author
Printed on acid-free paper.
press.princeton.edu
Printed in the United States of America
10 9 8 7 6 5 4 3 2 1

This book is dedicated to Morris and Henrietta Malena, Gerson Gintis, and Flora and Melvin Greisler. We get along without you very well. Of course we do.

There is no sorrow so great that it does not find its background in joy.

Niels Bohr

Contents

Preface

> The eternal mystery of the world is its comprehensibility.
>
> Albert Einstein

> Mathematics without natural history is sterile, but natural history without mathematics is muddled.
>
> John Maynard Smith

Game theory is central to understanding the dynamics of life forms in general, and humans in particular. Living creatures not only play games but also dynamically transform the games they play and have thereby evolved their unique identities. For this reason, the material in this book is foundational to all the behavioral sciences, from biology, psychology, and economics to anthropology, sociology, and political science. Disciplines that slight game theory are the worse—indeed, much worse—for it.

We humans have a completely stunning capacity to reason and to apply the fruits of reason to the transformation of our social existence. Social interactions in a vast array of species can be analyzed with game theory, yet only humans are capable of playing a game after being told its rules. This book is based on the appreciation that evolution and reason interact in constituting the social life and strategic interaction of humans.

Game theory, however, is not everything. This book systematically refutes one of the guiding prejudices of contemporary game theory. This is the notion that game theory is, insofar as human beings are rational, sufficient to explain all of human social existence. In fact, game theory is complementary to ideas developed and championed in all the behavioral disciplines. Behavioral scientists who have rejected game theory in reaction to the extravagant claims of some of its adherents may thus want to reconsider their positions, recognizing the fact that, just as game theory without broader social theory is merely technical bravado, so social theory without game theory is a handicapped enterprise.

The reigning culture in game theory asserts the sufficiency of game theory, allowing game theorists to do social theory without regard for either the facts or the theoretical contributions of the other social sciences. Only the feudal structure of the behavioral disciplines could possibly permit the persistence of such a manifestly absurd notion in a group of intelligent and open-minded scientists. Game theorists act like the proverbial "man with a hammer" for whom "all problems look like nails." I have explicitly started this volume with a broad array of social facts drawn from behavioral decision theory and behavioral game theory to disabuse the reader of this crippling notion. Game theory is a wonderful hammer, indeed a magical hammer. But, it is only a hammer and not the only occupant of the social scientist's toolbox.

The most fundamental failure of game theory is its lack of a theory of when and how rational agents share mental constructs. The assumption that humans are rational is an excellent first approximation. But, the Bayesian rational actors favored by contemporary game theory live in a universe of subjectivity and instead of constructing a truly social epistemology, game theorists have developed a variety of subterfuges that make it appear that rational agents may enjoy a commonality of belief (common priors, common knowledge), but all are failures. Humans have a *social epistemology*, meaning that we have reasoning processes that afford us forms of knowledge and understanding, especially the understanding and sharing of the content of other minds, that are unavailable to merely "rational" creatures. This social epistemology characterizes our species. The bounds of reason are thus not the irrational, but the social.

That game theory does not stand alone entails denying *methodological individualism*, a philosophical position asserting that all social phenomena can be explained purely in terms of the characteristics of rational agents, the actions available to them, and the constraints that they face. This position is incorrect because, as we shall see, human society is a system with *emergent properties*, including social norms, that can no more be analytically derived from a model of interacting rational agents than the chemical and biological properties of matter can be analytically derived from our knowledge of the properties of fundamental particles.

Evolutionary game theory often succeeds where classical game theory fails (Gintis 2009). The evolutionary approach to strategic interaction helps us understand the emergence, transformation, and stabilization of behaviors. In evolutionary game theory, successful strategies diffuse across populations of players rather than being learned inductively by disembodied rational agents. Moreover, reasoning is costly, so rational agents often do not even attempt to learn optimal strategies for complicated games but rather copy the behavior of successful agents whom they encounter. Evolutionary game theory allows us to investigate the interaction of learning, mutation, and imitation in the spread of strategies when information processing is costly.

But evolutionary game theory cannot deal with unique events, such as strangers interacting in a novel environment or Middle East peace negotiations. Moreover, by assuming that agents have very low-level cognitive capacities, evolutionary game theory ignores one of the most important of human capacities, that of being able to reason. Human society is an evolved system, but human reason is one of the key evolutionary forces involved. This book champions a unified approach based on modal logic, epistemic game theory, and social epistemology as an alternative to classical and a supplement to evolutionary game theory.

This approach holds that human behavior is most fruitfully modeled as the interaction of rational agents with a social epistemology, in the context of social norms that act as correlating devices that *choreograph* social interaction. This approach challenges contemporary sociology, which rejects the rational actor model. My response to the sociologists is that this rejection is the reason sociological theory has atrophied since the death of Talcott Parsons in 1979. This approach also challenges contemporary social psychology, which not only rejects the rational actor model but also generally delights in uncovering human "irrationalities." My response to the social psychologists is that this rejection accounts for the absence of a firm analytical base for the discipline, which must content itself with a host of nanomodels that illuminate highly specific aspects of human functioning with no analytical linkages among them.

The self-conceptions and dividing lines among the behavioral disciplines make no scientific sense. How can there be three separate fields, sociology, anthropology, and social psychology, for instance, studying social behavior and organization? How can the basic conceptual frameworks for the three fields, as outlined by their respective Great Masters and as taught to Ph.D. candidates, have almost nothing in common? In the name of science, these arbitrarities must be abolished. I propose, in the final chapter, a conceptual integration of the behavioral sciences that is analytically and empirically defensible and could be implemented now were it not for the virtually impassible feudal organization of the behavior disciplines in the contemporary university system, the structure of research funding agencies that mirror this feudal organization and interdisciplinary ethics that value comfort and tradition over the struggle for truth.

Game theory is a tool for investigating the world. By allowing us to specify carefully the conditions of social interaction (player characteristics, rules, informational assumptions, payoffs), its predictions can be *tested* and the results can be replicated in different laboratory settings. For this reason, *behavioral game theory* has become increasingly influential in setting research priorities. This aspect of game theory cannot be overstressed because the behavioral sciences currently consist of some fields where theory has evolved virtually without regard for the facts and others where facts abound and theory is absent.

Economic theory has been particularly compromised by its neglect of the facts concerning human behavior. This situation became clear to me in the summer of 2001, when I happened to be reading a popular introductory graduate text on quantum mechanics, as well as a leading graduate text on microeconomics. The physics text began with the anomaly of blackbody radiation, which could not be explained using the standard tools of electromagnetic theory. In 1900, Max Planck derived a formula that fit the data perfectly, assuming that radiation was discrete rather than continuous. In 1905, Albert Einstein explained another anomaly of classical electromagnetic theory, the photoelectric effect, using Planck's trick. The text continued, page after page, with new anomalies (Compton scattering, the spectral lines of elements of low atomic number, etc.) and new, partially successful models explaining the anomalies. In about 1925, this culminated with Heisenberg's wave mechanics and Schrödinger's equation, which fully unified the field.

By contrast, the microeconomics text, despite its beauty, did not contain a single fact in the whole thousand-page volume. Rather, the authors built economic theory in axiomatic fashion, making assumptions on the basis of their intuitive plausibility, their incorporation of the "stylized facts" of everyday life, or their appeal to the principles of rational thought. A bounty of excellent economic theory was developed in the twentieth century in this manner. But, the well has run dry. We will see that empirical evidence challenges the very foundations of both classical game theory and neoclassical economics. Future advances in economics will require that model-building dialogue with empirical testing, behavioral data-gathering, and agent-based models.

A simple generalization can be made: decision theory has developed valid algorithms by which people can best attain their objectives. Given these objectives, when people have the informational prerequisites of decision theory, yet fail to act as predicted, the theory is generally correct and the observed behavior faulty. Indeed, when deviations from theoretical predictions are pointed out to intelligent individuals, they generally agree that they have erred. By contrast, the extension of decision theory to the *strategic interaction* of Bayesian decision makers has led to a limited array of useful principles and when behavior differs from prediction, people generally stand by their behavior.

Most users of game theory remain unaware of this fact. Rather, the contemporary culture of game theory (as measured by what is accepted without complaint in a journal article) is to act as if epistemic game theory, which has flourished in the past two decades, did not exist. Thus, it is virtually universal to assume that rational agents play mixed strategies, use backward induction and more generally, play a Nash equilibrium. When people do not conform to these expectations, their rationality is called into question, whereas in fact, none of these assumptions can be successfully defended. Rational agents just do not behave the way classical game theory predicts, except in certain settings such as anonymous market interactions.

The reason for the inability of decision theory to extend to strategic interaction is quite simple. Decision theory shows that when a few plausible axioms hold, we can model agents as having beliefs (subjective priors) and a utility function over outcomes such that the agents' choices maximize the expected utility of the outcomes. In strategic interaction, nothing guarantees that all interacting parties have mutually consistent beliefs. Yet, as we shall see, a high degree of intersubjective belief consistency is required to ensure that agents play appropriately coordinated strategies.

The behavioral sciences have yet to adopt a serious commitment to linking basic theory and empirical research. Indeed, the various behavioral disciplines hold distinct and incompatible models of human behavior, yet their leading theoreticians make no attempt to adjudicate these differences (see chapter 12). Within economics there have been stunning advances in both theory and empirical data in the past few decades, yet theoreticians and experimentalists retain a hostile attitude to each other's work. This bizarre state of affairs must end.

It is often said that the mathematical rigor of contemporary economic theory is due to the economists' "physics envy." In fact, physicists generally judge models according to their ability to account for the facts, not their mathematical rigor. Physicists generally believe that rigor is the enemy of creative physical insight and they leave rigorous formulations to the mathematicians. The economic theorists' overvaluation of rigor is a symptom of their undervaluation of explanatory power. The truth is its own justification and needs no help from rigor.

Game theory can be used very profitably by researchers who do not know or care about mathematical intricacies but rather treat mathematics as but one of several tools deployed in the search for truth. I assert then that my arguments are correct and logically argued. I will leave rigor to the mathematicians.

In a companion volume, *Game Theory Evolving* (2009), I stress that understanding game theory requires solving lots of problems. I also stress therein that many of the weaknesses of classical game theory have beautiful remedies in evolutionary game theory. Neither of these considerations is dealt with in *The Bounds of Reason*, so I invite the reader to treat *Game Theory Evolving* as a complementary treatise.

The intellectual environments of the Santa Fe Institute, the Central European University (Budapest), and the University of Siena afforded me the time, resources, and research atmosphere to complete *The Bounds of Reason*. I would also like to thank Robert Aumann, Robert Axtell, Kent Bach, Kaushik Basu, Pierpaolo Battigalli, Larry Blume, Cristina Bicchieri, Ken Binmore, Samuel Bowles, Robert Boyd, Adam Brandenburger, Songlin Cai, Colin Camerer, Graciela Chichilnisky, Cristiano Castelfranchi, Rosaria Conte, Catherine Eckel, Jon Elster, Armin Falk, Ernst Fehr, Alex Field, Urs Fischbacher, Daniel Gintis, Jack Hirshleifer, Sung Ha Hwang, David Laibson, Michael Mandler, Stephen Morris, Larry Samuelson, Rajiv Sethi, Giacomo Sillari, E. Somanathan, Lones Smith, Roy A. Sorensen, Peter Vanderschraaf, Muhamet Yildiz, and Eduardo Zambrano for helping me with particular points. Thanks especially to Sean Brocklebank and Yusuke Narita, who read and corrected the entire manuscript. I am grateful to Tim Sullivan, Seth Ditchik, and Peter Dougherty, my editors at Princeton University Press, who persevered with me in making this volume possible.

1

Decision Theory and Human Behavior

People are not logical. They are *psycho*logical.
Anonymous

People often make mistakes in their maths. This does not mean that we should abandon arithmetic.
Jack Hirshleifer

Decision theory is the analysis of the behavior of an individual facing nonstrategic uncertainty—that is, uncertainty that is due to what we term "Nature" (a stochastic natural event such as a coin flip, seasonal crop loss, personal illness, and the like) or, if other individuals are involved, their behavior is treated as a statistical distribution known to the decision maker. Decision theory depends on probability theory, which was developed in the seventeenth and eighteenth centuries by such notables as Blaise Pascal, Daniel Bernoulli, and Thomas Bayes.

A *rational actor* is an individual with *consistent preferences* (§1.1). A rational actor need not be selfish. Indeed, if rationality implied selfishness, the only rational individuals would be sociopaths. Beliefs, called *subjective priors* in decision theory, logically stand between choices and payoffs. Beliefs are primitive data for the rational actor model. In fact, beliefs are the product of social processes and are shared among individuals. To stress the importance of beliefs in modeling choice, I often describe the rational actor model as the *beliefs, preferences and constraints* model, or the *BPC model*. The BPC terminology has the added attraction of avoiding the confusing and value-laden term "rational."

The BPC model requires only preference consistency, which can be defended on basic evolutionary grounds. While there are eminent critics of preference consistency, their claims are valid in only a few narrow areas. Because preference consistency does not presuppose unlimited information-processing capacities and perfect knowledge, even *bounded rationality* (Simon 1982) is consistent with the BPC model.[1] Because one cannot do behavioral game theory, by which I mean the application of game theory to the experimental study of human behavior, without assuming preference consistency, we must accept this axiom to avoid the analytical weaknesses of the behavioral disciplines that reject the BPC model, including psychology, anthropology, and sociology (see chapter 12).

Behavioral decision theorists have argued that there are important areas in which individuals appear to have inconsistent preferences. Except when individuals do not know their own preferences, this is a conceptual error based on a misspecification of the decision maker's preference function. We show in this chapter that, assuming individuals know their preferences, adding information concerning the current state of the individual to the choice space eliminates preference inconsistency. Moreover, this addition is completely reasonable because preference functions do not make any sense unless we include information about the decision maker's current state. When we are hungry, scared, sleepy, or sexually deprived, our preference ordering adjusts accordingly. The idea that we should have a utility function that does not depend on our current wealth, the current time, or our current strategic circumstances is also not plausible. Traditional decision theory ignores the individual's current state, but this is just an oversight that behavioral decision theory has brought to our attention.

Compelling experiments in behavioral decision theory show that humans violate the principle of expected utility in systematic ways (§1.7). Again, is must be stressed that this does *not* imply that humans violate preference consistency over the appropriate choice space but rather that they have incorrect beliefs deriving from what might be termed "folk probability theory" and make systematic *performance errors* in important cases (Levy 2008).

To understand why this is so, we begin by noting that, with the exception of hyperbolic discounting when time is involved (§1.4), there are no reported failures of the expected utility theorem in nonhumans, and there are some extremely beautiful examples of its satisfaction (Real 1991). Moreover, territoriality in many species is an indication of loss aversion (Chapter 11). The difference between humans and other animals is that the latter are tested in *real life*, or in elaborate simulations of real life, as in Leslie Real's work with bumblebees (1991), where subject bumblebees are re-

[1] Indeed, it can be shown (Zambrano 2005) that every boundedly rational individual is a fully rational individual subject to an appropriate set of Bayesian priors concerning the state of nature.

leased into elaborate spatial models of flowerbeds. Humans, by contrast, are tested using imperfect *analytical models* of real-life lotteries. While it is important to know how humans choose in such situations, there is certainly no guarantee they will make the same choices in the real-life situation and in the situation analytically generated to represent it. Evolutionary game theory is based on the observation that individuals are more likely to adopt behaviors that appear to be successful for others. A heuristic that says "adopt risk profiles that appear to have been successful to others" may lead to preference consistency even when individuals are incapable of evaluating analytically presented lotteries in the laboratory.

In addition to the explanatory success of theories based on the BPC model, supporting evidence from contemporary neuroscience suggests that expected utility maximization is not simply an "as if" story. In fact, the brain's neural circuitry actually makes choices by internally representing the payoffs of various alternatives as neural firing rates and choosing a maximal such rate (Shizgal 1999; Glimcher 2003; Glimcher and Rustichini 2004; Glimcher, Dorris, and Bayer 2005). Neuroscientists increasingly find that an aggregate decision making process in the brain synthesizes all available information into a single unitary value (Parker and Newsome 1998; Schall and Thompson 1999). Indeed, when animals are tested in a repeated trial setting with variable rewards, dopamine neurons appear to encode the difference between the reward that the animal expected to receive and the reward that the animal actually received on a particular trial (Schultz, Dayan, and Montague 1997; Sutton and Barto 2000), an evaluation mechanism that enhances the environmental sensitivity of the animal's decision making system. This error prediction mechanism has the drawback of seeking only local optima (Sugrue, Corrado, and Newsome 2005). Montague and Berns (2002) address this problem, showing that the orbitofrontal cortex and striatum contain a mechanism for more global predictions that include risk assessment and discounting of future rewards. Their data suggest a decision-making model that is analogous to the famous Black-Scholes options-pricing equation (Black and Scholes 1973).

The existence of an integrated decision-making apparatus in the human brain itself is predicted by evolutionary theory. The fitness of an organism depends on how effectively it make choices in an uncertain and varying environment. Effective choice must be a function of the organism's state of knowledge, which consists of the information supplied by the sensory inputs that monitor the organism's internal states and its external environment. In relatively simple organisms, the choice environment is primitive and is distributed in a decentralized manner over sensory inputs. But in three separate groups of animals, craniates (vertebrates and related creatures), arthropods (including insects, spiders, and crustaceans), and cephalopods (squid, octopuses, and other mollusks), a central nervous system with a brain (a centrally located decision-making and control apparatus) evolved. The phylogenetic tree of vertebrates exhibits increasing complexity through time and increasing metabolic and morphological costs of maintaining brain activity. Thus, *the brain evolved because larger and more complex brains, despite their costs, enhanced the fitness of their carriers.* Brains therefore are ineluctably structured to make consistent choices in the face of the various constellations of sensory inputs their bearers commonly experience.

Before the contributions of Bernoulli, Savage, von Neumann, and other experts, no creature on Earth knew how to value a lottery. The fact that people do not know how to evaluate abstract lotteries does not mean that they lack consistent preferences over the lotteries that they face in their daily lives.

Despite these provisos, experimental evidence on choice under uncertainty is still of great importance because in the modern world we are increasingly called upon to make such "unnatural" choices based on scientific evidence concerning payoffs and their probabilities.

1.1 Beliefs, Preferences, and Constraints

In this section we develop a set of behavioral properties, among which consistency is the most prominent, that together ensure that we can model agents as maximizers of preferences.

A *binary relation* $\odot_A$ on a set A is a subset of $A \times A$. We usually write the proposition $(x, y) \in \odot_A$ as $x \odot_A y$. For instance, the arithmetical operator "less than" ($<$) is a binary relation, where $(x, y) \in <$ is normally written $x < y$.[2] A *preference ordering* $\succeq_A$ on A is a binary relation with the following three properties, which must hold for all $x, y, z \in A$ and any set B:

1. **Complete**: $x \succeq_A y$ or $y \succeq_A x$;
2. **Transitive**: $x \succeq_A y$ and $y \succeq_A z$ imply $x \succeq_A z$;

[2]See chapter 14 for the basic mathematical notation used in this book. Additional binary relations over the set $\mathbf{R}$ of real numbers include $>$, $<$, $\leq$, $=$, $\geq$, and $\neq$, but $+$ is not a binary relation because $x + y$ is not a proposition.

3. **Independent of irrelevant alternatives**: For $x, y \in B$, $x \succeq_B y$ if and only if $x \succeq_A y$.

Because of the third property, we need not specify the choice set and can simply write $x \succeq y$. We also make the behavioral assumption that given any choice set A, the individual chooses an element $x \in A$ such that for all $y \in A$, $x \succeq y$. When $x \succeq y$, we say "x is weakly preferred to y."

The first condition is *completeness*, which implies that any member of A is weakly preferred to itself (for any x in A, $x \succeq x$). In general, we say a binary relation $\odot$ is *reflexive* if, for all x, $x \odot x$. Thus, completeness implies reflexivity. We refer to $\succeq$ as "weak preference" in contrast with "strong preference" $\succ$. We define $x \succ y$ to mean "it is false that $y \succeq x$." We say x and y are *equivalent* if $x \succeq y$ and $y \succeq x$, and we write $x \simeq y$. As an exercise, you may use elementary logic to prove that if $\succeq$ satisfies the completeness condition, then $\succ$ satisfies the following *exclusion* condition: if $x \succ y$, then it is false that $y \succ x$.

The second condition is *transitivity*, which says that $x \succeq y$ and $y \succeq z$ imply $x \succeq z$. It is hard to see how this condition could fail for anything we might like to call a preference ordering.[3] As a exercise, you may show that $x \succ y$ and $y \succeq z$ imply $x \succ z$, and $x \succeq y$ and $y \succ z$ imply $x \succ z$. Similarly, you may use elementary logic to prove that if $\succeq$ satisfies the completeness condition, then $\simeq$ is transitive (i.e., satisfies the transitivity condition).

The third condition, *independence of irrelevant alternatives* (IIA) means that the relative attractiveness of two choices does not depend upon the other choices available to the individual. For instance, suppose an individual generally prefers meat to fish when eating out, but if the restaurant serves lobster, the individual believes the restaurant serves superior fish, and hence prefers fish to meat, even though he never chooses lobster; thus, IIA fails. When IIA fails, it can be restored by suitably refining the choice set. For instance, we can specify two qualities of fish instead of one, in the preceding example. More generally, if the desirability of an outcome x depends on the set A from which it is chosen, we can form a new choice space Ω^*, elements of which are ordered pairs (A, x), where $x \in A \subseteq \Omega$, and restrict choice sets in Ω^* to be subsets of Ω^* all of whose first elements are equal. In this new choice space, IIA is trivially satisfied.

[3]The only plausible model of intransitivity with some empirical support is *regret theory* (Loomes 1988; Sugden 1993). Their analysis applies, however, only to a narrow range of choice situations.

When the preference relation $\succeq$ is complete, transitive, and independent of irrelevant alternatives, we term it *consistent*. If $\succeq$ is a consistent preference relation, then there will always exist a preference function such that the individual behaves as if maximizing this preference function over the set A from which he or she is constrained to choose. Formally, we say that a preference function $u: A \rightarrow \mathbf{R}$ *represents* a binary relation $\succeq$ if, for all $x, y \in A$, $u(x) \geq u(y)$ if and only if $x \succeq y$. We have the following theorem.

THEOREM 1.1 *A binary relation $\succeq$ on the finite set A of payoffs can be represented by a preference function $u: A \rightarrow \mathbf{R}$ if and only if $\succeq$ is consistent.*

It is clear that $u(\cdot)$ is not unique, and indeed, we have the following theorem.

THEOREM 1.2 *If $u(\cdot)$ represents the preference relation $\succeq$ and $f(\cdot)$ is a strictly increasing function, then $v(\cdot) = f(u(\cdot))$ also represents $\succeq$. Conversely, if both $u(\cdot)$ and $v(\cdot)$ represent $\succeq$, then there is an increasing function $f(\cdot)$ such that $v(\cdot) = f(u(\cdot))$.*

The first half of the theorem is true because if f is strictly increasing, then $u(x) > u(y)$ implies $v(x) = f(u(x)) > f(u(y)) = v(y)$, and conversely. For the second half, suppose $u(\cdot)$ and $v(\cdot)$ both represent $\succeq$, and for any $y \in \mathbf{R}$ such that $v(x) = y$ for some $x \in X$, let $f(y) = u(v^{-1}(y))$, which is possible because v is an increasing function. Then $f(\cdot)$ is increasing (because it is the composition of two increasing functions) and $f(v(x)) = u(v^{-1}(v(x))) = u(x)$, which proves the theorem. ∎

1.2 The Meaning of Rational Action

The origins of the BPC model lie in the eighteenth century research of Jeremy Bentham and Cesare Beccaria. In his *Foundations of Economic Analysis* (1947), economist Paul Samuelson removed the hedonistic assumptions of utility maximization by arguing, as we have in the previous section, that utility maximization presupposes nothing more than transitivity and some harmless technical conditions akin to those specified above.

Rational does not imply self-interested. There is nothing irrational about caring for others, believing in fairness, or sacrificing for a social ideal. Nor do such preferences contradict decision theory. For instance, suppose a man with \$100 is considering how much to consume himself and how much to give to charity. Suppose he faces a tax or subsidy such that for each \$1 he contributes to charity, he is obliged to pay p dollars. Thus, $p > 1$ represents a tax, while $0 < p < 1$ represents a subsidy. We can then treat p as the *price* of a unit contribution to charity and model the individual as maximizing his utility for personal consumption x and contributions to charity y, say $u(x, y)$ subject to the budget constraint $x + py = 100$. Clearly, it is perfectly rational for him to choose $y > 0$. Indeed, Andreoni and Miller (2002) have shown that in making choices of this type, consumers behave in the same way as they do when choosing among personal consumption goods; i.e., they satisfy the generalized axiom of revealed preference.

Decision theory does not presuppose that the choices people make are welfare-improving. In fact, people are often slaves to such passions as smoking cigarettes, eating junk food, and engaging in unsafe sex. These behaviors in no way violate preference consistency.

If humans fail to behave as prescribed by decision theory, we need not conclude that they are irrational. In fact, they may simply be ignorant or misinformed. However, if human subjects consistently make intransitive choices over lotteries (e.g., §1.7), then either they do not satisfy the axioms of expected utility theory or they do not know how to evaluate lotteries. The latter is often called *performance error*. Performance error can be reduced or eliminated by formal instruction, so that the experts that society relies upon to make efficient decisions may behave quite rationally even in cases where the average individual violates preference consistency.

1.3 Why Are Preferences Consistent?

Preference consistency flows from evolutionary biology (Robson 1995). Decision theory often applies extremely well to nonhuman species, including insects and plants (Real 1991; Alcock 1993; Kagel, Battalio, and Green 1995). Biologists define the *fitness* of an organism as its expected number of offspring. Assume, for simplicity, asexual reproduction. A maximally fit individual will then produce the maximal expected number of offspring, each of which will inherit the genes for maximal fitness. Thus, fitness maximization is a precondition for evolutionary survival. If organisms maximized fitness directly, the conditions of decision theory would be directly satisfied because we could simply represent the organism's utility function as its fitness.

However, organisms do *not* directly maximize fitness. For instance, moths fly into flames and humans voluntarily limit family size. Rather, organisms have preference orderings that are themselves subject to selection according to their ability to promote fitness (Darwin 1872). We can expect preferences to satisfy the completeness condition because an organism must be able to make a consistent choice in any situation it habitually faces or it will be outcompeted by another whose preference ordering can make such a choice.

Of course, unless the current environment of choice is the same as the historical environment under which the individual's preference system evolved, we would not expect an individual's choices to be fitness-maximizing, or even necessarily welfare-improving.

This biological explanation also suggests how preference consistency might fail in an imperfectly integrated organism. Suppose the organism has three decision centers in its brain, and for any pair of choices, majority rule determines which the organism prefers. Suppose the available choices are A, B, and C and the three decision centers have preferences $A \succ B \succ C$, $B \succ C \succ A$, and $C \succ A \succ B$, respectively. Then when offered A or B, the individual chooses A, when offered B or C, the individual chooses B, and when offered A and C, the individual chooses C. Thus $A \succ B \succ C \succ A$, and we have intransitivity. Of course, if an objective fitness is associated with each of these choices, Darwinian selection will favor a mutant who suppresses two of the three decision centers or, better yet, integrates them.

1.4 Time Inconsistency

Several human behavior patterns appear to exhibit *weakness of will*, in the sense that if there is a long time period between choosing and experiencing the costs and benefits of the choice, individuals can choose wisely, but when costs or benefits are immediate, people make poor choices, longrun payoffs being sacrificed in favor of immediate payoffs. For instance, smokers may know that their habit will harm them in the long run, but cannot bear to sacrifice the present urge to indulge in favor of the far-off reward of a healthy future. Similarly, a couple in the midst of sexual passion may appreciate that they may well regret their inadequate safety precautions at some point in the future, but they cannot control their present urges. We call this behavior *time-inconsistent*.[4]

[4]For an excellent survey of empirical results in this area, see Frederick, Loewenstein, and O'Donoghue (2002).

Are people time-consistent? Take, for instance, impulsive behavior. Economists are wont to argue that what appears to be impulsive—cigarette smoking, drug use, unsafe sex, overeating, dropping out of school, punching out your boss, and the like—may in fact be welfare-maximizing for people who have high time discount rates or who prefer acts that happen to have high future costs. Controlled experiments in the laboratory cast doubt on this explanation, indicating that people exhibit a *systematic* tendency to discount the near future at a higher rate than the distant future (Chung and Herrnstein 1967; Loewenstein and Prelec 1992; Herrnstein and Prelec 1992; Fehr and Zych 1994; Kirby and Herrnstein 1995; McClure et al. 2004).

For instance, consider the following experiment conducted by Ainslie and Haslam (1992). Subjects were offered a choice between $10 on the day of the experiment or $11 a week later. Many chose to take the $10 without delay. However, when the same subjects were offered $10 to be delivered a year from the day of the experiment or $11 to be delivered a year and a week from the day of the experiment, many of those who could not wait a week *right now* for an extra 10%, preferred to wait a week for an extra 10%, provided the agreed-upon wait was one year in the future.

It is instructive to see exactly where the consistency conditions are violated in this example. Let x mean "$10 at some time t" and let y mean "$11 at time $t+7$," where time t is measured in days. Then the present-oriented subjects display $x \succ y$ when $t = 0$, and $y \succ x$ when $t = 365$. Thus the exclusion condition for $\succ$ is violated, and because the completeness condition for $\succeq$ implies the exclusion condition for $\succ$, the completeness condition must be violated as well.

However, time inconsistency *disappears* if we model the individuals as choosing over a slightly more complicated choice space in which the distance between the time of choice and the time of delivery of the object chosen is explicitly included in the object of choice. For instance, we may write x_0 to mean "$10 delivered immediately" and x_{365} to mean "$10 delivered a year from today," and similarly for y_7 and y_{372}. Then the observation that $x_0 \succ y_7$ and $y_{372} \succ x_{365}$ is no contradiction.

Of course, if you are not time-consistent and if you know this, you should not expect that your will carry out your plans for the future when the time comes. Thus, you may be willing to *precommit* yourself to making these future choices, even at a cost. For instance, if you are saving in year 1 for a purchase in year 3, but you know you will be tempted to spend the money in year 2, you can put it in a bank account that cannot be accessed until the year after next. My teacher Leo Hurwicz called this the "piggy bank effect."

The central theorem on choice over time is that time consistency results from assuming that *utility is additive across time periods and that the instantaneous utility function is the same in all time periods, with future utilities discounted to the present at a fixed rate* (Strotz 1955). This is called *exponential discounting* and is widely assumed in economic models. For instance, suppose an individual can choose between two consumption streams $x = x_0, x_1, \ldots$ or $y = y_0, y_1, \ldots$. According to exponential discounting, he has a utility function $u(x)$ and a constant $\delta \in (0, 1)$ such that the total utility of stream x is given by[5]

$$U(x_0, x_1, \ldots) = \sum_{k=0}^{\infty} \delta^k u(x_k). \tag{1.1}$$

We call δ the individual's *discount factor*. Often we write $\delta = e^{-r}$ where we interpret $r > 0$ as the individual's one-period continuously compounded *interest rate*, in which case (1.1) becomes

$$U(x_0, x_1, \ldots) = \sum_{k=0}^{\infty} e^{-rk} u(x_k). \tag{1.2}$$

This form clarifies why we call this "exponential" discounting. The individual strictly prefers consumption stream x over stream y if and only if $U(x) > U(y)$. In the simple compounding case, where the interest accrues at the end of the period, we write $\delta = 1/(1+r)$, and (1.2) becomes

$$U(x_0, x_1, \ldots) = \sum_{k=0}^{\infty} \frac{u(x_k)}{(1+r)^k}. \tag{1.3}$$

Despite the elegance of exponential discounting, observed intertemporal choice for humans appears to fit more closely the model of *hyperbolic discounting* (Ainslie and Haslam 1992; Ainslie 1975; Laibson 1997), first observed by Richard Herrnstein in studying animal behavior (Herrnstein, Laibson, and Rachlin 1997) and reconfirmed many times since (Green et al. 2004). For instance, continuing the previous example, let z_t mean

[5]Throughout this text, we write $x \in (a,b)$ for $a < x < b$, $x \in [a,b)$ for $a \leq x < b$, $x \in (a,b]$ for $a < x \leq b$, and $x \in [a,b]$ for $a \leq x \leq b$.

"amount of money delivered t days from today." Then let the utility of z_t be $u(z_t) = z/(t+1)$. The value of x_0 is thus $u(x_0) = u(10_0) = 10/1 = 10$, and the value of y_7 is $u(y_7) = u(11_7) = 11/8 = 1.375$, so $x_0 \succ y_7$. But $u(x_{365}) = 10/366 = 0.027$ while $u(y_{372}) = 11/373 = 0.029$, so $y_{372} \succ x_{365}$.

There is also evidence that people have different rates of discount for different types of outcomes (Loewenstein 1987; Loewenstein and Sicherman 1991). This would be irrational for outcomes that could be bought and sold in perfect markets, because all such outcomes should be discounted at the market interest rate in equilibrium. But, of course, there are many things that people care about that cannot be bought and sold in perfect markets.

Neurological research suggests that balancing current and future payoffs involves adjudication among structurally distinct and spatially separated modules that arose in different stages in the evolution of *H. sapiens* (Tooby and Cosmides 1992; Sloman 2002; McClure et al. 2004). The long-term decision-making capacity is localized in specific neural structures in the prefrontal lobes and functions improperly when these areas are damaged, despite the fact that subjects with such damage appear to be otherwise completely normal in brain functioning (Damasio 1994). *H. sapiens* may be structurally predisposed, in terms of brain architecture, to exhibit a systematic present orientation.

In sum, time inconsistency doubtless exists and is important in modeling human behavior, but this does not imply that people are irrational in the weak sense of preference consistency. Indeed, we can model the behavior of time-inconsistent rational individuals by assuming they maximize their time-dependent preference functions (O'Donoghue and Rabin, 1999a,b, 2000, 2001). For axiomatic treatment of time-dependent preferences, see Ahlbrecht and Weber (1995) and Ok and Masatlioglu (2003). In fact, humans are much closer to time consistency and have much longer time horizons than any other species, probably by several orders of magnitude (Stephens, McLinn, and Stevens 2002; Hammerstein 2003). We do not know why biological evolution so little values time consistency and long time horizons even in long-lived creatures.

1.5 Bayesian Rationality and Subjective Priors

Consider decisions in which a stochastic event determines the payoffs to the players. Let X be a set of prizes. A *lottery* with payoffs in X is a function $p{:}X \rightarrow [0, 1]$ such that $\sum_{x \in X} p(x) = 1$. We interpret $p(x)$ as the probability that the payoff is $x \in X$. If $X = \{x_1, \ldots, x_n\}$ for some finite number n, we write $p(x_i) = p_i$.

The *expected value* of a lottery is the sum of the payoffs, where each payoff is weighted by the probability that the payoff will occur. If the lottery l has payoffs $x_1, \ldots, x_n$ with probabilities $p_1, \ldots, p_n$, then the expected value $\mathbf{E}[l]$ of the lottery l is given by

$$\mathbf{E}[l] = \sum_{i=1}^{n} p_i x_i.$$

The expected value is important because of the law of large numbers (Feller 1950), which states that as the number of times a lottery is played goes to infinity, the average payoff converges to the expected value of the lottery with probability 1.

Consider the lottery l_1 in figure 1.1(a), where p is the probability of winning amount a and $1-p$ is the probability of winning amount b. The expected value of the lottery is then $\mathbf{E}[l_1] = pa + (1 - p)b$. Note that we model a lottery a lot like an extensive form game—except that there is only one player.

Consider the lottery l_2 with the three payoffs shown in figure1.1(b). Here p is the probability of winning amount a, q is the probability of winning amount b, and $1-p-q$ is the probability of winning amount c. The expected value of the lottery is $\mathbf{E}[l_2] = pa + qb + (1 - p - q)c$.

A lottery with n payoffs is given in figure 1.1(c). The prizes are now $a_1, \ldots, a_n$ with probabilities $p_1, \ldots, p_n$, respectively. The expected value of the lottery is now $\mathbf{E}[l_3] = p_1a_1 + p_2a_2 + \cdots + p_na_n$.

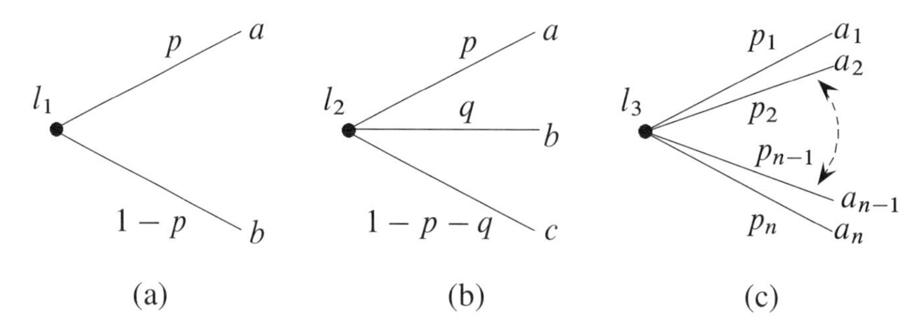

Figure 1.1. Lotteries with two, three, and n potential outcomes.

In this section we generalize the previous argument, developing a set of behavioral properties that yield both a utility function over outcomes and a probability distribution over states of nature, such that the expected utility principle holds. Von Neumann and Morgenstern (1944), Friedman and Savage (1948), Savage (1954), and Anscombe and Aumann (1963) showed that the expected utility principle can be derived from the assumption that individuals have consistent preferences over an appropriate set of lotteries. We outline here Savage's classic analysis of this problem.

For the rest of this section, we assume $\succeq$ is a preference relation (§1.1). To ensure that the analysis is not trivial, we also assume that $x \succeq y$ is false for at least some $x, y \in X$. Savage's accomplishment was to show that if the individual has a preference relation over *lotteries* that has some plausible properties, then not only can the individual's preferences be represented by a utility function, but also we can infer the probabilities the individual implicitly places on various events, and the expected utility principle holds for these probabilities.

Let Ω be a finite set of *states of nature*. We call $A \subseteq \Omega$ *events*. Let $\mathcal{L}$ be a set of lotteries, where a *lottery* is a function $\pi : \Omega \rightarrow X$ that associates with each state of nature $\omega \in \Omega$ a payoff $\pi(\omega) \in X$. Note that this concept of a lottery does not include a probability distribution over the states of nature. Rather, the Savage axioms allow us to associate a subjective prior over each state of nature ω, expressing the decision maker's personal assessment of the probability that ω will occur. We suppose that the individual chooses among lotteries without knowing the state of nature, after which Nature chooses the state $\omega \in \Omega$ that obtains, so that if the individual chose lottery $\pi \in \mathcal{L}$, his payoff is $\pi(\omega)$.

Now suppose the individual has a preference relation $\succ$ over $\mathcal{L}$ (we use the same symbol $\succ$ for preferences over both outcomes and lotteries). We seek a set of plausible properties of $\succ$ over lotteries that together allow us to deduce (a) a utility function $u : X \rightarrow \mathbf{R}$ corresponding to the preference relation $\succ$ over outcomes in X; (b) a probability distribution $p : \Omega \rightarrow \mathbf{R}$ such that the expected utility principle holds with respect to the preference relation $\succ$ over lotteries and the utility function $u(\cdot)$; i.e., if we define

$$\mathbf{E}_\pi[u; p] = \sum_{\omega \in \Omega} p(\omega) u(\pi(\omega)), \tag{1.4}$$

then for any $\pi, \rho \in \mathcal{L}$,

$$\pi \succ \rho \Longleftrightarrow \mathbf{E}_\pi[u; p] > \mathbf{E}_\rho[u; p].$$

Our first condition is that $\pi \succ \rho$ depends only on states of nature where π and ρ have different outcomes. We state this more formally as follows.

A1. For any $\pi, \rho, \pi', \rho' \in \mathcal{L}$, let $A = \{\omega \in \Omega | \pi(\omega) \neq \rho(\omega)\}$. Suppose we also have $A = \{\omega \in \Omega | \pi'(\omega) \neq \rho'(\omega)\}$. Suppose also that $\pi(\omega) = \pi'(\omega)$ and $\rho(\omega) = \rho'(\omega)$ for $\omega \in A$. Then $\pi \succ \rho \Leftrightarrow \pi' \succ \rho'$.

This axiom says, reasonably enough, that the relative desirability of two lotteries does not depend on the payoffs where the two lotteries agree. The axiom allows us to define a *conditional preference* $\pi \succ_A \rho$, where $A \subseteq \Omega$, which we interpret as "π is strictly preferred to ρ, conditional on event A," as follows. We say $\pi \succ_A \rho$ if, for some $\pi', \rho' \in \mathcal{L}$, $\pi(\omega) = \pi'(\omega)$ and $\rho(\omega) = \rho'(\omega)$ for $\omega \in A$, $\pi'(\omega) = \rho'(\omega)$ for $\omega \notin A$, and $\pi' \succ \rho'$. Because of A1, this is well defined (i.e., $\pi \succ_A \rho$ does not depend on the particular $\pi', \rho' \in \mathcal{L}$). This allows us to define $\succeq_A$ and $\sim_A$ in a similar manner. We then define an event $A \subseteq \Omega$ to be *null* if $\pi \sim_A \rho$ for all $\pi, \rho \in \mathcal{L}$.

Our second condition is then the following, where we write $\pi = x|A$ to mean $\pi(\omega) = x$ for all $\omega \in A$ (i.e., $\pi = x|A$ means π is a lottery that pays x when A occurs).

A2. If $A \subseteq \Omega$ is not null, then for all $x, y \in X$, $\pi = x|A \succ_A \pi = y|A \Leftrightarrow x \succ y$.

This axiom says that a natural relationship between outcomes and lotteries holds: if π pays x given event A and ρ pays y given event A, and if $x \succ y$, then $\pi \succ_A \rho$, and conversely.

Our third condition asserts that the probability that a state of nature occurs is independent of the outcome one receives when the state occurs. The difficulty in stating this axiom is that the individual cannot choose probabilities but only lotteries. But, if the individual prefers x to y, and if $A, B \subseteq \Omega$ are events, then the individual treats A as more probable than B if and only if a lottery that pays x when A occurs and y when A does not occur is preferred to a lottery that pays x when B occurs and y when B does not. However, this must be true for any $x, y \in X$ such that $x \succ y$, or the individual's notion of probability is incoherent (i.e., it depends on what particular payoffs we are talking about—for instance, wishful thinking, where if the prize associated with an event increases, the individual thinks it is more likely to occur). More formally, we have the following, where we write $\pi = x, y|A$ to mean "$\pi(\omega) = x$ for $\omega \in A$ and $\pi(\omega) = y$ for $\omega \notin A$."

A3. Suppose $x \succ y$, $x' \succ y'$, $\pi, \rho, \pi', \rho' \in \mathcal{L}$, and $A, B \subseteq \Omega$. Suppose that $\pi = x, y|A$, $\rho = x', y'|A$, $\pi' = x, y|B$, and $\rho' = x', y'|B$. Then $\pi \succ \pi' \Leftrightarrow \rho \succ \rho'$.

The fourth condition is a weak version of *first-order stochastic dominance*, which says that if one lottery has a higher payoff than another for any event, then the first is preferred to the second.

A4. For any event A, if $x \succ \rho(\omega)$ for all $\omega \in A$, then $\pi = x|A \succ_A \rho$. Also, for any event A, if $\rho(\omega) \succ x$ for all $\omega \in A$, then $\rho \succ_A \pi = x|A$.

In other words, if for any event A, $\pi = x$ on A pays more than the best ρ can pay on A, then $\pi \succ_A \rho$, and conversely.

Finally, we need a technical property to show that a preference relation can be represented by a utility function. We say nonempty sets $A_1, \ldots, A_n$ form a *partition* of set X if the A_i are mutually disjoint ($A_i \cap A_j = \emptyset$ for $i \neq j$) and their union is X (i.e., $A_1 \cup \cdots \cup A_n = X$). The technical condition says that for any $\pi, \rho \in \mathcal{L}$, and any $x \in X$, there is a partition $A_1, \ldots, A_n$ of Ω such that, for each A_i, if we change π so that its payoff is x on A_i, then π is still preferred to ρ, and similarly, for each A_i, if we change ρ so that its payoff is x on A_i, then π is still preferred to ρ. This means that no payoff is "supergood," so that no matter how unlikely an event A is, a lottery with that payoff when A occurs is always preferred to a lottery with a different payoff when A occurs. Similarly, no payoff can be "superbad." The condition is formally as follows.

A5. For all $\pi, \pi', \rho, \rho' \in \mathcal{L}$ with $\pi \succ \rho$, and for all $x \in X$, there are disjoint subsets $A_1, \ldots, A_n$ of Ω such that $\cup_i A_i = \Omega$ and for any A_i (a) if $\pi'(\omega) = x$ for $\omega \in A_i$ and $\pi'(\omega) = \pi(\omega)$ for $\omega \notin A_i$, then $\pi' \succ \rho$, and (b) if $\rho'(\omega) = x$ for $\omega \in A_i$ and $\rho'(\omega) = \rho(\omega)$ for $s \notin A_i$, then $\pi \succ \rho'$.

We then have Savage's theorem.

THEOREM 1.3 *Suppose A1–A5 hold. Then there is a probability function p on Ω and a utility function $u : X \to \mathbf{R}$ such that for any $\pi, \rho \in \mathcal{L}$, $\pi \succ \rho$ if and only if $\mathbf{E}_\pi[u; p] > \mathbf{E}_\rho[u; p]$.*

The proof of this theorem is somewhat tedious; it is sketched in Kreps 1988.

We call the probability p the individual's *Bayesian prior*, or *subjective prior* and say that A1–A5 imply *Bayesian rationality*, because the they together imply Bayesian probability updating.

1.6 The Biological Basis for Expected Utility

Suppose an organism must choose from action set X under certain conditions. There is always uncertainty as to the degree of success of the various options in X, which means essentially that each $x \in X$ determines a lottery that pays i offspring with probability $p_i(x)$ for $i = 0, 1, \ldots, n$. Then the expected number of offspring from this lottery is $\psi(x) = \sum_{j=1}^{n} jp_j(x)$. Let L be a lottery on X that delivers $x_i \in X$ with probability q_i for $i = 1, \ldots, k$. The probability of j offspring given L is then $\sum_{i=1}^{k} q_i p_j(x_i)$, so the expected number of offspring given L is

$$\sum_{j=1}^{n} j \sum_{i=1}^{k} q_i p_j(x_i) = \sum_{i=1}^{k} q_i \sum_{i=1}^{k} jp_j(x_i) = \sum_{i=1}^{k} q_i \psi(x_i), \tag{1.5}$$

which is the expected value theorem with utility function $\psi(\cdot)$. See also Cooper (1987).

1.7 The Allais and Ellsberg Paradoxes

Although most decision theorists consider the expected utility principle acceptably accurate as a basis of modeling behavior, there are certainly well established situations in which individuals violate it. Machina (1987) reviews this body of evidence and presents models to deal with them. We sketch here the most famous of these anomalies, the *Allais paradox* and the *Ellsberg paradox*. They are, of course, not paradoxes at all but simply empirical regularities that do not fit the expected utility principle.

Maurice Allais (1953) offered the following scenario. There are two choice situations in a game with prizes $x =$ \$2,500,000, $y =$ \$500,000, and $z =$ \$0. The first is a choice between lotteries $\pi = y$ and $\pi' = 0.1x + 0.89y + 0.01z$. The second is a choice between $\rho = 0.11y + 0.89z$ and $\rho' = 0.1x + 0.9z$. Most people, when faced with these two choice situations, choose $\pi \succ \pi'$ and $\rho' \succ \rho$. Which would you choose?

This pair of choices is not consistent with the expected utility principle. To see this, let us write $u_h = u(2500000)$, $u_m = u(500000)$, and $u_l =$

$u(0)$. Then if the expected utility principle holds, $\pi \succ \pi'$ implies $u_m > 0.1u_h + 0.89u_m + 0.01u_l$, so $0.11u_m > 0.10u_h + 0.01u_l$, which implies (adding $0.89u_l$ to both sides) $0.11u_m + 0.89u_l > 0.10u_h + 0.9u_l$, which says $\rho \succ \rho'$.

Why do people make this mistake? Perhaps because of *regret*, which does not mesh well with the expected utility principle (Loomes 1988; Sugden 1993). If you choose π' in the first case and you end up getting nothing, you will feel really foolish, whereas in the second case you are probably going to get nothing anyway (not your fault), so increasing the chances of getting nothing a tiny bit (0.01) gives you a good chance (0.10) of winning the really big prize. Or perhaps because of *loss aversion* (§1.9), because in the first case, the anchor point (the most likely outcome) is $500,000, while in the second case the anchor is $0. Loss-averse individuals then shun π', which gives a positive probability of loss whereas in the second case, neither lottery involves a loss, from the standpoint of the most likely outcome.

The Allais paradox is an excellent illustration of problems that can arise when a lottery is consciously chosen by an act of will and one *knows* that one has made such a choice. The regret in the first case arises because if one chose the risky lottery and the payoff was zero, one knows for certain that one made a poor choice, at least ex post. In the second case, if one received a zero payoff, the odds are that it had nothing to do with one's choice. Hence, there is no regret in the second case. But in the real world, most of the lotteries we experience are chosen by default, not by acts of will. Thus, if the outcome of such a lottery is poor, we feel bad because of the poor outcome but not because we made a poor choice.

Another classic violation of the expected utility principle was suggested by Daniel Ellsberg (1961). Consider two urns. Urn A has 51 red balls and 49 white balls. Urn B also has 100 red and white balls, but the fraction of red balls is unknown. One ball is chosen from each urn but remains hidden from sight. Subjects are asked to choose in two situations. First, a subject can choose the ball from urn A or urn B, and if the ball is red, the subject wins $10. In the second situation, the subject can choose the ball from urn A or urn B, and if the ball is white, the subject wins $10. Many subjects choose the ball from urn A in both cases. This violates the expected utility principle no matter what probability the subject places on the probability p that the ball from urn B is white. For in the first situation, the payoff from choosing urn A is $0.51u(10)+0.49u(0)$ and the payoff from choosing urn B

is $(1-p)u(10)+pu(0)$, so strictly preferring urn A means $p > 0.49$. In the second situation, the payoff from choosing urn A is $0.49u(10) + 0.51u(0)$ and the payoff from choosing urn B is $pu(10) + (1 - p)u(0)$, so strictly preferring urn A means $p < 0.49$. This shows that the expected utility principle does not hold.

Whereas the other proposed anomalies of classical decision theory can be interpreted as the failure of linearity in probabilities, regret, loss aversion, and epistemological ambiguities, the Ellsberg paradox strikes even more deeply because it implies that humans systematically violate the following principle of first-order stochastic dominance (FOSD).

> Let $p(x)$ and $q(x)$ be the probabilities of winning x or more in lotteries A and B, respectively. If $p(x) \geq q(x)$ for all x, then $A \succeq B$.

The usual explanation of this behavior is that the subject *knows* the probabilities associated with the first urn, while the probabilities associated with the second urn are *unknown*, and hence there appears to be an added degree of risk associated with choosing from the second urn rather than the first. If decision makers are risk-averse and if they perceive that the second urn is considerably riskier than the first, they will prefer the first urn. Of course, with some relatively sophisticated probability theory, we are assured that there is in fact no such additional risk, it is hardly a failure of rationality for subjects to come to the opposite conclusion. The Ellsberg paradox is thus a case of performance error on the part of subjects rather than a failure of rationality.

1.8 Risk and the Shape of the Utility Function

If $\succeq$ is defined over X, we can say nothing about the *shape* of a utility function $u(\cdot)$ representing $\succeq$ because, by theorem 1.2, any increasing function of $u(\cdot)$ also represents $\succeq$. However, if $\succeq$ is represented by a utility function $u(x)$ satisfying the expected utility principle, then $u(\cdot)$ is determined up to an arbitrary constant and unit of measure.[6]

[6]Because of this theorem, the difference between two utilities means nothing. We thus say utilities over outcomes are *ordinal*, meaning we can say that one bundle is preferred to another, but we cannot say by how much. By contrast, the next theorem shows that utilities over lotteries are *cardinal*, in the sense that, up to an arbitrary constant and an arbitrary positive choice of units, utility is numerically uniquely defined.

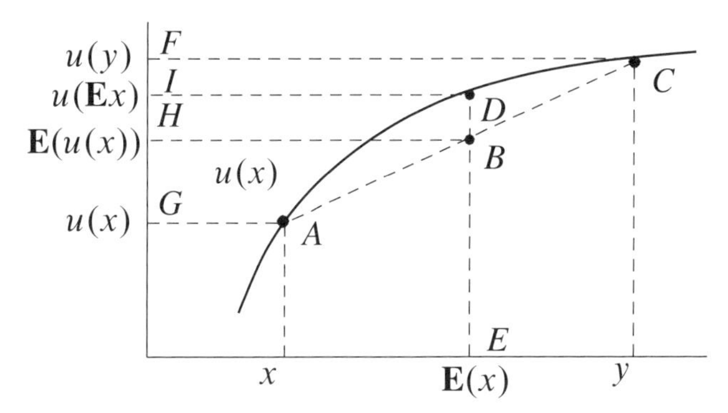

Figure 1.2. A concave utility function

THEOREM 1.4 *Suppose the utility function $u(\cdot)$ represents the preference relation $\succeq$ and satisfies the expected utility principle. If $v(\cdot)$ is another utility function representing $\succeq$, then there are constants $a, b \in \mathbf{R}$ with $a > 0$ such that $v(x) = au(x) + b$ for all $x \in X$.*

For a proof of this theorem, see Mas-Collel, Whinston, and Green (1995, p. 173).

If $X = \mathbf{R}$, so the payoffs can be considered to be money, and utility satisfies the expected utility principle, what shape do such utility functions have? It would be nice if they were linear in money, in which case expected utility and expected value would be the same thing (why?). But generally utility is *strictly concave*, as illustrated in figure 1.2. We say a function $u: X \to \mathbf{R}$ is strictly concave if, for any $x, y \in X$ and any $p \in (0,1)$, we have $pu(x) + (1-p)u(y) < u(px + (1-p)y)$. We say $u(x)$ is *weakly concave*, or simply *concave*, if $u(x)$ is either strictly concave or linear, in which case the above inequality is replaced by $pu(x) + (1-p)u(y) = u(px + (1-p)y)$.

If we define the lottery π as paying x with probability p and y with probability $1-p$, then the condition for strict concavity says that *the expected utility of the lottery is less than the utility of the expected value of the lottery*, as depicted in figure 1.2. To see this, note that the expected value of the lottery is $E = px + (1-p)y$, which divides the line segment between x and y into two segments, the segment xE having length $(px + (1-p)y) - x = (1-p)(y-x)$ and the segment Ey having length $y - (px + (1-p)y) = p(y-x)$. Thus, E divides $[x, y]$ into two segments whose lengths have the ratio $(1-p)/p$. From elementary geometry, it follows that B divides segment $[A, C]$ into two segments whose lengths have the same ratio. By the same reasoning, point H divides segments $[F, G]$ into segments with the same ratio of lengths. This means that point H has the coordinate value $pu(x) + (1 - p)u(y)$, which is the expected utility of the lottery. But by definition, the utility of the expected value of the lottery is at D, which lies above H. This proves that the utility of the expected value is greater than the expected value of the lottery for a strictly concave utility function. This is know as *Jensen's inequality*.

What *are* good candidates for $u(x)$? It is easy to see that strict concavity means $u''(x) < 0$, providing $u(x)$ is twice differentiable (which we assume). But there are lots of functions with this property. According to the famous *Weber-Fechner law* of psychophysics, for a wide range of sensory stimuli and over a wide range of levels of stimulation, a just noticeable change in a stimulus is a constant fraction of the original stimulus. If this holds for money, then the utility function is logarithmic.

We say an individual is *risk-averse* if the individual prefers the expected value of a lottery to the lottery itself (provided, of course, the lottery does not offer a single payoff with probability 1, which we call a sure thing). We know, then, that an individual with utility function $u(\cdot)$ is risk-averse if and only if $u(\cdot)$ is concave.[7] Similarly, we say an individual is *risk-loving* if he prefers any lottery to the expected value of the lottery, and *risk-neutral* if he is indifferent between a lottery and its expected value. Clearly, an individual is risk-neutral if and only if he has linear utility.

Does there exist a measure of risk aversion that allows us to say when one individual is more risk-averse than another, or how an individual's risk aversion changes with changing wealth? We may define individual A to be *more risk-averse* than individual B if whenever A prefers a lottery to an amount of money x, B will also prefer the lottery to x. We say A is *strictly* more risk-averse than B if he is more risk-averse and there is some lottery that B prefers to an amount of money x but such that A prefers x to the lottery.

Clearly, the degree of risk aversion depends on the curvature of the utility function (by definition the *curvature* of $u(x)$ at x is $u''(x)$), but because $u(x)$ and $v(x) = au(x) + b$ $(a > 0)$ describe the same behavior, although $v(x)$ has curvature a times that of $u(x)$, we need something more sophis-

[7]One may ask why people play government-sponsored lotteries or spend money at gambling casinos if they are generally risk-averse. The most plausible explanation is that people enjoy the act of gambling. The same woman who will have insurance on her home and car, both of which presume risk aversion, will gamble small amounts of money for recreation. An excessive love for gambling, of course, leads an individual either to personal destruction or to wealth and fame (usually the former).

ticated. The obvious candidate is $\lambda_u(x) = -u''(x)/u'(x)$, which does not depend on scaling factors. This is called the *Arrow-Pratt coefficient of absolute risk aversion*, and it is exactly the measure that we need. We have the following theorem.

THEOREM 1.5 *An individual with utility function $u(x)$ is more risk-averse than an individual with utility function $v(x)$ if and only if $\lambda_u(x) > \lambda_v(x)$ for all x.*

For example, the logarithmic utility function $u(x) = \ln(x)$ has Arrow-Pratt measure $\lambda_u(x) = 1/x$, which decreases with x; i.e., as the individual becomes wealthier, he becomes less risk-averse. Studies show that this property, called *decreasing absolute risk aversion*, holds rather widely (Rosenzweig and Wolpin 1993; Saha, Shumway, and Talpaz 1994; Nerlove and Soedjiana 1996). Another increasing concave function is $u(x) = x^a$ for $a \in (0, 1)$, for which $\lambda_u(x) = (1-a)/x$, which also exhibits decreasing absolute risk aversion. Similarly, $u(x) = 1 - x^{-a}$ $(a > 0)$ is increasing and concave, with $\lambda_u(x) = -(a+1)/x$, which again exhibits decreasing absolute risk aversion. This utility has the additional attractive property that *utility is bounded:* no matter how rich you are, $u(x) < 1$.[8] Yet another candidate for a utility function is $u(x) = 1 - e^{-ax}$ for some $a > 0$. In this case $\lambda_u(x) = a$, which we call *constant absolute risk aversion.*

Another commonly used term is *coefficient of relative risk aversion*, $\mu_u(x) = \lambda_u(x)/x$. Note that for any of the utility functions $u(x) = \ln(x)$, $u(x) = x^a$ for $a \in (0, 1)$, and $u(x) = 1 - x^{-a}$ $(a > 0)$, $\mu_u(x)$ is constant, which we call *constant relative risk aversion*. For $u(x) = 1 - e^{-ax}$ $(a > 0)$, we have $\mu_u(x) = a/x$, so we have *decreasing relative risk aversion.*

1.9 Prospect Theory

A large body of experimental evidence indicates that people value payoffs according to whether they are *gains* or *losses* compared to their current status quo position. This is related to the notion that individuals adjust to an accustomed level of income, so that subjective well-being is associated more with *changes* in income rather than with the *level* of income. See, for instance, Helson (1964), Easterlin (1974, 1995), Lane (1991, 1993), and

[8] If utility is unbounded, it is easy to show that there is a lottery that you would be willing to give all your wealth to play no matter how rich you are. This is not plausible behavior.

Oswald (1997). Indeed, people appear to be about twice as averse to taking losses as to enjoying an equal level of gains (Kahneman, Knetsch, and Thaler 1990; Tversky and Kahneman 1981b). This means, for instance, that an individual may attach zero value to a lottery that offers an equal chance of winning \$1000 and losing \$500. This also implies that people are *risk-loving over losses* while they remain risk-averse over gains (§1.8 explains the concept of risk aversion). For instance, many individuals choose a 25% probability of losing \$2000 rather than a 50% chance of losing \$1000 (both have the same expected value, of course, but the former is riskier).

More formally, suppose an individual has utility function $v(x-r)$, where r is the status quo (his current position), and x represents a change from the status quo. *Prospect theory*, developed by Daniel Kahneman and Amos Tversky, asserts that (a) there is a "kink" in $v(x-r)$ such that the slope of $v(\cdot)$ is two to three times as great just to the left of $x=r$ as to the right; (b) that the curvature of $v(\cdot)$ is positive for positive values and negative for negative values; and (c) the curvature goes to zero for large positive and negative values. In other words, individuals are two to three times more sensitive to small losses than they are to small gains, they exhibit declining marginal utility over gains and declining absolute marginal utility over losses, and they are very insensitive to change when all alternatives involve either large gains or large losses. This utility function is exhibited in figure 1.3.

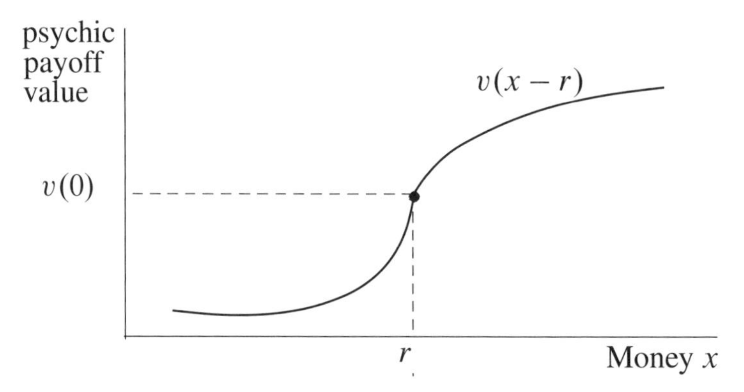

Figure 1.3. Loss aversion according to prospect theory

Experimental economists have long known that the degree of risk aversion exhibited in the laboratory over small gambles cannot be explained by standard expected utility theory, according to which risk aversion is measured by the curvature of the utility function (§1.8). The problem is that for small gambles the utility function should be almost flat. This issue has been formalized by Rabin (2000). Consider a lottery that imposes a \$100 loss and offers a \$125 gain with equal probability $p = 1/2$. Most subjects in the laboratory reject this lottery. Rabin shows that if this is true for all expected lifetime wealth levels less than \$300,000, then in order to induce a subject to sustain a loss of \$600 with probability 1/2, you would have to offer him a gain of at least \$36,000,000,000 with probability 1/2. This is, of course, quite absurd.

There are many regularities in empirical data on human behavior that fit prospect theory very well (Kahneman and Tversky 2000). For instance, returns on stocks in the United States have exceeded the returns on bonds by about 8 percentage points, averaged over the past 100 years. Assuming investors are capable of correctly estimating the shape of the return schedule, if this were due to risk aversion alone, then the average individual would be indifferent between a sure \$51,209 and a lottery that paid \$50,000 with probability 1/2 and paid \$100,000 with probability 1/2. It is, of course, quite implausible that more than a few individuals would be this risk-averse. However, a loss aversion coefficient (the ratio of the slope of the utility function over losses at the kink to the slope over gains) of 2.25 is sufficient to explain this phenomenon. This loss aversion coefficient is very plausible based on experiments.

In a similar vein, people tend to sell stocks when they are doing well but hold onto stocks when they are doing poorly. A kindred phenomenon holds for housing sales: homeowners are extremely averse to selling at a loss and sustain operating, tax, and mortgage costs for long periods of time in the hope of obtaining a favorable selling price.

One of the earliest examples of loss aversion is the *ratchet effect* discovered by James Duesenberry, who noticed that over the business cycle, when times are good, people spend all their additional income, but when times start to go bad, people incur debt rather than curb consumption. As a result, there is a tendency for the fraction of income saved to decline over time. For instance, in one study unionized teachers consumed more when next year's income was going to increase (through wage bargaining) but did not consume less when next year's income was going to decrease. We can explain this behavior with a simple loss aversion model. A teacher's utility can be written as $u(c_t - r_t) + s_t(1 + \rho)$, where c_t is consumption in period t, s_t is savings in period t, ρ is the rate of interest on savings, and r_t is the reference point (status quo point) in period t. This assumes that the marginal utility of savings is constant, which is a very good approximation. Now suppose the reference point changes as follows: $r_{t+1} = \alpha r_t + (1 - \alpha)c_t$, where $\alpha \in [0, 1]$ is an adjustment parameter ($\alpha = 1$ means no adjustment and $\alpha = 0$ means complete adjustment to last year's consumption). Note that when consumption in one period rises, the reference point in the next period rises, and conversely.

Now, dropping the time subscripts and assuming the individual has income M, so $c + s = M$, the individual chooses c to maximize

$$u(c - r) + (M - c)(1 + \rho).$$

This gives the first order condition $u'(c - r) = 1 + \rho$. Because this must hold for all r, we can differentiate totally with respect to r, getting

$$u''(c - r)\frac{dc}{dr} = u''(c - r).$$

This shows that $dc/dr = 1 > 0$, so when the individual's reference point rises, his consumption rises an equal amount.

One general implication of prospect theory is a *status quo bias*, according to which people often prefer the status quo over any of the alternatives but if one of the alternatives becomes the status quo, that too is preferred to any of the alternatives (Kahneman, Knetsch, and Thaler 1991). Status quo bias makes sense if we recognize that any change can involve a loss, and because on the average gains do not offset losses, it is possible that any one of a number of alternatives might be preferred if it is the status quo. For instance, if employers make joining a 401k savings plan the default position, almost all employees join. If *not* joining is made the default position, most employees do *not* join. Similarly, if the state automobile insurance commission declares one type of policy the default option and insurance companies ask individual policyholders how they would like to vary from the default, the policyholders tend not to vary, no matter what the default is (Camerer 2000).

Another implication of prospect theory is the *endowment effect* (Kahneman, Knetsch, and Thaler 1991), according to which people place a higher value on what they possess than they place on the same things when they do not possess them. For instance, if you win a bottle of wine that you could sell for \$200, you may drink it rather than sell it, but you would never think of buying a \$200 bottle of wine. A famous experimental result exhibiting the endowment effect was the "mug" experiment described by Kahneman, Knetsch and Thaler (1990). College student subjects given coffee mugs with the school logo on them demand a price two to three times as high to *sell* the mugs as those without mugs are willing to pay to *buy* the mugs. There is evidence that people underestimate the endowment effect and hence cannot appropriately *correct* for it in their choice behavior (Loewenstein and Adler 1995).

Yet another implication of prospect theory is the existence of a *framing effect*, whereby one form of a lottery is strictly preferred to another even though they have the same payoffs with the same probabilities (Tversky and Kahneman 1981a). For instance, people prefer a price of $10 plus a $1 discount to a price of $8 plus a $1 surcharge. Framing is, of course, closely associated with the endowment effect because framing usually involves privileging the initial state from which movements are assessed.

The framing effect can seriously distort effective decision making. In particular, when it is not clear what the appropriate reference point is, decision makers can exhibit serious inconsistencies in their choices. Kahneman and Tversky give a dramatic example from health care policy. Suppose we face a flu epidemic in which we expect 600 people to die if nothing is done. If program A is adopted, 200 people will be saved, while if program B is adopted, there is a 1/3 probability 600 will be saved and a 2/3 probability no one will be saved. In one experiment, 72% of a sample of respondents preferred A to B. Now suppose that if program C is adopted, 400 people will die, while if program D is adopted there is a 1/3 probability nobody will die and a 2/3 probability 600 people will die. Now, 78% of respondents preferred D to C, even though A and C are equivalent in terms of the probability of each final state, and B and D are similarly equivalent. However, in the choice between A and B, alternatives are over gains, whereas in the choice between C and D, the alternatives are over losses, and people are loss-averse. The inconsistency stems from the fact that there is no natural reference point for the decision maker, because the gains and losses are experienced by others, not by the decision maker himself.

The brilliant experiments by Kahneman, Tversky, and their coworkers clearly show that humans exhibit systematic biases in the way they make decisions. However, it should be clear that none of the above examples illustrates preference inconsistency once the appropriate parameter (current time, current position, status quo point) is admitted into the preference function. This point is formally demonstrated in Sugden (2003). Sugden considers a preference relation of the form $f \succeq g|h$, which means "lottery f is weakly preferred to lottery g when one's status quo position is lottery h." Sugden shows that if several conditions on this preference relation, most of which are direct generalizations of the Savage conditions (§1.5), obtain, then there is a utility function $u(x, z)$ such that $f \succeq g|h$ if and only if $\mathbf{E}[u(f, h)] \geq \mathbf{E}[u(g, h)]$, where the expectation is taken over the probability of events derived from the preference relation.

1.10 Heuristics and Biases in Decision Making

Laboratory testing of the standard economic model of choice under uncertainty was initiated by the psychologists Daniel Kahneman and Amos Tversky. In a famous article in the journal *Science*, Tversky and Kahneman (1974) summarized their early research as follows:

> How do people assess the probability of an uncertain event or the value of an uncertain quantity? ...people rely on a limited number of heuristic principles which reduce the complex tasks of assessing probabilities and predicting values to simpler judgmental operations. In general, these heuristics are quite useful, but sometimes they lead to severe and systematic errors.

Subsequent research has strongly supported this assessment (Kahneman, Slovic, and Tversky 1982; Shafir and Tversky 1992; Shafir and Tversky 1995). Although we still do not have adequate models of these heuristics, we can make certain generalizations.

First, in judging whether an event A or object A belongs to a class or process B, one heuristic that people use is to consider whether A is *representative* of B but consider no other relevant facts, such as the frequency of B. For instance, if informed that an individual has a good sense of humor and likes to entertain friends and family, and asked if the individual is a professional comic or a clerical worker, people are more likely to say the former. This is despite the fact that a randomly chosen person is much more likely to be a clerical worker than a professional comic, and many people have a good sense of humor, so there are many more clerical workers satisfying the description than professional comics.

A particularly pointed example of this heuristic is the famous Linda the Bank Teller problem (Tversky and Kahneman 1983). Subjects are given the following description of a hypothetical person named Linda:

> Linda is 31 years old, single, outspoken, and very bright. She majored in philosophy. As a student, she was deeply concerned with issues of discrimination and social justice and also participated in antinuclear demonstrations.

The subjects were then asked to rank-order eight statements about Linda according to their probabilities. The statements included the following two:

> Linda is a bank teller.
> Linda is a bank teller and is active in the feminist movement.

More than 80% of the subjects—graduate and medical school students with statistical training and doctoral students in the decision science program at Stanford University's business school—ranked the second statement as more probable than the first. This seems like a simple logical error because every bank teller feminist is also a bank teller. It appears, once again, that subjects measure probability by *representativeness* and ignore baseline frequencies.

However, there is another interpretation according to which the subjects are correct in their judgments. Let p and q be properties that every member of a population either has or does not have. The standard definition of "the probability that member x is p" is the fraction of the population for which p is true. But an equally reasonable definition is "the probability that x is a member of a random sample of the subset of the population for which p is true." According to the standard definition, the probability of p and q cannot be greater than the probability of p. But, according to the second, the opposite inequality can hold: x might be more likely to appear in a random sample of individuals who are both p and q than in a random sample of the same size of individuals who are p. In other words, the probability that a randomly chosen bank teller is Linda is probably much lower than the probability that a randomly chosen feminist bank teller is Linda. Another way of expressing this point is that the probability that a randomly chosen member of the set "is a feminist bank teller" may be linda is greater than the probability that a randomly chosen member of the set "is a bank teller," is Linda.

A second heuristic is that in assessing the frequency of an event, people take excessive account of information that is easily *available* or highly *salient*, even though a selective bias is obviously involved. For this reason, people tend to overestimate the probability of rare events because such events are highly newsworthy while nonoccurrences are not reported. Thus, people worry much more about dying in an accident while flying than they do while driving, even though air travel is much safer than automobile travel.

A third heuristic in problem solving is to start from an initial guess, chosen for its representativeness or salience, and adjust upward or downward toward a final figure. This is called *anchoring* because there is a tendency to underadjust, so the result is too close to the initial guess. Probably as a result of anchoring, people tend to overestimate the probability of conjunctions (p and q) and underestimate the probability of disjunctions (p or q).

For an instance of the former, a person who knows an event occurs with 95% probability may overestimate the probability that the event occurs 10 times in a row, suggesting a probability of 90%. The actual probability is about 60%. In this case the individual starts with 95% and does not adjust downward sufficiently. Similarly, if a daily event has a failure one time in a thousand, people will underestimate the probability that a failure occurs at least once in a year, suggesting a figure of 5%. The actual probability is 30.5%. Again, the individual starts with 0.1% and doesn't adjust upward enough.

A fourth heuristic is that people prefer objective probability distributions to subjective distributions derived from applying probabilistic principles, such as the principle of insufficient reason, which says that if you are completely ignorant as to which of several outcomes will occur, you should treat them as equally probable. For example, if you give a subject a prize for drawing a red ball from an urn containing red and white balls, the subject will pay to have the urn contain 50% red balls rather than contain an indeterminate percentage of red balls. This is the famous Ellsberg paradox, analyzed in §1.7.

Choice theorists often express dismay over the failure of people to apply the laws of probability and conform to normative decision theory. Yet, people may be applying rules that serve them well in daily life. It takes many years of study to feel at home with the laws of probability, the understanding of which is the product of the last couple of hundred years of scientific research. Moreover, it is costly, in terms of time and effort, to apply these laws even if we know them. Of course, if the stakes are high enough, it is worthwhile to make the effort or engage an expert who will do it for you. But generally, as Kahneman and Tversky suggest, we apply a set of heuristics that more or less get the job done. Among the most prominent heuristics is simply *imitation*: decide what class of phenomenon is involved, find out what people normally do in that situation, and do it. If there is some mechanism leading to the survival and growth of relatively successful behaviors, and if the problem in question recurs with sufficient regularity, the choice-theoretic solution will describe the winner of a dynamic social process of trial, error, and imitation.

Should we expect people to conform to the axioms of choice theory—transitivity, independence from irrelevant alternatives, the sure-thing principle, and the like? Where we know that individuals are really optimizing, and have expertise in decision theory, we doubtless should. But this applies only to a highly restricted range of actions. In more general settings we should not. We might have recourse to Darwinian analysis, demonstrating that under the appropriate conditions individuals who are genetically constituted to obey the axioms of choice theory are better fit to solve general decision-theoretic problems and hence will emerge triumphant through an evolutionary dynamic. But human beings did not evolve facing general decision-theoretic problems. Rather, they faced a few specific decision-theoretic problems associated with survival in small social groups. We may have to settle for modeling these specific choice contexts to discover how our genetic constitution and cultural tools interact in determining choice under uncertainty.

2

Game Theory: Basic Concepts

> High-rationality solution concepts in game theory can emerge in a world populated by low-rationality agents.
>
> Young (1998)

> The philosophers kick up the dust and then complain that they cannot see.
>
> Bishop Berkeley

2.1 The Extensive Form

An *extensive form game* $\mathcal{G}$ consists of a number of *players*, a *game tree*, and a set of *payoffs*. A game tree consists of a number of *nodes* connected by *branches*. Each branch connects a *head node* to a distinct *tail node*. If b is a branch of the game tree, we denote the head node of b by b^h, and the tail node of b by b^t.

A *path* from node a to node a' in the game tree is a connected sequence of branches starting at a and ending at a'.[1] If there is a path from node a to node a', we say a is an *ancestor* of a', and a' is a *successor* to a. We call the number of branches between a and a' the *length* of the path. If a path from a to a' has length 1, we call a the *parent* of a', and a' is a *child* of a.

We require that the game tree have a unique node r, called the *root node*, that has no parent, and a set T of nodes, called *terminal nodes* or *leaf nodes*, that have no children. We associate with each terminal node $t \in T$ ($\in$ means "is an element of"), and each player i, a *payoff* $\pi_i(t) \in \mathbf{R}$ ($\mathbf{R}$ is the set of real numbers). We say the game is *finite* if it has a finite number of nodes. We assume all games are finite unless otherwise stated.

We also require that the graph of $\mathcal{G}$ have the following *tree property*. There must be *exactly one path* from the root node to any given terminal

[1]Technically, a path is a sequence $b_1, \ldots, b_k$ of branches such that $b_1^h = a$, $b_i^t = b_{i+1}^h$ for $i = 1, \ldots, k-1$, and $b_k^t = a'$; i.e., the path starts at a, the tail of each branch is the head of the next branch, and the path ends at a'. The length of the path is k.

node in the game tree. Equivalently, *every node except the root node has exactly one parent.*

Players relate to the game tree as follows. Each nonterminal node is assigned to a player who moves at that node. Each branch b with head node b^h represents a particular *action* that the player assigned to that node can take there, and hence determines either a terminal node or the next point of play in the game—the particular child node b^t to be visited next.[2]

If a stochastic event occurs at a node a (for instance, the weather is good or bad, or your partner is nice or nasty), we assign the fictitious player Nature to that node, the actions Nature takes representing the possible outcomes of the stochastic event, and we attach a *probability* to each branch of which a is the head node, representing the probability that Nature chooses that branch (we assume all such probabilities are strictly positive).

The tree property thus means that there is a *unique* sequence of moves by the players (including Nature) leading from the root node to any specific node of the game tree, and for any two nodes there is *at most one* sequence of player moves leading from the first to the second.

A player may know the exact node in the game tree when it is his turn to move, or he may know only that he is at one of several possible nodes. We call such a collection of nodes an *information set*. For a set of nodes to form an information set, the same player must be assigned to move at each of the nodes in the set and have the same array of possible actions at each node.

We also require that if two nodes a and a' are in the same information set for a player, the moves that player made up to a and a' must be the same. This criterion is called *perfect recall*, because if a player never forgets his moves, he cannot make two different choices that subsequently land him in the same information set.[3]

A *strategy* s_i for player i is a choice of an action at every information set assigned to i. Suppose each player $i = 1, \ldots, n$ chooses strategy s_i. We call $s = (s_1, \ldots, s_n)$ a *strategy profile* for the game, and we define the *payoff* to player i, given strategy profile s, as follows. If there are no moves

[2]Thus, if $\mathbf{p} = (b_1, \ldots, b_k)$ is a path from a to a', then starting from a, if the actions associated with b_j are taken by the various players, the game moves to a'.

[3]Another way to describe perfect recall is to note that the information sets $\mathcal{N}_i$ for player i are the nodes of a graph in which the children of an information set $\nu \in \mathcal{N}_i$ are the $\nu' \in \mathcal{N}_i$ that can be reached by one move of player i, plus some combination of moves of the other players and Nature. Perfect recall means that this graph has the tree property.

by Nature, then s determines a unique path through the game tree and hence a unique terminal node $t \in T$. The payoff $\pi_i(s)$ to player i under strategy profile s is then defined to be simply $\pi_i(t)$.

Suppose there are moves by Nature, by which we mean that at one or more nodes in the game tree, there is a lottery over the various branches emanating from that node rather than a player choosing at that node. For every terminal node $t \in T$, there is a unique path $\mathbf{p}_t$ in the game tree from the root node to t. We say $\mathbf{p}_t$ is *compatible* with strategy profile s if, for every branch b on $\mathbf{p}_t$, if player i moves at b^h (the head node of b), then s_i chooses action b at b^h. If $\mathbf{p}_t$ is not compatible with s, we write $p(s,t) = 0$. If $\mathbf{p}_t$ is compatible with s, we define $p(s,t)$ to be the product of all the probabilities associated with the nodes of $\mathbf{p}_t$ at which Nature moves along $\mathbf{p}_t$, or 1 if Nature makes no moves along $\mathbf{p}_t$. We now define the payoff to player i as

$$\pi_i(s) = \sum_{t \in T} p(s,t)\pi_i(t). \tag{2.1}$$

Note that this is the expected payoff to player i given strategy profile s, assuming that Nature's choices are independent, so that $p(s,t)$ is just the probability that path $\mathbf{p}_t$ is followed, given strategy profile s. We generally assume in game theory that players attempt to maximize their expected payoffs, as defined in (2.1).

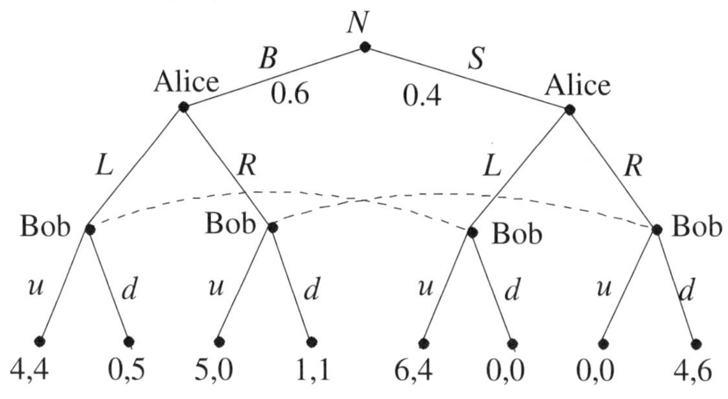

Figure 2.1. Evaluating payoffs when Nature moves

For example, consider the game depicted in figure 2.1. Here, Nature moves first, and with probability $p_l = 0.6$ chooses B where the game between Alice and Bob is known as the Prisoner's Dilemma (§2.10), and with probability $p_l = 0.4$ chooses S, where the game between Alice and Bob is known as the Battle of the Sexes (§2.8). Note that Alice knows Nature's move, because she has separate information sets on the two branches where Nature moves, but Bob does not, because when he moves, he does not know whether he is on the left- or right-hand branch.

Alice's strategies can be written LL, LR, RL, and RR, where LL means choose L whatever Nature chooses, RR means choose R whatever Nature chooses, LR means chose L when Nature chooses B, and choose R when Nature chooses S, and finally, RL means choose R when Nature chooses B and choose L when Nature chooses S. Similarly we can write Bob's choices as uu, ud, du, and dd, where uu means choose u whatever Alice chooses, dd means choose d whatever Alice chooses, ud means chose u when Alice chooses L, and choose d when Alice chooses R, and finally, and du means choose d when Alice chooses L and choose u when Alice chooses R.

Let us write, $\pi_A(x, y, z)$ and $\pi_B(x, y, z)$ for the payoffs to Alice and Bob, respectively, when Alice plays $x \in \{LL, LR, RL, RR\}$, Bob plays $y \in \{uu, ud, du, dd\}$ and Nature plays $z \in \{B, S\}$. Then, using the above parameter values, (2.1) gives the following equations.

$$\begin{aligned}
\pi_A(LL, uu) &= p_u \pi_A(LL, uu, B) + p_r \pi_A(LL, uu, S) \\
&= 0.6(4) + 0.4(6) = 4.8; \\
\pi_B(LL, uu) &= p_u \pi_B(LL, uu, B) + p_r \pi_B(LL, uu, S) \\
&= 0.6(4) + 0.4(4) = 4.0;
\end{aligned}$$

The reader should fill in the payoffs at the remaining nodes.

2.2 The Normal Form

The *strategic form* or *normal form* game consists of a number of players, a set of strategies for each of the players, and a payoff function that associates a payoff to each player with a choice of strategies by each player. More formally, an n-player normal form game consists of

a. A set of *players* $i = 1, \ldots, n$.
b. A set S_i of *strategies* for player $i = 1, \ldots, n$. We call $s = (s_1, \ldots, s_n)$, where $s_i \in S_i$ for $i = 1, \ldots, n$, a *strategy profile* for the game.[4]
c. A function $\pi_i : S \to \mathbf{R}$ for player $i = 1, \ldots, n$, where S is the set of strategy profiles, so $\pi_i(s)$ is player i's payoff when strategy profile s is chosen.

[4]Technically, these are *pure strategies* because in §2.3 we will consider *mixed strategies* that are probabilistic combinations of pure strategies.

Two extensive form games are said to be *equivalent* if they correspond to the same normal form game, except perhaps for the labeling of the actions and the naming of the players. But given an extensive form game, how exactly do we form the corresponding normal form game? First, the players in the normal form are the same as the players in the extensive form. Second, for each player i, let S_i be the set of strategies for that player, each strategy consisting of the choice of an action at each information set where i moves. Finally, the payoff functions are given by equation (2.1). If there are only two players and a finite number of strategies, we can write the payoff function in the form of a matrix.

As an exercise, you should work out the normal form matrix for the game depicted in figure 2.1.

2.3 Mixed Strategies

Suppose a player has pure strategies $s_1, \ldots, s_k$ in a normal form game. A *mixed strategy* for the player is a probability distribution over $s_1, \ldots, s_k$; i.e., a mixed strategy has the form

$$\sigma = p_1 s_1 + \cdots + p_k s_k,$$

where $p_1, \ldots, p_k$ are all nonnegative and $\sum_1^n p_j = 1$. By this we mean that the player chooses s_j with probability p_j, for $j = 1, \ldots, k$. We call p_j the *weight* of s_j in σ. If all the p_j's are zero except one, say $p_l = 1$, we say σ is a *pure strategy*, and we write $\sigma = s_l$. We say that pure strategy s_j is *used* in mixed strategy σ if $p_j > 0$. We say a strategy is *strictly mixed* if it is not pure, and we say that it is *completely mixed* if all pure strategies are used in it. We call the set of pure strategies used in a mixed strategy σ_i the *support* of σ_i.

In an n-player normal form game where player i has pure-strategy set S_i for $i = 1, \ldots, n$, a *mixed-strategy profile* $\sigma = (\sigma_1, \ldots, \sigma_n)$ is the choice of a mixed strategy σ_i by each player. We define the *payoffs* to σ as follows. Let $\pi_i(s_1, \ldots, s_n)$ be the payoff to player i when players use the pure strategy profile $(s_1, \ldots, s_n)$, and if s is a pure strategy for player i, let p_s be the weight of s in σ_i. Then we define

$$\pi_i(\sigma) = \sum_{s_1 \in S_1} \cdots \sum_{s_n \in S_n} p_{s_1} p_{s_2} \cdots p_{s_n} \pi_i(s_1, \ldots, s_n).$$

This is a formidable expression, but the idea behind it is simple. We assume the players' choices are made independently, so the probability that the particular pure strategies $s_1 \in S_1, \ldots, s_n \in S_n$ will be used is simply the product $p_{s_1} \cdots p_{s_n}$ of their weights, and the payoff to player i in this case is just $\pi_i(s_1, \ldots, s_n)$. We get the expected payoff by multiplying and adding up over all n-tuples of mixed strategies.

2.4 Nash Equilibrium

The concept of a Nash equilibrium of a game is formulated most easily in terms of the normal form. Suppose the game has n players, with strategy sets S_i and payoff functions $\pi_i : S \rightarrow \mathbf{R}$, for $i = 1, \ldots, n$, where S is the set of strategy profiles. We use the following very useful notation. Let ΔS_i be the set of mixed strategies for player i and let $\Delta^* S = \prod_{i=1}^n \Delta S_i$, be the mixed-strategy profiles for the game. If $\sigma \in \Delta^* S$, we write σ_i for the ith component of σ (i.e., σ_i is player i's mixed strategy in σ). If $\sigma \in \Delta^* S$ and $\tau_i \in \Delta S_i$, we write

$$(\sigma_{-i}, \tau_i) = (\tau_i, \sigma_{-i}) = \begin{cases} (\tau_1, \sigma_2, \ldots, \sigma_n) & \text{if } i = 1 \\ (\sigma_1, \ldots, \sigma_{i-1}, \tau_i, \sigma_{i+1}, \ldots, \sigma_n) & \text{if } 1 < i < n. \\ (\sigma_1, \ldots, \sigma_{n-1}, \tau_n) & \text{if } i = n \end{cases}$$

In other words, (σ_{-i}, τ_i) is the strategy profile obtained by replacing σ_i with τ_i for player i.

We say a strategy profile $\sigma^* = (\sigma_1^*, \ldots, \sigma_n^*) \in \Delta^* S$ is a *Nash equilibrium* if, for every player $i = 1, \ldots, n$ and every $\sigma_i \in \Delta S_i$, we have $\pi_i(\sigma^*) \geq \pi_i(\sigma_{-i}^*, \sigma_i)$; i.e., choosing σ_i^* is at least as good for player i as choosing any other σ_i given that the other players choose σ_{-i}^*. Note that in a Nash equilibrium, the strategy of each player is a *best response* to the strategies chosen by all the other players. Finally, notice that a player could have responses that are *equally good* as the one chosen in the Nash equilibrium—there just cannot be a strategy that is strictly better.

The Nash equilibrium concept is important because in many cases we can accurately (or reasonably accurately) predict how people will play a game by assuming they will choose strategies that implement a Nash equilibrium. In dynamic games that model an evolutionary process whereby successful strategies drive out unsuccessful ones over time, stable stationary states are always Nash equilibria. Conversely, Nash equilibria that seem implausible are often *unstable* equilibria of an evolutionary process, so we would not expect to see them in the real world (Gintis 2009). Where people appear to deviate systematically from implementing Nash equilibria, we sometimes find that they do not understand the game, or that we have misspecified the game they are playing or the payoffs we attribute to them. But, in important cases, as we shall see in later chapters, people simply do not play Nash equilibria at all.

2.5 The Fundamental Theorem of Game Theory

John Nash showed that every finite game has a Nash equilibrium in mixed strategies (Nash 1950). More concretely, we have the following theorem.

THEOREM 2.1 Nash Existence Theorem. If each player in an n-player game has a finite number of pure strategies, then the game has a (not necessarily unique) Nash equilibrium in (possibly) mixed strategies.

The following fundamental theorem of mixed-strategy equilibrium develops the principles for finding Nash equilibria. Let $\sigma = (\sigma_1, \ldots, \sigma_n)$ be a mixed-strategy profile for an n-player game. For any player $i = 1, \ldots, n$, let σ_{-i} represent the mixed strategies used by all the players other than player i. The *fundamental theorem of mixed-strategy Nash equilibrium* says that σ is a Nash equilibrium if and only if, for any player $i = 1, \ldots, n$ with pure-strategy set S_i,

a. If $s, s' \in S_i$ occur with positive probability in σ_i, then the payoffs to s and s' when played against σ_{-i}, are equal.
b. If s occurs with positive probability in σ_i and s' occurs with zero probability in σ_i, then the payoff to s' is less than or equal to the payoff to s, when played against σ_{-i}.

The proof of the fundamental theorem is straightforward. Suppose σ is the player's mixed strategy in a Nash equilibrium that uses s with probability $p > 0$ and s' with probability $p' > 0$. If s has a higher payoff than s' when played against σ_{-i}, then i's mixed strategy that uses s with probability $(p + p')$, does not use s', and assigns the same probabilities to the other pure strategies as does σ has a higher payoff than σ, so σ is not a best response to σ_{-i}. This is a contradiction, which proves the assertion. The rest of the proof is similar.

2.6 Solving for Mixed-Strategy Nash Equilibria

This problem asks you to apply the general method of finding mixed-strategy equilibria in normal form games. Consider the game at the right. First, of course, you should check for pure-strategy equilibria. To check for a completely mixed-strategy equilibrium, we use the fundamental theorem (§2.5). Suppose the column player uses the strategy $\sigma = \alpha L + (1-\alpha)R$ (i.e., plays L with probability α). Then, if the row player uses both U and D, they must both have the same payoff against σ. The payoff to U against σ is $\alpha a_1 + (1-\alpha)b_1$, and the payoff to D against σ is $\alpha c_1 + (1-\alpha)d_1$. Equating these two, we find

	L	R
U	a_1, a_2	b_1, b_2
D	c_1, c_2	d_1, d_2

$$\alpha = \frac{d_1 - b_1}{d_1 - b_1 + a_1 - c_1}.$$

For this to make sense, the denominator must be nonzero and the right-hand side must lie between zero and one. Note that the *column* player's strategy is determined by the requirement that the *row* player's two strategies be equal.

Now suppose the row player uses strategy $\tau = \beta U + (1-\beta)D$ (i.e., plays U with probability β). Then, if the column player uses both L and R, they must both have the same payoff against τ. The payoff to L against τ is $\beta a_2 + (1-\beta)c_2$, and the payoff to R against τ is $\beta b_2 + (1-\beta)d_2$. Equating these two, we find $\beta = (d_2 - c_2)/(d_2 - c_2 + a_2 - b_2)$.

Again, for this to make sense, the denominator must be nonzero and the right-hand side must lie between zero and one. Note that now the *row* player's strategy is determined by the requirement that the *column* player's two strategies be equal.

a. Suppose the above really is a mixed-strategy equilibrium. What are the payoffs to the two players?
b. Note that to solve a 2×2 game, we have checked for five different configurations of Nash equilibria—four pure and one mixed. But there are four more possible configurations, in which one player uses a pure strategy and the second player uses a mixed strategy. Show that if there is a Nash equilibrium in which the row player uses a pure strategy (say UU) and the column player uses a completely mixed strategy, then *any* strategy for the column player is a best response to UU.

c. How many different configurations are there to check for in a 2×3 game? In a 3×3 game? In an $n \times m$ game?

2.7 Throwing Fingers

Alice \ Bob	c_1	c_2
c_1	$1,-1$	$-1,1$
c_2	$-1,1$	$1,-1$

Alice and Bob each throws one (c_1) or two (c_2) fingers simultaneously. If they are the same, Alice wins; otherwise, Bob wins. The winner takes \$1 from the loser. The normal form of this game is depicted to the right. There are no pure-strategy equilibria, so suppose Bob uses the mixed strategy σ that consists of playing c_1 with probability α and c_2 with probability $1-\alpha$. We write this as $\sigma = \alpha c_1 + (1-\alpha)c_2$. If Alice uses both c_1 and c_2 with positive probability, they both must have the same payoff against σ, or else Alice should drop the lower-payoff strategy and use only the higher-payoff strategy. The payoff to c_1 against σ is $\alpha \cdot 1 + (1-\alpha)(-1) = 2\alpha - 1$, and the payoff to c_2 against σ is $\alpha(-1) + (1-\alpha)1 = 1-2\alpha$. If these are equal, then $\alpha = 1/2$. Similar reasoning shows that Alice chooses each strategy with probability 1/2. The expected payoff to Alice is then $2\alpha - 1 = 1 - 2\alpha = 0$, and the same is true for Bob.

2.8 The Battle of the Sexes

Alfredo \ Violetta	g	o
g	2,1	0,0
o	0,0	1,2

Violetta and Alfredo love each other so much that they would rather be together than apart. But Alfredo wants to go gambling, and Violetta wants to go to the opera. Their payoffs are described to the right. There are two pure-strategy equilibria and one mixed-strategy equilibrium for this game. We will show that Alfredo and Violetta would be better off if they stuck to either of their pure-strategy equilibria.

Let α be the probability of Alfredo going to the opera and let β be the probability of Violetta going to the opera. Because in a strictly mixed-strategy equilibrium the payoff from gambling and from going to the opera must be equal for Alfredo, we must have $\beta = 2(1-\beta)$, which implies $\beta = 2/3$. Because the payoff from gambling and from going to the opera must also be equal for Violetta, we must have $2\alpha = 1-\alpha$, so $\alpha = 1/3$.

The payoff of the game to each is then

$$\frac{2}{9}(1,2) + \frac{5}{9}(0,0) + \frac{2}{9}(2,1) = \left(\frac{2}{3},\frac{2}{3}\right)$$

because both go gambling $(1/3)(2/3) = 2/9$ of the time, both go to the opera $(1/3)(2/3) = 2/9$ of the time, and otherwise they miss each other.

Both players do better if they can coordinate because (2,1) and (1,2) are both better than (2/3,2/3).

2.9 The Hawk-Dove Game

	H	D
H	z, z	v, v
D	0,0	$v/2, v/2$

Consider a population of birds that fight over valuable territory. There are two possible strategies. The hawk (H) strategy is to escalate battle until injured or your opponent retreats. The dove (D) strategy is to display hostility but retreat before sustaining injury if your opponent escalates. The payoff matrix is given in the figure, where $v > 0$ is the value of territory, $w > v$ is the cost of injury, and $z = (v - w)/2$ is the payoff when two hawks meet. The birds can play mixed strategies, but they cannot condition their play on whether they are player 1 or player 2, and hence both players must use the same mixed strategy.

As an exercise, explain the entries in the payoff matrix and show that there are no symmetric pure-strategy Nash equilibria. The pure strategy pairs (H,D) and (D,H) are Nash equilibria, but they are not symmetric, so cannot be attained assuming, as we do, that the birds cannot which is player 1 and which is player 2. There is only one symmetric Nash equilibrium, in which players do not condition their behaviors on whether they are player 1 or player 2. This is the game's unique mixed-strategy Nash equilibrium, which we will now analyze.

Let α be the probability of playing hawk. The payoff to playing hawk is then $\pi_h = \alpha(v - w)/2 + (1 - \alpha)v$, and the payoff to playing dove is $\pi_d = \alpha(0) + (1 - \alpha)v/2$. These two are equal when $\alpha^* = v/w$, so the unique symmetric Nash equilibrium occurs when $\alpha = \alpha^*$. The payoff to each player is thus

$$\pi_d = (1 - \alpha)\frac{v}{2} = \frac{v}{2}\left(\frac{w - v}{w}\right).$$

Note that when w is close to v, almost all the value of the territory is dissipated in fighting, while for very high w, very little value is lost. This is known as "mutually assured destruction" in military parlance. Of course, if there is some possibility of error, where each player plays hawk by mistake with positive probability, then you can easily show that mutually assured destruction may have a very poor payoff.

2.10 The Prisoner's Dilemma

	C	D
C	R,R	S,T
D	T,S	P,P

Alice and Bob can each earn a profit R if they both cooperate (C). However, either can defect by working secretly on private jobs (D), earning $T > R$, but the other player will earn only $S < R$. If both defect, however, they will each earn P, where $S < P < R$. Each must decide independently of the other whether to choose C or D. The game tree is depicted in the figure on the right. The payoff T stands for "temptation" (to defect on a partner), S stands for "sucker" (for cooperating when your partner defected), P stands for "punishment" (for both defecting), and R stands for "reward" (for both cooperating). We usually assume also that $S + T < 2R$, so there is no gain from taking turns playing C and D.

Let α be the probability of playing C if you are Alice and let β be the probability of playing C if you are Bob. To simplify the algebra, we assume $P = 1$, $R = 0$, $T = 1 + t$, and $S = -s$, where $s, t > 0$. It is easy to see that these assumptions involve no loss of generality because adding a constant to all payoffs or multiplying all payoffs by a positive constant does not change the Nash equilibria of the game. The payoffs to Alice and Bob are now

$$\pi_A = \alpha\beta + \alpha(1-\beta)(-s) + (1-\alpha)\beta(1+t) + (1-\alpha)(1-\beta)(0),$$
$$\pi_B = \alpha\beta + \alpha(1-\beta)(1+t) + (1-\alpha)\beta(-s) + (1-\alpha)(1-\beta)(0),$$

which simplify to

$$\pi_A = \beta(1+t) - \alpha(s(1-\beta) + \beta t),$$
$$\pi_B = \alpha(1+t) - \beta(s(1-\alpha) + \alpha t).$$

It is clear from these equations that π_A is maximized by choosing $\alpha = 0$ no matter what Bob does, and similarly π_B is maximized by choosing $\beta = 0$, no matter what Alice does. This is the mutual defect equilibrium.

As we shall see in §3.5, this is not the way many people play this game in the experimental laboratory. Rather, people very often prefer to cooperate, provided their partners cooperate as well. We can capture this phenomenon by assuming that there is a psychic gain $\lambda_A > 0$ for Alice and $\lambda_B > 0$ for Bob when both players cooperate, above the temptation payoff $T = 1 + t$. If we rewrite the payoffs using this assumption, we get

$$\begin{aligned}
\pi_A &= \alpha\beta(1 + t + \lambda_A) + \alpha(1 - \beta)(-s) \\
&\quad + (1 - \alpha)\beta(1 + t) + (1 - \alpha)(1 - \beta)(0) \\
\pi_B &= \alpha\beta(1 + t + \lambda_B) + \alpha(1 - \beta)(1 + t) \\
&\quad + (1 - \alpha)\beta(-s) + (1 - \alpha)(1 - \beta)(0),
\end{aligned}$$

which simplify to

$$\begin{aligned}
\pi_A &= \beta(1 + t) - \alpha(s - \beta(s + \lambda_A)) \\
\pi_B &= \alpha(1 + t) - \beta(s - \alpha(s + \lambda_B)).
\end{aligned}$$

The first equation shows that if $\beta > s/(s + \lambda_A)$, then Alice plays C, and if $\alpha > s/(s + \lambda_B)$, then Bob plays C. If the opposite equalities hold, then both play D.

2.11 Alice, Bob, and the Choreographer

Consider the game played by Alice and Bob, with the normal form matrix shown to the right. There are two Pareto-efficient pure-strategy equilibria: (1,5) and (5,1). There is also a mixed-strategy equilibrium with payoffs (2.5,2.5), in which Alice plays u with probability 0.5 and Bob plays l with probability 0.5.

Alice \ Bob	l	r
u	5,1	0,0
d	4,4	1,5

If the players can jointly observe a choreographer who signals ul and dr, each with probability 1/2, Alice and Bob can then achieve the payoff (3,3) by obeying the choreographer; i.e. by playing (u, l) if they see ul and playing (d, r) if they see dr. Note that this is a Nash equilibrium of a larger game in which the choreographer moves first and acts as Nature. This is a Nash equilibrium because each players chooses a best response to the move of the other, assuming the other carries out the choreographer's directive. This situation is termed a *correlated equilibrium* of the original game (Aumann 1974). The commonly observable event on which their behavior is conditioned is called a *correlating device*.

A more general correlated equilibrium for this game can be constructed as follows. Consider a choreographer who would like to direct Alice to play d and Bob to play l so the joint payoff $(4,4)$ could be realized. The problem is that if Alice obeys the choreographer, then Bob has an incentive to choose r, giving him a payoff of 5 instead of 4. Similarly, if Bob obeys the choreographer, then Alice has an incentive to choose u, giving her a payoff of 5 instead of 4. The choreographer must therefore be more sophisticated.

Suppose the choreographer has three states. In ω_1, which occurs with probability α_1, he advises Alice to play u and Bob to play l. In ω_2, which occurs with probability α_2, the choreographer advises Alice to play d and Bob to play l. In ω_3, which occurs with probability α_3, the choreographer advises Alice to play d and Bob to play r. We assume Alice and Bob know α_1, α_2, and $\alpha_3 = 1 - \alpha_1 - \alpha_2$, and it is common knowledge that both have a *normative predisposition* (see chapter 7) to obey the choreographer unless they can do better by deviating. However, neither Alice nor Bob can observe the state ω of the choreographer, and each hears only what the choreographer tells them, not what the choreographer tells the other player. We will find the values of α_1, α_2, and α_3 for which the resulting game has a Pareto-efficient correlated equilibrium.

Note that Alice has knowledge partition $[\{\omega_1\},\{\omega_2,\omega_3\}]$ (§1.5), meaning that she knows when ω_1 occurs but cannot tell whether the state is ω_2 or ω_3. This is because she is told to move u only in state ω_1 but to move d in both states ω_2 and ω_3. The conditional probability of ω_2 for Alice given $\{\omega_2,\omega_3\}$ is $p_A(\omega_2) = \alpha_2/(\alpha_2+\alpha_3)$, and similarly $p_A(\omega_3) = \alpha_3/(\alpha_2+\alpha_3)$. Note also that Bob has knowledge partition $[\{\omega_3\},\{\omega_1,\omega_2\}]$ because he is told to move r only at ω_3 but to move l at both ω_1 and ω_2. The conditional probability of ω_1 for Bob given $\{\omega_1,\omega_2\}$ is $p_B(\omega_1) = \alpha_1/(\alpha_1+\alpha_2)$, and similarly $p_B(\omega_2) = \alpha_2/(\alpha_1+\alpha_3)$.

When ω_1 occurs, Alice knows that Bob plays l, to which Alice's best response is u. When ω_2 or ω_3 occurs, Alice knows that Bob is told l by the choreographer with probability $p_A(\omega_2)$ and is told r with probability $p_A(\omega_3)$. Thus, despite the fact that Bob plays only pure strategies, Alice knows she effectively faces the mixed strategy l played with probability $\alpha_2/(\alpha_2+\alpha_3)$ and r played with probability $\alpha_3/(\alpha_2+\alpha_3)$. The payoff to u in this case is $5\alpha_2/(\alpha_2+\alpha_3)$, and the payoff to d is $4\alpha_2/(\alpha_2+\alpha_3)+\alpha_3/(\alpha_2+\alpha_3)$. If d is to be a best response, we must thus have $\alpha_1 + 2\alpha_2 \leq 1$.

Turning to the conditions for Bob, when ω_3 occurs, Alice plays d so Bob's best response is r. When ω_1 or ω_2 occurs, Alice plays u with probability $p_B(\omega_1)$ and d with probability $p_B(\omega_2)$. Bob chooses l when $\alpha_1 + 4\alpha_2 \geq 5\alpha_2$. Thus, any α_1 and α_2 that satisfy $1 \geq \alpha_1 + 2\alpha_2$ and $\alpha_1 \geq \alpha_2$ permit a correlated equilibrium. Another characterization is $1 - 2\alpha_2 \geq \alpha_1 \geq \alpha_2 \geq 0$.

What are the Pareto-optimal choices of α_1 and α_2? Because the correlated equilibrium involves $\omega_1 \rightarrow (u, l)$, $\omega_2 \rightarrow (d, l)$, and $\omega_3 \rightarrow (d, r)$, the payoffs to (α_1, α_2) are $\alpha_1(5, 1) + \alpha_2(4, 4) + (1 - \alpha_1 - \alpha_2)(1, 5)$, which simplifies to $(1 + 4\alpha_1 + 3\alpha_2, 5 - 4\alpha_1 - \alpha_2)$, where $1 - 2\alpha_2 \geq \alpha_1 \geq \alpha_2 \geq 0$. This is a linear programming problem. It is easy to see that either $\alpha_1 = 1 - 2\alpha_2$ or $\alpha_1 = \alpha_2$ and $0 \leq \alpha_2 \leq 1/3$. The solution is shown in figure 2.2.

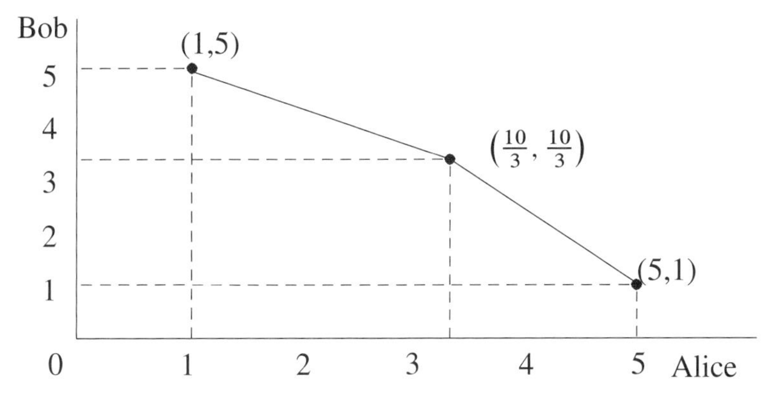

Figure 2.2. Alice, Bob, and the choreographer

The pair of straight lines connecting (1,5) to (10/3,10/3) to (5,1) is the set of Pareto-optimal points. Note that the symmetric point (10/3,10/3) corresponds to $\alpha_1 = \alpha_2 = \alpha_3 = 1/3$.

2.12 An Efficiency-Enhancing Choreographer

Consider an n-player game in which each player can choose an integer in the range $k = 1, \ldots, 10$. Nature chooses an integer k in this range, and if all n players also choose k, each has payoff 1. Otherwise, each has payoff 0. Nature also supplies to any agent who inquires (one sample per agent) a noisy signal that equals the true value with probability $p > 0.10$. A best response for each player is to sample the signal and choose a number equal to the signal received. The payoff is p^n. For a correlated equilibrium, suppose there is a social rule that obligates the youngest player to reveal his choice. There is then a correlated equilibrium in which each player follows the youngest player's choice, and the payoff to each player is now p. For instance, when $p = 90\%$ and $n = 25$, the Nash equilibrium payoff is 0.071, which is only 8% of the value of the correlated equilibrium payoff.

This example shows that there may be huge gains to groups that manage to find an appropriate correlating device.

2.13 The Correlated Equilibrium Solution Concept

The correlated equilibrium concept will be studied further in chapter 7. This solution concept has been neglected in classical game theory, although we will show that it is a more natural solution concept than the Nash equilibrium. This is because the correlated equilibrium directly addresses the central weaknesses of the Nash equilibrium concept: its lack of a mechanism for choosing among various equally plausible alternatives, for coordinating the behaviors of players who are indifferent among several pure strategies, and for providing incentives for players to follow the suggested strategy even when they may have private payoffs that would lead self-regarding agents to do otherwise (§6.3, 6.4).

Game theorists have not embraced the correlated equilibrium concept because it appears to require an active social agency, in the form of the choreographer, that cannot be accounted for within game theory. We will argue that therein lies the true power of the correlated equilibrium concept: it points away from game theory to a larger, complementary, social epistemology, that will be explored in chapters 7 and 12.

3

Game Theory and Human Behavior

God is crafty, but He is not malicious.
Albert Einstein

My motive for doing what I am going to do is simply personal revenge. I do not expect to accomplish anything by it.
Theodore Kaczynski (the Unabomber)

Game theory is multiplayer decision theory where the choices of each player affect the payoffs to other players, and the players take this into account in their choice behavior. In this chapter we address the contribution of game theory to the design of experiments aimed at understanding the behavior of individuals engaged in strategic interaction. We call this *behavioral game theory.*

Game theory is a general lexicon that applies to all life forms. Strategic interaction neatly separates living from nonliving entities and defines life itself. Strategic interaction is the sole concept commonly used in the analysis of living systems that has no counterpart in physics or chemistry.

Game theory provides the conceptual and procedural tools for studying social interaction, including the characteristics of the players, the rules of the game, the informational structure, and the payoffs associated with particular strategic interactions. The various behavioral disciplines (economics, psychology, sociology, politics, anthropology, and biology) are currently based on distinct principles and rely on distinct types of data. However, game theory fosters a unified analytical framework available to *all* the behavioral disciplines. This facilitates cross-disciplinary information exchange that may eventually culminate in a degree of unity within the behavioral sciences now enjoyed only by the natural sciences (see chapter 12). Moreover, because behavioral game-theoretic predictions can be systematically tested, the results can be replicated by different laboratories (Plott 1979; Smith 1982; Sally 1995). This turns social science into true science.

Behavioral game theory presumes the BPC model, as developed in §1.1. Experiments subject individuals to a variety of game settings, including diverse payoffs, informational conditions, and constraints on action, and deduce their underlying preferences from their behavior. This would be impossible if the individuals were not maximizing consistent preferences. Chapter 1 showed that the deviations that human subjects exhibit from the prescriptions of normative decision theory, while important, are compatible with preference consistency plus performance error.

3.1 Self- and Other-Regarding Preferences

This chapter deals with the interplay of self-regarding and other-regarding behavior. By a *self-regarding* actor we mean a player i in a game $\mathcal{G}$ who maximizes his own payoff π_i as defined in §2.1. A self-regarding actor thus cares about the behavior of and payoffs to the other players only insofar as these impact his own payoff π_i. The term "self-regarding" is more accurate than "self-interested" because an other-regarding individual is still acting to maximize utility and so can be described as self-interested. For instance, if I get great pleasure from your consumption, my gift to you may be self-interested, even though it is surely other-regarding. We can avoid confusion (and much pseudophilosophical discussion) by employing the self-regarding/other-regarding terminology.

One major result of behavioral game theory is that *when modeling market processes with well-specified contracts, such as double auctions (supply and demand) and oligopoly, game-theoretic predictions assuming self-regarding actors are accurate under a wide variety of social settings* (Kachelmaier and Shehata 1992; Davis and Holt 1993). In such market settings behavioral game theory sheds much new light, particularly in dealing with price dynamics and their relationship to buyer and seller expectations (Smith and Williams 1992).

The fact that self-regarding behavior explains market dynamics lends credence to the practice in neoclassical economics of assuming that individuals are self-regarding. However, it by no means justifies "Homo economicus" because many economic transactions do *not* involve anonymous exchange. This includes employer-employee, creditor-debtor, and firm-client relationships. Nor does this result apply to the welfare implications of economic outcomes (e.g., people may care about the overall degree of economic inequality and/or their positions in the income and wealth distribution), to modeling the behavior of taxpayers (e.g., they may be more or less honest than a self-regarding individual, and they may prefer to transfer resources toward or away from other individuals even at an expense to themselves) or to important aspects of economic policy (e.g., dealing with corruption, fraud, and other breaches of fiduciary responsibility).

A second major result is that *when contracts are incomplete and individuals can engage in strategic interaction, with the power to reward and punish the behavior of other individuals, game-theoretic predictions based on the self-regarding actor model generally fail.* In such situations, the *character virtues* (including, honesty, promise keeping, trustworthiness, and decency), as well as both *altruistic cooperation* (helping others at a cost to oneself) and *altruistic punishment* (hurting others at a cost to oneself) are often observed. These behaviors are particularly common in a *social dilemma*, which is an n-player Prisoner's Dilemma—a situation in which all gain when all cooperate but each has a personal incentive to defect, gaining at the expense of the others (see, for instance, §3.9).

Other-regarding preferences were virtually ignored until recently in both economics and biology, although they are standard fare in anthropology, sociology, and social psychology. In economics, the notion that enlightened self-interest allows individuals to cooperate in large groups goes back to Bernard Mandeville's "private vices, public virtues" (1924 [1705]) and Adam Smith's "invisible hand" (2000 [1759]). The great Francis Ysidro Edgeworth considered self-interest "the first principle of pure economics" (Edgeworth 1925, p. 173). In biology, the selfishness principle has been touted as a central implication of rigorous evolutionary modeling. In *The Selfish Gene* (1976), for instance, Richard Dawkins asserts "We are survival machines—robot vehicles blindly programmed to preserve the selfish molecules known as genes. ...Let us try to teach generosity and altruism, because we are born selfish." Similarly, in *The Biology of Moral Systems* (1987, p. 3), R. D. Alexander asserts that "ethics, morality, human conduct, and the human psyche are to be understood only if societies are seen as collections of individuals seeking their own self-interest." More poetically, Michael Ghiselin (1974) writes: "No hint of genuine charity ameliorates our vision of society, once sentimentalism has been laid aside. What passes for cooperation turns out to be a mixture of opportunism and exploitation. ...Scratch an altruist, and watch a hypocrite bleed."

The Darwinian struggle for existence may explain why the concept of virtue does not add to our understanding of animal behavior in general, but by all available evidence, it is a central aspect of human behavior. The reasons for this are the subject of some speculation (Gintis 2003a,2006b), but they come down to the plausible insight that human social life is so complex, and the rewards for prosocial behavior so distant and indistinct, that adherence to general rules of propriety, including the strict control of such deadly sins as anger, avarice, gluttony, and lust, is individually fitness-enhancing (Simon 1990; Gintis 2003a).

One salient behavior in social dilemmas revealed by behavioral game theory is *strong reciprocity*. Strong reciprocators come to a social dilemma with a propensity to cooperate (*altruistic cooperation*), respond to cooperative behavior by maintaining or increasing their level of cooperation, and respond to noncooperative behavior by punishing the "offenders," even at a cost to themselves and even when they cannot reasonably expect future personal gains to flow therefrom (*altruistic punishment*). When other forms of punishment are not available, the strong reciprocator responds to defection with defection.

The strong reciprocator is thus neither the selfless altruist of utopian theory, nor the self-regarding individual of traditional economics. Rather, he is a conditional cooperator whose penchant for reciprocity can be elicited under circumstances in which self-regard would dictate otherwise. The positive aspect of strong reciprocity is commonly known as gift exchange, in which one individual behaves more kindly than required toward another with the hope and expectation that the other will treat him kindly as well (Akerlof 1982). For instance, in a laboratory-simulated work situation in which employers can pay higher than market-clearing wages in hopes that workers will reciprocate by supplying a high level of effort (§3.7), the generosity of employers was generally amply rewarded by their workers.

A second salient behavior in social dilemmas revealed by behavioral game theory is *inequality aversion*. The inequality-averse individual is willing to reduce his own payoff to increase the degree of equality in the group (whence widespread support for charity and social welfare programs). But he is especially displeased when placed on the *losing side* of an unequal relationship. The inequality-averse individual is willing to reduce his own payoff if that reduces the payoff of relatively favored individuals even more. In short, an inequality-averse individual generally exhibits a *weak* urge to reduce inequality when he is the beneficiary and a *strong* urge to reduce inequality when he is the victim (Loewenstein, Thompson, and Bazerman 1989). Inequality aversion differs from strong reciprocity in that the inequality-averse individual cares only about the distribution of final payoffs and not at all about the role of other players in bringing about this distribution. The strong reciprocator, by contrast, does not begrudge others their payoffs but is sensitive to how fairly he is treated by others.

Self-regarding agents are in common parlance called *sociopaths*. A sociopath (e.g., a sexual predator, a recreational cannibal, or a professional killer) treats others instrumentally, caring only about what he derives from an interaction, whatever the cost to the other party. In fact, for most people, interpersonal relations are guided as much by empathy (and hostility) as by self-regard. The principle of *sympathy* is the guiding theme of Adam Smith's great book, *The Theory of Moral Sentiments*, despite the fact that his self-regarding principle of the "invisible hand" is one of the central insights of economic theory.

We conclude from behavioral game theory that one must treat individuals' objectives as a matter of *fact*, not *logic*. We can just as well build models of honesty, promise keeping, regret, strong reciprocity, vindictiveness, status seeking, shame, guilt, and addiction as of choosing a bundle of consumption goods subject to a budget constraint (§12.8), (Gintis 1972a,b, 1974, 1975; Becker and Murphy 1988; Bowles and Gintis 1993; Becker 1996; Becker and Mulligan 1997). ,

3.2 Methodological Issues in Behavioral Game Theory

Vernon Smith, who was awarded the Nobel prize in 2002, began running laboratory experiments of market exchange in 1956 at Purdue and Stanford universities. Until the 1980s, aside from Smith, whose results supported the traditional theory of market exchange, virtually the only behavioral discipline to use laboratory experiments with humans as a basis for modeling human behavior was social psychology. Despite the many insights afforded by experimental social psychology, its experimental design was weak. For instance, the BPC model was virtually ignored and game theory was rarely used, so observed behavior could not be analytically modeled, and experiments rarely used incentive mechanisms (such as monetary rewards and penalties) designed to reveal the real, underlying preferences of subjects. As a result, social psychological findings that were at variance with the assumptions of other behavioral sciences were widely ignored.

The results of the *Ultimatum Game* (Güth, Schmittberger, and Schwarze 1982) changed all that (§3.6), showing that in one-shot games that preserved the anonymity of subjects, people were quite willing to reject monetary rewards that they considered unfair. This, and a barrage of succeeding experiments, some of which are analyzed below, did directly challenge the widely used assumption that individuals are self-regarding. Not surprisingly, the first reaction within the disciplines was to criticize the experiments rather than to question their theoretical preconceptions. This is a valuable reaction to new data, so we shall outline the various objections to these findings.

First, the behavior of subjects in simple games under controlled circumstances may bear no implications for their behavior in the complex, rich, temporally extended social relationships into which people enter in daily life. We discuss the *external validity* of laboratory experiments in §3.15.

Second, games in the laboratory are unusual, so people do not know how best to behave in these games. They therefore simply play as they would in daily life, in which interactions are repeated rather than one-shot, and take place among acquaintances rather than being anonymous. For instance, critics suggest that strong reciprocity is just a confused carryover into the laboratory of the subject's extensive experience with the value of building a reputation for honesty and willingness to punish defectors, both of which benefit the self-regarding actor. However, when opportunities for reputation building are incorporated into a game, subjects make predictable strategic adjustments compared to a series of one-shot games without reputation building, indicating that subjects are capable of distinguishing between the two settings (Fehr and Gächter 2000). Postgame interviews indicate that subjects clearly comprehend the one-shot aspect of the games.

Moreover, one-shot, anonymous interactions are not rare. We face them frequently in daily life. Members of advanced market societies are engaged in one-shot games with very high frequency—virtually every interaction we have with strangers is of this form. Major rare events in people's lives (fending off an attacker, battling hand to hand in wartime, experiencing a natural disaster or major illness) are one-shots in which people appear to exhibit strong reciprocity much as in the laboratory. While members of the small-scale societies we describe below may have fewer interactions with strangers, they are no less subject to one-shots for the other reasons mentioned. Indeed, in these societies, greater exposure to market exchange led to stronger, not weaker, deviations from self-regarding behavior (Henrich et al. 2004).

Another indication that the other-regarding behavior observed in the laboratory is not simply confusion on the part of the subjects is that when experimenters point out that subjects could have earned more money by behaving differently, the subjects generally respond that of course they knew that but preferred to behave in an ethically or emotionally satisfying manner rather than simply maximize their material gain. This, by the way, contrasts sharply with the experiments in behavioral decision theory described in chapter 1 where subjects generally admitted their errors.

Recent neuroscientific evidence supports the notion that subjects punish those who are unfair to them simply because this gives them pleasure. de-Quervain et al. (2004) used positron emission tomography to examine the neural basis for the altruistic punishment of defectors in an economic exchange. The experimenters scanned the subjects' brains while they learned about the defector's abuse of trust and determined the punishment. Punishment activated the *dorsal striatum*, which has been implicated in the processing of rewards that accrue as a result of goal-directed actions. Moreover, subjects with stronger activations in the dorsal striatum were willing to incur greater costs in order to punish. This finding supports the hypothesis that people derive satisfaction from punishing norm violations and that the activation in the dorsal striatum reflects the anticipated satisfaction from punishing defectors.

Third, it may be that subjects really do not believe that conditions of anonymity will be respected, and they behave altruistically because they fear their selfish behavior will be revealed to others. There are several problems with this argument. First, one of the strict rules of behavioral game research is that *subjects are never told untruths or otherwise misled*, and they are generally informed of this fact by experimenters. Thus, revealing the identity of participants would be a violation of scientific integrity. Second, there are generally no penalties that could be attached to being discovered behaving in a selfish manner. Third, an exaggerated fear of being discovered cheating is *itself* a part of the strong reciprocity syndrome—it is a psychological characteristic that induces us to behave prosocially even when we are most attentive to our selfish needs. For instance, subjects might feel embarrassed and humiliated were their behavior revealed, but shame and embarrassment are themselves *other-regarding emotions* that contribute to prosocial behavior in humans (Bowles and Gintis 2004; Carpenter et al. 2009). In short, the tendency of subjects to overestimate the probability of detection and the costs of being detected are prosocial mental processes (H. L. Mencken once defined "conscience" as "the little voice that warns us that someone may be looking"). Fourth, and perhaps most telling, in tightly controlled experiments designed to test the hypothesis that subject-experimenter anonymity is important in fostering altruistic behavior, it is found that subjects behave similarly regardless of the experimenter's knowledge of their behavior (Bolton and Zwick 1995; Bolton, Katok, and Zwick 1998).

A final argument is that while a game may be one-shot and the players may be anonymous to one another, they nonetheless *remember* how they played a game, and they may derive great pleasure from recalling their generosity or their willingness to incur the costs of punishing others for being selfish. This is quite correct and probably explains a good deal of non-self-regarding behavior in experimental games.[1] But this does not contradict the fact that our behavior is other-regarding! Rather, it affirms that it may be in one's personal interest to engage in other-regarding acts. Only for sociopaths are the set of self-regarding acts and the set of self-interested acts the same.

In all the games described below, unless otherwise stated, subjects were college students who were anonymous to one another, were paid real money, were not deceived or misled by the experimenters, and they were instructed to the point where they fully understood the rules and the payoffs before playing for real.

3.3 An Anonymous Market Exchange

By *neoclassical economics* I mean the standard fare of microeconomics courses, including the Walrasian general equilibrium model, as developed by Kenneth Arrow, Gérard Debreu, Frank Hahn, Tjalling Koopmans, and others (Arrow 1951; Arrow and Hahn 1971; Koopmans 1957). Neoclassical economic theory holds that in a market for a product, the equilibrium price is at the intersection of the supply and demand curves for the good. It is easy to see that at any other point a self-regarding seller could gain by asking a higher price, or a self-regarding buyer could gain by offering a lower price. This situation was among the first to be simulated experimentally, the neoclassical prediction virtually always receiving strong support (Holt 1995). Here is a particularly dramatic example, provided by Holt, Langan, and Villamil (1986) (reported by Charles Holt in Kagel and Roth, 1995).

[1]William Shakespeare understands this well when he has Henry V use the following words to urge his soldiers to fight for victory against a much larger French army: "Whoever lives past today ... will rouse himself every year on this day, show his neighbor his scars, and tell embellished stories of all their great feats of battle. These stories he will teach his son and from this day until the end of the world we shall be remembered."

In the Holt-Langan-Villamil experiment there are four "buyers" and four "sellers." The good is a chip that the seller can redeem for \$5.70 (unless it is sold) but a buyer can redeem for \$6.80 at the end of the game. In analyzing the game, we assume throughout that buyers and sellers are self-regarding. In each of the first five rounds, each buyer was informed, privately, that he could redeem up to 4 chips, while 11 chips were distributed to sellers (three sellers were given 3 chips each, and the fourth was given 2 chips). Each player knew only the number of chips in his possession, the number he could redeem, and their redemption value, and did not know the value of the chips to others or how many they possessed or were permitted to redeem. Buyers should be willing to pay up to \$6.80 per chip for up to 4 chips each, and sellers should be willing to sell a chip for any amount at or above \$5.70. Total demand is thus 16 for all prices at or below \$6.80, and total supply is 11 chips at or above \$5.70. Because there is an excess demand for chips at every price between \$5.70 and \$6.80, the only point of intersection of the demand and supply curves is at the price $p = \$6.80$. The subjects in the game, however, have absolutely no knowledge of aggregate demand and supply because each knows only his own supply of or demand for chips.

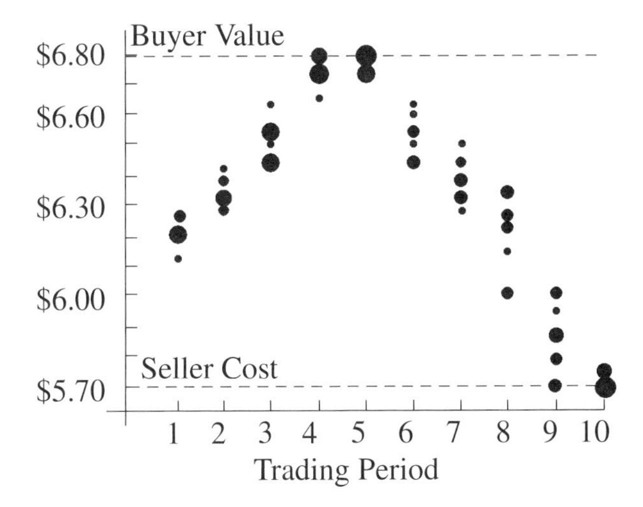

Figure 3.1. The double auction. The size of the circle is proportional to the number of trades that occurred at the stated price.

The rules of the game are that at any time a seller can call out an asking price for a chip, and a buyer can call out an offer price for a chip. This price remains "on the table" until it is accepted by another player, or a lower asking price is called out, or a higher offer price is called out. When a deal is made, the result is recorded and that chip is removed from the game. As seen in figure 3.1, in the first period of play, actual prices were about midway between \$5.70 and \$6.80. Over the succeeding four rounds the average price increased until in period 5 prices were very close to the equilibrium price predicted by neoclassical theory.

In period 6 and each of the succeeding four periods, buyers were given the right to redeem a total of 11 chips, and each seller was given 4 chips. In this new situation, it is clear (to observers who know these facts, though not the subjects in the experiment) that there is now an *excess supply* of chips at each price between \$5.70 and \$6.80, so supply and demand intersect precisely at \$5.70. While sellers, who previously made a profit of about \$1.10 per chip in each period, must have been delighted with their additional supplies of chips, succeeding periods witnessed a steady fall in price until in the tenth period the price is close to the neoclassical prediction, and now buyers are earning about \$1.10 per chip. We see that even when agents are completely ignorant of macroeconomics conditions of supply and demand, they can move quickly to a market-clearing equilibrium under the appropriate conditions.

3.4 The Rationality of Altruistic Giving

There is nothing irrational about caring for others. But do preferences for altruistic acts entail transitive preferences as required by the notion of rationality in decision theory? Andreoni and Miller (2002) showed that in the case of the Dictator Game, they do. Moreover, there are no known counterexamples.

In the Dictator Game, first studied by Forsythe et al. (1994), the experimenter gives a subject, called the Dictator, a certain amount of money and instructs him to give any portion of it he desires to a second, anonymous, subject, called the Receiver. The Dictator keeps whatever he does not choose to give to the Receiver. Obviously, a self-regarding Dictator will give nothing to the Receiver. Suppose the experimenter gives the Dictator m points (exchangeable at the end of the session for real money) and tells him that the price of giving some of these points to the Receiver is p, meaning that each point the Receiver gets costs the giver p points. For instance, if $p=4$, then it costs the Dictator 4 points for each point that he transfers to the Receiver. The Dictator's choices must then satisfy the budget constraint $\pi_s + p\pi_o = m$, where π_s is the amount the Dictator keeps and π_o is the amount the Receiver gets. The question, then, is simply, is there a preference function $u(\pi_s, \pi_o)$ that the Dictator maximizes subject to the budget constraint $\pi_s + p\pi_o = m$? If so, then it is just as rational, from a behavioral standpoint, to care about giving to the Receiver as to care about consuming marketed commodities.

Varian (1982) showed that the following generalized axiom of revealed preference (GARP) is sufficient to ensure not only rationality but that individuals have nonsatiated, continuous, monotone, and concave utility functions—the sort expected in traditional consumer demand theory. To define GARP, suppose the individual purchases bundle $x(p)$ when prices are p. We say consumption bundle $x(p_s)$ is *directly revealed to be preferred* to bundle $x(p_t)$ if $p_s x(p_t) \leq p_s x(p_s)$; i.e., $x(p_t)$ could have been purchased when $x(p_s)$ was purchased. We say $x(p_s)$ is *indirectly revealed to be preferred* to $x(p_t)$ if there is a sequence $x(p_s) = x(p_1), x(p_2), \ldots, x(p_k) = x(p_t)$, where each $x(p_i)$ is directly revealed preferred to $x(p_{i+1})$ for $i = 1, \ldots, k-1$. GARP then is the following condition: if $x(p_s)$ is indirectly revealed to be preferred to $x(p_t)$, then $p_t x(p_t) \leq p_t x(p_s)$; i.e., $x(p_s)$ does not cost less than $x(p_t)$ when $x(p_s)$ is purchased.

Andreoni and Miller (2002) worked with 176 students in an elementary economics class and had them play the Dictator Game multiple times each, with the price p taking on the values $p = 0.25, 0.33, 0.5, 1, 2, 3$, and 4, with amounts of tokens equaling $m = 40, 60, 75, 80$, and 100. They found that only 18 of the 176 subjects violated GARP at least once and that of these violations, only four were at all significant. By contrast, if choices were randomly generated, we would expect that between 78% and 95% of subjects would have violated GARP.

As to the degree of altruistic giving in this experiment, Andreoni and Miller found that 22.7% of subjects were perfectly selfish, 14.2% were perfectly egalitarian at all prices, and 6.2% always allocated all the money so as to maximize the total amount won (i.e., when $p > 1$, they kept all the money, and when $p < 1$, they gave all the money to the Receiver).

We conclude from this study that, at least in some cases, and perhaps in all, we can treat altruistic preferences in a manner perfectly parallel to the way we treat money and private goods in individual preference functions. We use this approach in the rest of the problems in this chapter.

3.5 Conditional Altruistic Cooperation

Both strong reciprocity and inequality aversion imply *conditional altruistic cooperation* in the form of a predisposition to cooperate in a social dilemma as long as the other players also cooperate, although they have different reasons: the strong reciprocator believes in returning good for good, whatever the distributional implications, whereas the inequality-averse individual simply does not want to create unequal outcomes by making some parties bear a disproportionate share of the costs of cooperation.

Social psychologist Toshio Yamagishi and his coworkers used the Prisoner's Dilemma (§2.10) to show that a majority of subjects (college students in Japan and the United States) positively value altruistic cooperation. In this game, let CC stand for "both players cooperate," let DD stand for "both players defect," let CD stand for "I cooperate but my partner defects," and let DC stand for "I defect and my partner cooperates." A self-regarding individual will exhibit $DC \succ CC \succ DD \succ CD$ (check it), while an altruistic cooperator will exhibit $CC \succ DC \succ DD \succ CD$ (for notation, see §1.1); i.e. the self-regarding individual prefers to defect no matter what his partner does, whereas the conditional altruistic cooperator prefers to cooperate so long as his partner cooperates. Watabe et al. (1996), using 148 Japanese subjects, found that the average desirability of the four outcomes conformed to the altruistic cooperator preferences ordering. The experimenters also asked 23 of the subjects if they would cooperate if they already knew that their partner was going to cooperate, and 87% (20) said they would. Hayashi et al. (1999) ran the same experiment with U.S. students with similar results. In this case, all the subjects said they would cooperate if their partners were already committed to cooperating.

While many individuals appear to value conditional altruistic cooperation, the above studies did not use real monetary payoffs, so it is unclear how strongly these values are held, or if they are held at all, because subjects might simply be paying lip service to altruistic values that they in fact do not hold. To address this issue, Kiyonari, Tanida and Yamagishi (2000) ran an experiment with real monetary payoffs using 149 Japanese university students. The experimenters ran three distinct treatments, with about equal numbers of subjects in each treatment. The first treatment was a standard "simultaneous" Prisoner's Dilemma, the second was a "second-player" situation in which the subject was told that the first player in the Prisoner's Dilemma had already chosen to cooperate, and the third was a "first-player" treatment in which the subject was told that his decision to cooperate or defect would be made known to the second player before the latter made his own choice. The experimenters found that 38% of the subjects cooperated in the simultaneous treatment, 62% cooperated in the second player treatment, and 59% cooperated in the first-player treatment. The decision to cooperate in each treatment cost the subject about $5 (600 yen). This shows unambiguously that a majority of subjects were conditional altruistic cooperators (62%). Almost as many were not only cooperators, but were also willing to bet that their partners would be (59%), provided the latter were assured of not being defected upon, although under standard conditions, without this assurance, only 38% would in fact cooperate.

3.6 Altruistic Punishment

Both strong reciprocity and inequality aversion imply *altruistic punishment* in the form of a predisposition to punish those who fail to cooperate in a social dilemma. The source of this behavior is different in the two cases: the strong reciprocator believes in returning harm for harm, whatever the distributional implications, whereas the inequality-averse individual wants to create a more equal distribution of outcomes even at the cost of lower outcomes for himself and others. The simplest game exhibiting altruistic punishment is the *Ultimatum Game* (Güth, Schmittberger, and Schwarze 1982). Under conditions of anonymity, two player are shown a sum of money, say $10. One of the players, called the Proposer, is instructed to offer any number of dollars, from $1 to $10, to the second player, who is called the Responder. The Proposer can make only one offer and the Responder can either accept or reject this offer. If the Responder accepts the offer, the money is shared accordingly. If the Responder rejects the offer, both players receive nothing. The two players do not face each other again.

There is only *one* Responder strategy that is a best response for a self-regarding individual: accept anything you are offered. Knowing this, a self-regarding Proposer who believes he faces a self-regarding Responder, offers the minimum possible amount, $1, and this is accepted.

However, when actually played, the self-regarding outcome is almost never attained or even approximated. In fact, as many replications of this experiment have documented, under varying conditions and with varying amounts of money, Proposers routinely offer Responders very substantial amounts (50% of the total generally being the modal offer) and Responders frequently reject offers below 30% (Güth and Tietz 1990; Camerer and Thaler 1995). Are these results culturally dependent? Do they have a strong genetic component or do all successful cultures transmit similar values of reciprocity to individuals? Roth et al. (1991) conducted the Ultimatum Game in four different countries (United States, Yugoslavia, Japan, and Israel) and found that while the level of offers differed a small but significant amount in different countries, the probability of an offer being rejected did not. This indicates that both Proposers and Responders share the same notion of what is considered fair in that society and that Proposers adjust their offers to reflect this common notion. The differences in level of offers across countries, by the way, were relatively small. When a much greater degree of cultural diversity is studied, however, large differences in behavior are found, reflecting different standards of what it means to be fair in different types of societies (Henrich et al. 2004).

Behavior in the Ultimatum Game thus conforms to the strong reciprocity model: fair behavior in the Ultimatum Game for college students is a 50–50 split. Responders reject offers under 40% as a form of altruistic punishment of the norm-violating Proposer. Proposers offer 50% because they are altruistic cooperators, or 40% because they fear rejection. To support this interpretation, we note that if the offers in an Ultimatum Game are generated by a computer rather than by the Proposer, and if Responders know this, low offers are rarely rejected (Blount 1995). This suggests that players are motivated by *reciprocity*, reacting to a violation of behavioral norms (Greenberg and Frisch 1972). Moreover, in a variant of the game in which a Responder rejection leads to the Responder getting nothing but allows the Proposer to keep the share he suggested for himself, Responders never reject offers, and proposers make considerably smaller (but still positive) offers (Bolton and Zwick 1995). As a final indication that strong reciprocity motives are operative in this game, after the game is over, when asked why they offered more than the lowest possible amount, Proposers commonly said that they were afraid that Responders will consider low offers unfair and reject them. When Responders rejected offers, they usually claimed they want to punish unfair behavior. In all of the above experiments a significant fraction of subjects (about a quarter, typically) conformed to self-regarding preferences.

3.7 Strong Reciprocity in the Labor Market

Gintis (1976) and Akerlof (1982) suggested that, in general, employers pay their employees higher wages than necessary in the expectation that workers will respond by providing higher effort than necessary. Fehr, Gächter, and Kirchsteiger (1997) (see also Fehr and Gächter 1998) performed an experiment to validate this *legitimation* or *gift exchange* model of the labor market.

The experimenters divided a group of 141 subjects (college students who had agreed to participate in order to earn money) into "employers" and "employees." The rules of the game are as follows. If an employer hires an employee who provides effort e and receives a wage w, his profit is $\pi = 100e - w$. The wage must be between 1 and 100, and the effort is between 0.1 and 1. The payoff to the employee is then $u = w - c(e)$, where $c(e)$ is the cost of effort function shown in figure 3.2. All payoffs involve real money that the subjects are paid at the end of the experimental session. We call this the *Experimental Labor Market Game*.

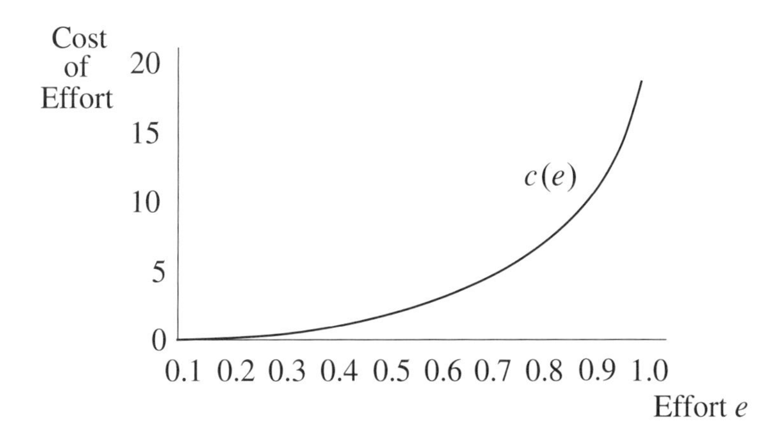

Figure 3.2. The Cost-of-effort schedule in Fehr, Gächter, and Kirchsteiger (1997).

The sequence of actions is as follows. The employer first offers a "contract" specifying a wage w and a desired amount of effort e^*. A contract is made with the first employee who agrees to these terms. An employer can make a contract (w, e^*) with at most one employee. The employee who agrees to these terms receives the wage w and supplies an effort level e that *need not equal the contracted effort* e^*. In effect, there is no penalty if the employee does not keep his promise, so the employee can choose any effort level, $e \in [0.1, 1]$, with impunity. Although subjects may play this game several times with different partners, each employer-employee interaction is a one-shot (nonrepeated) event. Moreover, the identity of the interacting partners is never revealed.

If employees are self-regarding, they will choose the zero-cost effort level, $e = 0.1$, no matter what wage is offered them. Knowing this, employers will never pay more than the minimum necessary to get the employee to accept a contract, which is 1 (assuming only integer wage offers are permitted).[2] The employee will accept this offer and will set $e = 0.1$. Because $c(0.1) = 0$, the employee's payoff is $u = 1$. The employer's payoff is $\pi = 0.1 \times 100 - 1 = 9$.

In fact, however, this self-regarding outcome rarely occurred in this experiment. The average net payoff to employees was $u = 35$, and the more generous the employer's wage offer to the employee, the higher the effort provided. In effect, employers presumed the strong reciprocity predispositions of the employees, making quite generous wage offers and receiving higher effort, as a means to increase both their own and the employee's payoff, as depicted in figure 3.3. Similar results have been observed in Fehr, Kirchsteiger, and Riedl (1993, 1998).

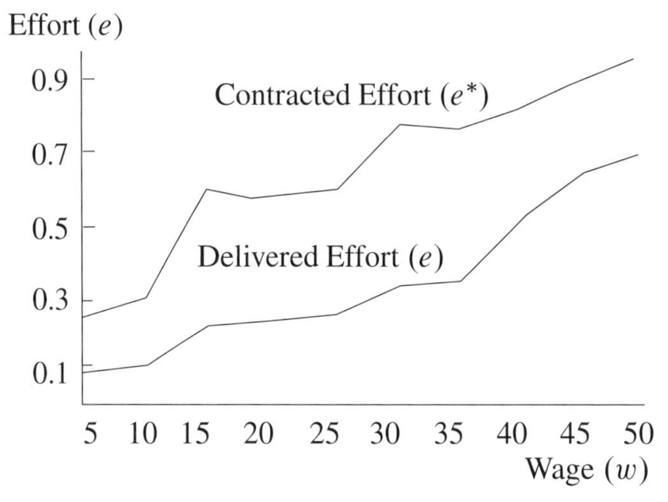

Figure 3.3. Relation of contracted and delivered effort to worker wage (141 subjects). From Fehr, Gächter, and Kirchsteiger (1997).

Figure 3.3 also shows that, though most employees are strong reciprocators, at any wage rate there still is a significant gap between the amount of

[2]This is because the experimenters created more employees than employers, thus ensuring an excess supply of employees.

effort agreed upon and the amount actually delivered. This is not because there are a few "bad apples" among the set of employees but because only 26% of the employees delivered the level of effort they promised! We conclude that strong reciprocators are inclined to compromise their morality to some extent.

To see if employers are also strong reciprocators, the authors extended the game by allowing the employers to respond reciprocally to the *actual effort choices* of their workers. At a cost of 1, an employer could *increase* or *decrease* his employee's payoff by 2.5. If employers were self-regarding, they would of course do neither because they would not (knowingly) interact with the same worker a second time. However, 68% of the time, employers punished employees who did not fulfill their contracts, and 70% of the time, employers rewarded employees who overfulfilled their contracts. Employers rewarded 41% of employees who *exactly* fulfilled their contracts. Moreover, employees *expected* this behavior on the part of their employers, as shown by the fact that their effort levels *increased significantly* when their bosses gained the power to punish and reward them. Underfulfilling contracts dropped from 71% to 26% of the exchanges, and overfulfilled contracts rose from 3% to 38% of the total. Finally, allowing employers to reward and punish led to a 40% increase in the net payoffs to all subjects, even when the payoff reductions resulting from employer punishment of employees are taken into account.

We conclude from this study that subjects who assume the role of employee conform to internalized standards of reciprocity even when they are certain there are no material repercussions from behaving in a self-regarding manner. Moreover, subjects who assume the role of employer expect this behavior and are rewarded for acting accordingly. Finally, employers reward good behavior and punish bad behavior when they are allowed, and employees expect this behavior and adjust their own effort levels accordingly. In general, then, subjects follow an internalized norm not because it is prudent or useful to do so, or because they will suffer some material loss if they do not, but rather because they desire to do this *for its own sake*.

3.8 Altruistic Third-Party Punishment

Prosocial behavior in human society occurs not only because those directly helped and harmed by an individual's actions are likely to reciprocate in kind but also because there are general *social norms* that foster prosocial behavior and many people are willing to bestow favors on someone who conforms to social norms, and to punish someone who does not, even if they are not personally helped or hurt by the individual's actions. In everyday life, third parties who are not the beneficiaries of an individual's prosocial act, help the individual and his family in times of need, preferentially trade favors with the individual, and otherwise reward the individual in ways that are not costly but are nonetheless of great benefit to the cooperator. Similarly, third parties who have not been personally harmed by the selfish behavior of an individual refuse aid even when it is not costly to do so, shun the offender, and approve of the offender's ostracism from beneficial group activities, again at low cost to the third party but at high cost to the offender.

It is hard to conceive of human societies operating at a high level of efficiency in the absence of such third-party reward and punishment. Yet, self-regarding actors will never engage in such behavior if it is at all costly. Fehr and Fischbacher (2004) addressed this question by conducting a series of third-party punishment experiments using the Prisoner's Dilemma (§2.10) and the Dictator Game (§3.4). The experimenters implemented four experimental treatments in each of which subjects were grouped into threes. In each group, in stage 1, subject A played a Prisoner's Dilemma or the Dictator Game with subject B as the Receiver, and subject C was an outsider whose payoff was not affected by A's decision. Then, in stage two, subject C was endowed with 50 points and allowed to deduct points from subject A such that every 3 points deducted from A's score cost C 1 point. In the first treatment, TP-DG, the game was the Dictator Game, in which A was endowed with 100 points, and could give 0, 10, 20, 30, 40, or 50 points to B, who had no endowment.

The second treatment (TP-PD) was the same, except that the game was the Prisoner's Dilemma. Subjects A and B were each endowed with 10 points, and each could either keep the 10 points or transfer them to the other subject, in which case the points were tripled by the experimenter. Thus, if both cooperated, each earned 30 points, and if both defected, each earned 10 points. If one cooperated and one defected, however, the cooperator earned 0 points and the defector earned 40 points. In the second stage, C was given an endowment of 40 points, and was allowed to deduct points from A and/or B, just as in the TP-DG treatment.

To compare the relative strengths of second- and third-party punishment in the Dictator Game, the experimenters implemented a third treatment,

S&P-DG. In this treatment, subjects were randomly assigned to player A and player B, and A-B pairs were randomly formed. In the first stage of this treatment, each A was endowed with 100 points and each B with none, and the A's played the Dictator Game as before. In the second stage of each treatment, each player was given an additional 50 points, and the B players were permitted to deduct points from A players on the same terms as in the first two treatments. S&P-DG also had two conditions. In the S condition, a B player could punish only his *own* Dictator, whereas in the T condition, a B player could punish only an A player *from another pair*, to which he was randomly assigned by the experimenters. In the T condition, each B player was informed of the behavior of the A player to which he was assigned.

To compare the relative strengths of second and third-party punishment in the Prisoner's Dilemma, the experimenters implemented a fourth treatment, S&P-PG. This was similar to the S&P-DG treatment, except that now they played the Prisoner's Dilemma.[3]

In the first two treatments, because subjects were randomly assigned to positions A, B, and C, the obvious fairness norm is that all should have equal payoffs (an *equality norm*). For instance, if A gave 50 points to B and C deducted no points from A, each subject would end up with 50 points. In the Dictator Game treatment, TP-DG, 60% of third parties (Cs) punished Dictators (As) who give less than 50% of the endowment to Receivers (Bs). Statistical analysis (ordinary least squares regression) showed that for every point an A kept for himself above the 50-50 split, he was punished an average 0.28 points by C's, leading to a total punishment of $3 \times 0.28 = 0.84$ points. Thus, a Dictator who kept the whole 100 points would have $0.84 \times 50 = 42$ points deducted by C's, leaving a meager gain of 8 points over equal sharing.

The results for the Prisoner's Dilemma treatment, TP-PD, was similar, with an interesting twist. If one partner in the A-B pair defected and the other cooperated, the defector would have on average 10.05 points deducted by Cs, but if both defected, the punished player lost only an average of 1.75 points. This shows that third parties (Cs) cared not only about the intentions of defectors but also about how much harm they caused and/or how unfair they turned out to be. Overall, 45.8% of third parties punished defectors

[3]The experimenters never used value-laden terms such as "punish" but rather used neutral terms, such as "deduct points."

whose partners cooperated, whereas only 20.8% of third parties punished defectors whose partners defected.

Turning to the third treatment (S&P-DG), second-party sanctions of selfish Dictators were found to be considerably stronger than third-party sanctions, although both were highly significant. On average, in the first condition, where Receivers could punish their own Dictators, they imposed a deduction of 1.36 points for each point the Dictator kept above the 50-50 split, whereas they imposed a deduction of only 0.62 point per point kept on third-party Dictators. In the final treatment, S&P-PD, defectors were severely punished by both second and third parties, but second-party punishment was again found to be much more severe than third-party punishment.. Thus, cooperating subjects deducted on average 8.4 points from a defecting partner, but only 3.09 points from a defecting third party.

This study confirms the general principle that punishing norm violators is very common but not universal, and that individuals are prone to be more harsh in punishing those who hurt them personally, as opposed to violating a social norm that hurts others than themselves.

3.9 Altruism and Cooperation in Groups

A *Public Goods Game* is an n-person game in which, by cooperating, each individual A adds more to the payoff of the other members than A's cost of cooperating, but A's share of the total gains he creates is less than his cost of cooperating. By not contributing, the individual incurs no personal cost and produces no benefit for the group. The Public Goods Game captures many social dilemmas, such as voluntary contribution to team and community goals. Researchers (Ledyard 1995; Yamagishi 1986; Ostrom, Walker, and Gardner 1992; Gächter and Fehr 1999) uniformly found that groups exhibit a much higher rate of cooperation than can be expected assuming the standard model of the self-regarding actor.

A typical Public Goods Game consists of a number of rounds, say 10. In each round, each subject is grouped with several other subjects—say 3 others. Each subject is then given a certain number of points, say 20, redeemable at the end of the experimental session for real money. Each subject then places some fraction of his points in a "common account" and the remainder in the subject's "private account." The experimenter then tells the subjects how many points were contributed to the common account and adds to the private account of *each* subject some fraction, say 40%, of the total amount in the common account. So if a subject contributes his whole 20 points to the common account, each of the 4 group members will receive 8 points at the end of the round. In effect, by putting the whole endowment into the common account, a player loses 12 points but the other 3 group members gain in total 24 (8 times 3) points. The players keep whatever is in their private accounts at the end of the round.

A self-regarding player contributes nothing to the common account. However, only a fraction of the subjects in fact conform to the self-regarding model. Subjects begin by contributing on average about half of their endowments to the public account. The level of contributions decays over the course of the 10 rounds until in the final rounds most players are behaving in a self-regarding manner. This is, of course, exactly what is predicted by the strong reciprocity model. Because they are altruistic contributors, strong reciprocators start out by contributing to the common pool, but in response to the norm violation of the self-regarding types, they begin to refrain from contributing themselves.

How do we know that the decay of cooperation in the Public Goods Game is due to cooperators punishing free riders by refusing to contribute themselves? Subjects often report this behavior retrospectively. More compelling, however, is the fact that when subjects are given a more constructive way of punishing defectors, they use it in a way that helps sustain cooperation (Orbell, Dawes, and Van de Kragt 1986, Sato 1987, and Yamagishi 1988a, 1988b, 1992).

For instance, in Ostrom, Walker, and Gardner (1992) subjects in a Public Goods Game, by paying a "fee," could impose costs on others by "fining" them. Because fining costs the individual who uses it but the benefits of increased compliance accrue to the group as a whole, the only subgame perfect Nash equilibrium in this game is for no player to pay the fee, so no player is ever punished for defecting, and all players defect by contributing nothing to the public account. However, the authors found a significant level of punishing behavior. The experiment was then repeated with subjects being allowed to communicate without being able to make binding agreements. In the framework of the self-regarding actor model, such communication is called *cheap talk* and cannot lead to a distinct subgame perfect equilibrium. But in fact such communication led to almost perfect cooperation (93%) with very little sanctioning (4%).

The design of the Ostrom-Walker-Gardner study allowed individuals to engage in strategic behavior because costly punishment of defectors could increase cooperation in future periods, yielding a positive net return for the punisher. What happens if we remove any possibility of punishment being strategic? This is exactly what Fehr and Gächter (2000) studied.

Fehr and Gächter (2000) set up an experimental situation in which the possibility of strategic punishment was removed. They used 6- and 10-round Public Goods Games with groups of size 4, and with costly punishment allowed at the end of each round, employing three different methods of assigning members to groups. There were sufficient subjects to run between 10 and 18 groups simultaneously. Under the Partner treatment, the four subjects remained in the same group for all 10 periods. Under the Stranger treatment, the subjects were randomly reassigned after each round. Finally, under the Perfect Stranger treatment, the subjects were randomly reassigned but assured that they would never meet the same subject more than once.

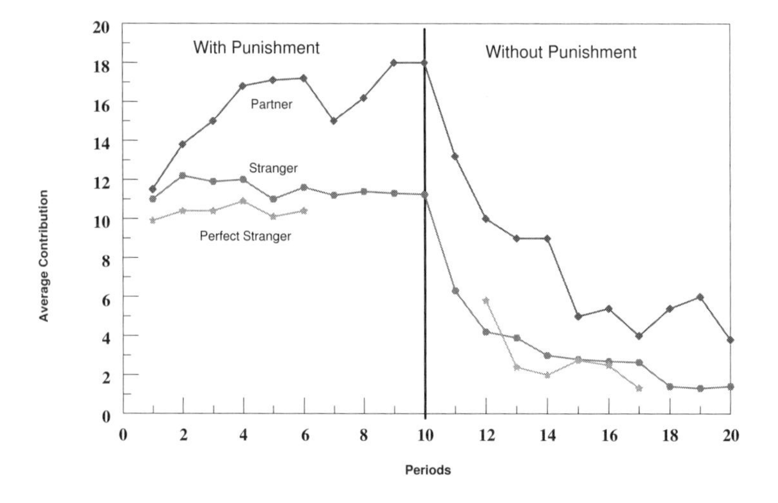

Figure 3.4. Average contributions over time in the Partner, Stranger, and Perfect Stranger Treatments when the punishment condition is played first (Fehr and Gächter 2000).

Fehr and Gächter (2000) performed their experiment for 10 rounds with punishment and 10 rounds without. Their results are illustrated in figure 3.4. We see that when costly punishment is permitted, cooperation does not deteriorate, and in the Partner game, despite strict anonymity, cooperation increases almost to full cooperation even in the final round. When punishment is not permitted, however, the same subjects experienced the deterioration of cooperation found in previous Public Goods Games. The contrast in cooperation rates between the Partner treatment and the two Stranger treatments is worth noting because the strength of punishment is roughly the same across all treatments. This suggests that the credibility of the punishment threat is greater in the Partner treatment because in this treatment the punished subjects are certain that, once they have been punished in previous rounds, the punishing subjects are in their group. The prosociality impact of strong reciprocity on cooperation is thus more strongly manifested, the more coherent and permanent the group in question.[4]

Many behavioral game theorists have found that, while altruistic punishment increases participation, it often leads to such a high level of punishment that overall average payoffs, net of punishment, are low (Carpenter and Matthews 2005; Page, Putterman, and Unel 2005; Casari and Luini 2007; Anderson and Putterman 2006; Nikiforakis 2008). Some have interpreted this as showing that strong reciprocity "could not have evolved," or "is not an adaptation." It is more likely, however, that the problem is with the experiments themselves. These experiments attempt to refute the standard "homo economicus" model of the self-regarding actor and do not attempt to produce realistic punishment scenarios in the laboratory. In fact, the motive for punishing norm violators is sufficiently strong as to lower overall payoffs when not subject to some social regulation. In real societies, there tends to be collective control over the meting out of punishment, and the excessive zeal of individual punishers is frowned upon and socially punished. Indeed, in one of the rare studies that allowed groups to regulate punishment, Ertan, Page, and Putterman (2005) found that groups that voted to permit only punishment of below-average or of average and below-average contributors achieved significantly higher earnings than groups not using punishment.

[4]In Fehr and Gächter (2002), the experimenters reverse the order of the rounds with and without punishment to be sure that the decay in the "without punishment" phase was not due to its occurring at the end rather than at the start of the game. It was not.

3.10 Inequality Aversion

The inequality-averse individual exhibits a *weak* urge to reduce inequality when on top and a *strong* urge to reduce inequality when on the bottom (Loewenstein, Thompson, and Bazerman 1989). Since the advent of hierarchical societies based on settled agriculture, societies have attempted to inculcate in their less fortunate members precisely the opposite values—subservience to and acceptance of the status quo. The widely observed distaste for relative deprivation is thus probably a genetically based behavioral characteristic of humans. Because small children spontaneously share (even the most sophisticated of nonhuman primates, such as chimpanzees, fail to do this), the urge of the fortunate to redistribute may also be part of human nature, though doubtless a weaker impulse in most of us.

Support for inequality aversion comes from the anthropological literature. *H. sapiens* evolved in small hunter-gatherer groups. Contemporary groups of this type, although widely dispersed throughout the world, display many common characteristics. This commonality probably reflects their common material conditions. From this and other considerations we may tentatively infer the social organization of early human society from that of these contemporary foraging societies (Woodburn 1982; Boehm 1982, 2000).

Such societies have no centralized structure of governance (state, judicial system, church, Big Man), so the enforcement of norms depends on the voluntary participation of peers. There are many unrelated individuals, so cooperation cannot be explained by kinship ties. Status differences are very circumscribed, monogamy is widely enforced,[5] members who attempt to acquire personal power are banished or killed, and there is widespread sharing of large game and other food sources that are subject to substantial stochasticity, independent of the skill and/or luck of the hunters. Such conditions are, of course, conducive to the emergence of inequality aversion.

We model inequality aversion following Fehr and Schmidt (1999). Suppose the monetary payoffs to n players are given by $\pi = (\pi_1, \ldots, \pi_n)$. We take the utility function of player i to be

$$u_i(\pi) = \pi_i - \frac{\alpha_i}{n-1} \sum_{\pi_j > \pi_i} (\pi_j - \pi_i) - \frac{\beta_i}{n-1} \sum_{\pi_j < \pi_i} (\pi_i - \pi_j). \tag{3.1}$$

[5]Monogamy in considered to be an extremely egalitarian institution for men because it ensures that virtually all adult males will have a wife.

A reasonable range of values for β_i is $0 \leq \beta_i < 1$. Note that when $n = 2$ and $\pi_i > \pi_j$, if $\beta_i = 0.5$, then i is willing to transfer income to j dollar for dollar until $\pi_i = \pi_j$, and if $\beta_i = 1$ and i has the highest payoff, then i is willing to throw away money (or give it to the other players) at least until $\pi_i = \pi_j$ for some player j. We also assume $\beta_i < \alpha_i$, reflecting the fact that people are more sensitive to inequality when on the bottom than when on the top.

We shall show that with these preferences we can reproduce some of the salient behaviors in the Ultimatum and Public Goods games, where fairness appears to matter, as well as in market games, where it does not.

Consider first the Ultimatum Game. Let y be the share the Proposer offers the Responder, so the Proposer gets $x = 1 - y$. Because $n = 2$, we can write the two utility functions as

$$u(x) = \begin{cases} x - \alpha_1(1 - 2x) & x \leq 0.5 \\ x - \beta_1(2x - 1) & x > 0.5 \end{cases} \tag{3.2}$$

$$v(y) = \begin{cases} y - \alpha_2(1 - 2y) & y \leq 0.5 \\ y - \beta_2(2y - 1) & y > 0.5 \end{cases} \tag{3.3}$$

We have the following theorem.

THEOREM 3.1 *Suppose the payoffs in the Ultimatum Game are given by (3.2) and (3.3) and α_2 is uniformly distributed on the interval $[0, \alpha^*]$. Writing $y^* = \alpha^*/(1 + 2\alpha^*)$, we have the following:*

a. If $\beta_1 > 0.5$, the Proposer offers $y = 0.5$.
b. If $\beta_1 = 0.5$, the Proposer offers $y \in [y^*, 0.5]$.
c. If $\beta_1 < 0.5$, the Proposer offers y^*.

In all cases the Responder accepts. We leave the proof, which is straightforward, to the reader.

Now suppose we have a Public Goods Game $\mathcal{G}$ with $n \geq 2$ players. Each player i is given an amount 1 and decides independently what share x_i to contribute to the public account, after which the public account is multiplied by a number a, with $1 > a > 1/n$, and shared equally among the players. Because $1 > a$, contributions are costly to the contributor, and because $na > 1$, the group benefits of contributing exceed the costs, so contributing is a public good. The monetary payoff for each player then becomes $\pi_i = 1 - x_i + a\sum_{j=1}^{n} x_j$, and the utility payoffs are given by (3.1). We then have the following theorem.

THEOREM 3.2 *In the n-player Public Goods Game $\mathcal{G}$,*

a. *If $\beta_i < 1 - a$ for player i, then contributing nothing to the public account is a dominant strategy for i (a strategy is* dominant *for player i if it is a best response to any strategy profile of the other players).*
b. *If there are $k > a(n-1)/2$ players with $\beta_i < 1 - a$, then the only Nash equilibrium is for all players to contribute nothing to the public account.*
c. *If there are $k < a(n-1)/2$ players with $\beta_i < 1 - a$ and if all players i with $\beta_i > 1 - a$ satisfy $k/(n-1) < (a + \beta_i - 1)/(\alpha_i + \beta_i)$, then there is a Nash equilibrium in which the latter players contribute all their money to the public account.*

Note that if a player has a high β and hence could possibly contribute, but also has a high α so the player strongly dislikes being below the mean, then condition $k/(n-1) < (a+\beta_i-1)/(\alpha_i+\beta_i)$ in part (c) of the theorem will fail. In other words, cooperation with defectors requires that contributors not be excessively sensitive to relative deprivation.

The proof of this theorem is a bit tedious but straightforward and will be left to the reader (or consult Fehr and Schmidt 1999). We prove only part (c). We know from part (a) that players i with $\beta_i < 1-a$ will not contribute. Suppose $\beta_i > 1 - a$ and assume all other players satisfying this inequality contribute all their money to the public account. By reducing his contribution by $\delta > 0$, player i saves $(1-a)\delta$ directly and receives $k\alpha_i\delta/(n-1)$ in utility from the higher returns compared to the noncontributors, minus $(n-k-1)\delta\beta_i/(n-1)$ in utility from the lower returns compared with the contributors. The sum must be nonpositive in a Nash equilibrium, which reduces to the inequality in part (c).

Despite the fact that players have egalitarian preferences given by (3.1) if the game played has sufficiently marketlike qualities, the unique Nash equilibrium may settle on the competitive equilibrium however unfair this appears to be to the participants. Consider the following theorem.

THEOREM 3.3 *Suppose preferences are given by (3.1) and that \$1 is to be shared between player 1 and one of the players $i = 2, \ldots, n$ who submit simultaneous bids y_i for the share they are willing to give to player 1. The highest bid wins, and among equal highest bids, the winner is drawn at random. Then, for any set of (α_i, β_i), in every subgame perfect Nash equilibrium player 1 receives the whole \$1.*

The proof is left to the reader. Show that at least two bidders will set their y_i's to 1, and the seller will accept this offer.

3.11 The Trust Game

In the Trust Game, first studied by Berg, Dickhaut, and McCabe (1995), subjects are each given a certain endowment, say \$10. Subjects are then randomly paired, and one subject in each pair, Alice, is told she can transfer any number of dollars, from 0 to 10, to her (anonymous) partner, Bob, and keep the remainder. The amount transferred will be tripled by the experimenter and given to Bob, who can then give any number of dollars back to Alice (this amount is not tripled). If Alice transfers a lot, she is called "trusting," and if Bob returns a lot to Alice, he is called "trustworthy." In the terminology of this chapter, a trustworthy player is a strong reciprocator, and a trusting player is an individual who expects his partner to be a strong reciprocator.

If all individuals have self-regarding preferences, and if Alice believes Bob has self-regarding preferences, she will give nothing to Bob. On the other hand, if Alice believes Bob can be trusted, she will transfer all \$10 to Bob, who will then have \$40. To avoid inequality, Bob will give \$20 back to Alice. A similar result will obtain if Alice believes Bob is a strong reciprocator. On the other hand, if Alice is altruistic, she may transfer some money to Bob, on the grounds that it is worth more to Bob (because it is tripled) than it is to her, even if she does not expect anything back. It follows that several distinct motivations can lead to a positive transfer of money from Alice to Bob and then back to Alice.

Berg, Dickhaut, and McCabe (1995) found that, on average, \$5.16 was transferred from Alices to Bobs and on average, \$4.66 was transferred back from Bobs to Alices. Furthermore, when the experimenters revealed this result to the subjects and had them play the game a second time, \$5.36 was transferred from Alices to Bobs, and \$6.46 was transferred back from Bobs to Alices. In both sets of games there was a great deal of variability: some Alices transferring everything and some transferring nothing, and some Bobs more than fully repaying their partners, and some giving back nothing.

Note that the term "trustworthy" applied to Bob is inaccurate because Bob never, either explicitly or implicitly, promised to behave in any particular manner, so there is nothing concrete that Alice might trust him to do. The

Trust Game is really a strong reciprocity game in which Alice believes with some probability that Bob is a sufficiently motivated strong reciprocator and Bob either does or does not fulfill this expectation. To turn this into a real Trust Game, the second player should be able to promise to return a certain fraction of the money passed to him. We investigate this case in §3.12.

To tease apart the motivations in the Trust Game, Cox (2004) implemented three treatments, the first of which, treatment A, was the Trust Game as described above. Treatment B was a Dictator Game (§3.8) exactly like treatment A, except that now Bob could not return anything to Alice. Treatment C differs from treatment A in that each Alice was matched one-to-one with an Alice in treatment A, and each Bob was matched one-to-one with a Bob in treatment A. Each player in treatment C was then given an endowment equal to the amount his corresponding player had after the A-to-B transfer, but before the B-to-A transfer in treatment A. In other words, in treatment C, the Alice group and the Bob group have exactly what they had under treatment A, except that Alice now had nothing to do with Bob's endowment, so nothing transferred from Bob to Alice could be accounted for by strong reciprocity.

In all treatments, the rules of the game and the payoffs were accurately revealed to the subjects. However, in order to rule out third-party altruism (§3.8), the subjects in treatment C were not told the reasoning behind the sizes of their endowments. There were about 30 pairs in each treatment, each treatment was played two times, and no subject participated in more than one treatment. The experiment was run double-blind (subjects were anonymous to one another and to the experimenter).

In treatment B, the Dictator Game counterpart to the Trust Game, Alice transferred on average \$3.63 to player B, as opposed to \$5.97 in treatment A. This shows that \$2.34 of the \$5.97 transferred to B in treatment A can be attributed to trust, and the remaining \$3.63 to some other motive. Because players A and B both have endowments of \$10 in treatment B this other motive cannot be inequality aversion. This transfer may well reflect a reciprocity motive of the form, "If someone can benefit his partner at a cost that is low compared to the benefit, he should do so, even if he is on the losing end of the proposition." But we cannot tell from the experiment exactly what the \$3.63 represents.

In treatment C, the player B Dictator Game counterpart to the Trust Game, player B returned an average of \$2.06, as compared with \$4.94 in treatment A. In other words, \$2.06 of the original \$4.94 can be interpreted as a reflection of inequality aversion, and the remaining \$2.88 is a reflection of strong reciprocity.

Several other experiments confirm that other-regarding preferences depend on the actions of individuals and not simply on the distribution of payoffs, as is the case with inequality aversion. Charness and Haruvy (2002), for instance, developed a version of the gift exchange labor market described in §3.7 capable of testing self-regarding preferences, pure altruism, inequality aversion, and strong reciprocity simultaneously. Strong reciprocity had by far the greatest explanatory value.

3.12 Character Virtues

Character virtues are ethically desirable behavioral regularities that individuals value for their own sake, while having the property of facilitating cooperation and enhancing social efficiency. Character virtues include *honesty*, *loyalty*, *trustworthiness*, *promise keeping*, and *fairness*. Unlike such other-regarding preferences as strong reciprocity and empathy, these character virtues operate without concern for the individuals with whom one interacts. An individual is honest in his transactions because this is a desired state of being, not because he has any particular regard for those with whom he transacts. Of course, the sociopath "Homo economicus" is honest only when it serves his material interests to be so, whereas the rest of us are at times honest even when it is costly to be so and even when no one but us could possibly detect a breach.

Common sense, as well as the experiments described below, indicate that honesty, fairness, and promise keeping are not absolutes. If the cost of virtue is sufficiently high, and the probability of detection of a breach of virtue is sufficiently small, many individuals will behave dishonestly. When one is aware that others are unvirtuous in a particular region of their lives (e.g., marriage, tax paying, obeying traffic rules, accepting bribes), one is more likely to allow one's own virtue to lapse. Finally, the more easily one can delude oneself into inaccurately classifying an unvirtuous act as virtuous, the more likely one is to allow oneself to carry out such an act.

One might be tempted to model honesty and other character virtues as *self-constituted constraints* on one's set of available actions in a game, but a more fruitful approach is to include the state of being virtuous in a certain way as an argument in one's preference function, to be traded off against other valuable objects of desire and personal goals. In this respect, character virtues are in the same category as ethical and religious preferences and are often considered subcategories of the latter.

Numerous experiments indicate that most subjects are willing to sacrifice material rewards to maintain a virtuous character even under conditions of anonymity. Sally (1995) undertook a meta-analysis of 137 experimental treatments, finding that face-to-face communication, in which subjects are capable of making verbal agreements and promises, was the strongest predictor of cooperation. Of course, face-to-face interaction violates anonymity and has other effects besides the ability to make promises. However, both Bochet, Page, and Putterman (2006) and Brosig, Ockenfels, and Weimann (2003) report that only the ability to exchange verbal information accounts for the increased cooperation.

A particularly clear example of such behavior is reported by Gneezy (2005), who studied 450 undergraduate participants paired off to play three games of the following form, all payoffs to which were of the form (b, a), where player 1, Bob, receives b and player 2, Alice, receives a. In all games, Bob was shown two pairs of payoffs, A:(x, y) and B:(z, w) where x, y, z, and w are amounts of money with $x < z$ and $y > w$, so in all cases B is better for Bob and A is better for Alice. Bob could then say to Alice, who could not see the amounts of money, either "Option A will earn you more money than option B," or "Option B will earn you more money than option A." The first game was A:(5,6) vs. B:(6,5) so Bob could gain 1 by lying and being believed while imposing a cost of 1 on Alice. The second game was A:(5,15) vs. B:(6,5), so Bob could gain 1 by lying and being believed, while still imposing a cost of 10 on Alice. The third game was A:(5,15) vs. B:(15,5), so Bob could gain 10 by lying and being believed, while imposing a cost of 10 on Alice.

Before starting play, Gneezy asked the various Bobs whether they expected their advice to be followed. He induced honest responses by promising to reward subjects whose guesses were correct. He found that 82% of Bobs expected their advice to be followed (the actual number was 78%). It follows from the Bobs' expectations that if they were self-regarding, they would always lie and recommend B to Alice.

The experimenters found that, in game 2, where lying was very costly to Alice and the gain from lying was small for Bob, only 17% of Bobs lied. In game 1, where the cost of lying to Alice was only 1 but the gain to Bob was the same as in game 2, 36% of Bobs lied. In other words, Bobs were loathe to lie but considerably more so when it was costly to Alices. In game 3, where the gain from lying was large for Bob and equal to the loss to Alice, fully 52% of Bobs lied. This shows that many subjects are willing to sacrifice material gain to avoid lying in a one-shot anonymous interaction, their willingness to lie increasing with an increased cost to them of truth telling, and decreasing with an increased cost to their partners of being deceived. Similar results were found by Boles, Croson, and Murnighan (2000) and Charness and Dufwenberg (2004). Gunnthorsdottir, McCabe, and Smith (2002) and Burks, Carpenter, and Verhoogen (2003) have shown that a socio-psychological measure of "Machiavellianism" predicts which subjects are likely to be trustworthy and trusting.

3.13 The Situational Character of Preferences

This chapter has deepened the rational actor model, allowing it to apply to situations of strategic interaction. We have found that preferences are other-regarding as well as self-regarding. Humans have social preferences that facilitate cooperation and exchange, as well as moral preferences for such personal character virtues as honesty and loyalty. These extended preferences doubtless contribute to longrun individual well-being (Konow and Earley 2008). However, social and moral preferences are certainly not merely instrumental, because individuals exercise these preferences even when no longrun benefits can accrue.

Despite this deepening of rational choice, we have conserved the notion that the individual has an immutable underlying preferences ordering that entails situationally specific behaviors, depending on the particular strategic interaction involved. Our analysis in §7.8, however, is predicated upon the denial of this immutability. Rather, we suggest that generally a social situation, which we call a *frame*, is imbued with a set of customary social norms that individuals often desire to follow simply because these norms are socially appropriate in the given frame. To the extent that this occurs, preferences themselves, and not just their behavioral implications, are situationally specific. The desire to conform to the moral and conventional standards that people associate with particular social frames thus represents a *meta-preference* that regulates revealed preferences in specific social situations.

We present two studies by Dana, Cain, and Dawes (2006) that illustrate the situational nature of preferences and the desire to conform to social norms (which we term *normative predisposition* in chapter 7). The first study used 80 Carnegie-Mellon University undergraduate subjects who were divided into 40 pairs to play the Dictator Game (§3.4), one member of each pair being randomly assigned to be the Dictator, the other to be the Receiver. Dictators were given $10, and asked to indicate how many dollars each wanted to give the Receiver, but the Receivers were not informed they were playing a Dictator Game. After making their choices, but before informing the Receivers about the game, the Dictators were presented with the option of accepting $9 rather than playing the game. They were told that if a Dictator took this option, the Receiver would never find out that the game was a possibility and would go home with their show-up fee alone.

Eleven of the 40 Dictators took this exit option, including 2 who had chosen to keep all of the $10 in the Dictator Game. Indeed, 46% of the Dictators who had chosen to give a positive amount to their Receivers took the exit option in which the Receiver got nothing. This behavior is not compatible with the concept of immutable preferences for a division of the $10 between the Dictator and the Receiver because individuals who would have given their Receiver a positive amount in the Dictator Game instead gave them nothing by avoiding playing the game, and individuals who would have kept the whole $10 in the Dictator Game were willing to take a $1 loss not to have to play the game.

To rule out other possible explanations of this behavior, the authors executed a second study in which the Dictator was told that the Receiver would never find out that a Dictator Game had been played. Thus, if the Dictator gave $5 to the Receivers, the latter would be given the $5 but would be given no reason why. In this new study, only 1 of 24 Dictators chose to take the $9 exit option. Note that in this new situation, the same social situation between Dictator and Receiver obtains both in the Dictator Game and in the exit option. Hence, there is no difference in the norms applying to the two options, and it does not make sense to forfeit $1 simply to have the game not called a Dictator Game.

The most plausible interpretation of these results is that many subjects felt obliged to behave according to certain norms when playing the Dictator Game, or violated these norms in an uncomfortable way, and were willing to pay simply not to be in a situation subject to these norms.

3.14 The Dark Side of Altruistic Cooperation

The human capacity to cooperate in large groups by virtue of prosocial preferences extends not only to exploiting nature but also to conquering other human groups as well. Indeed, even a slight hint that there may be a basis for inter-group competition induces individuals to exhibit insider loyalty and outsider hostility (Dawes, de Kragt, and Orbell 1988; Tajfel 1970; Tajfel et al. 1971; Turner 1984). Group members then show more generous treatment to in-group members than to out-group members even when the basis for group formation is arbitrary and trivial (Yamagishi, Jin, and Kiyonari 1999; Rabbie, Schot, and Visser 1989).

An experiment conducted by Abbink et al. (2007), using undergraduate students recruited at the University of Nottingham, is an especially dramatic example of the tendency for individuals willingly to escalate a conflict well beyond the point of serving their interests in terms of payoffs alone. Experimenters first had pairs of students $i = 1, 2$ play the following game. Each individual was given 1000 points and could spend any portion of it, x_i, on "armaments." The probability of player i winning was then set to $p_i = x_i/(x_1 + x_2)$.

We can find the Nash equilibrium of this game as follows. If player 1 spends x_1, then the expenditure of player 2 that maximizes the expected payoff is given by

$$x_2^* = \sqrt{1000x_1} - x_1.$$

The symmetric Nash equilibrium sets $x_1^* = x_2^*$, which gives $x_1^* = x_2^* = 250$. Indeed, if one player spends more than 250 points, the other player's best response is to spend less than 250 points.

Fourteen pairs of subjects played this game in pairs for 20 rounds, each with the same partner. The average per capita armament expenditure started at 250% of the Nash equilibrium in round 1 and showed some tendency to decline, reaching 160% of the Nash level after 20 rounds.

The experimenters also played the same game with 4 players on each team, where each player on the winning team received 1000 points. It is easy to show that now the Nash equilibrium has each team spending 250 points on armaments. To see this, we write player 1's expected payoff as

$$\frac{1000 \sum_{i=1}^{4} x_i}{\sum_{i=1}^{8}}.$$

Differentiating this expression, setting the result to zero, and solving for x_1 gives

$$x_1 = \sqrt{1000(x_5 + x_6 + x_7 + x_8)} - \sum_{i=2}^{8} x_i .$$

Now, equating all the x_i's to find the symmetric equilibrium, we find $x_i^* = 62.5 = 250/4$. In this case, however, the teams spent about 600% of the optimum in the first few periods, and this declined fairly steadily to 250% of the optimum in the final few periods.

This experiment showcases the tendency of subjects to overspend vastly for competitive purposes, although familiarity with the game strongly dampens this tendency, and had the participants played another 20 periods, we might have seen an approach to best response behavior.

However, the experimenters followed up the above treatments with another in which, after each round, players were allowed to punish other players based on the level of their contributions in the previous period. The punishment was costly, three tokens taken from the punishee costing the punisher one token. This, of course, mirrors the Public Goods Game with costly punishment (§3.9), and indeed this game does have a public goods aspect since the more one team member contributes, the less the best response contribution of the others, because the optimal total contribution of team members is 250, no matter how it is divided up among the members.

In this new situation, competition with punishment, spending started at 640% of the best response level, rose to a high of 1000% of this level, and settled at 900% of the best response level in period 7, showing no tendency to increase or decrease in the remaining 13 periods. This striking behavior shows that the internal dynamics of altruistic punishment are capable of sustaining extremely high levels of combat expenditure far in excess of the material payoff-maximizing level. While much more work in this area remains to be done, it appears that the same prosocial preferences that allow humans to cooperate in large groups of unrelated individuals are also turned into the goal of mutual self-destruction with great ease.

3.15 Norms of Cooperation: Cross-Cultural Variation

Experimental results in the laboratory would not be very interesting if they did not aid us in understanding and modeling real-life behavior. There are strong and consistent indications that the external validity of experimental results is high. For instance, Binswanger (1980) and Binswanger and

Sillers (1983) used survey questions concerning attitudes towards risk and experimental lotteries with real financial rewards to successfully predict the investment decisions of farmers. Glaeser et al. (2000) explored whether experimental subjects who trusted others in the Trust Game (§3.11) also behaved in a trusting manner with their own personal belongings. The authors found that experimental behavior was a quite good predictor of behavior outside the laboratory, while the usual measures of trust, based on survey questions, provided virtually no information. Genesove and Mayer (2001) showed that loss aversion determined seller behavior in the 1990s Boston housing market. Condominium owners subject to nominal losses set selling prices equal to the market rate plus 25% to 35% of the difference between their purchase price and the market price and sold at prices 3% to 18% of this difference. These findings show that loss aversion is not confined to the laboratory but affects behavior in a market in which very high financial gains and losses can occur.

Similarly, Karlan (2005) used the Trust Game and the Public Goods Game to predict the probability that loans made by a Peruvian microfinance lender would be repaid. He found that individuals who were trustworthy in the Trust Game were less likely to default. Also, Ashraf, Karlan, and Yin (2006) studied Phillipino women, identifying through a baseline survey those women exhibited a lower discount rate for future relative to current tradeoffs. These women were indeed significantly more likely to open a savings account, and after 12 months, average savings balances increased by 81 percentage points for those clients assigned to a treatment group based on their laboratory performance, relative to those assigned to the control group. In a similar vein, Fehr and Goette (2007) found that in a group of bicycle messengers in Zürich, those and only those who exhibited loss aversion in a laboratory survey also exhibited loss aversion when faced with real-life wage rate changes. For additional external validity studies, see Andreoni, Erard, and Feinstein (1998) on tax compliance (§3.4), Bewley (2000) on fairness in wage setting, and Fong, Bowles, and Gintis (2005) on support for income redistribution.

In one very important study, Herrmann, Thöni, and Gächter (2008) had subjects play the Public Goods Game with punishment (§3.9) with 16 subject pools in 15 different countries with highly varying social characteristics (one country, Switzerland, was represent by two subject pools, one in Zurich and one in St. Gallen). To minimize the social diversity among subject pools, they used university students in each country. The phenomenon they aimed to study was *antisocial punishment*.

The phenomenon itself was first noted by Cinyabuguma, Page, and Putterman (2004), who found that some free riders, when punished, responded not by increasing their contributions, but rather by punishing the high contributors! The ostensible explanation of this perverse behavior is that some free riders believe it is their personal right to free-ride if they so desire, and they respond to the "bullies" who punish them in a strongly reciprocal manner—they retaliate against their persecutors. The result, of course, is a sharp decline in the level of cooperation for the whole group.

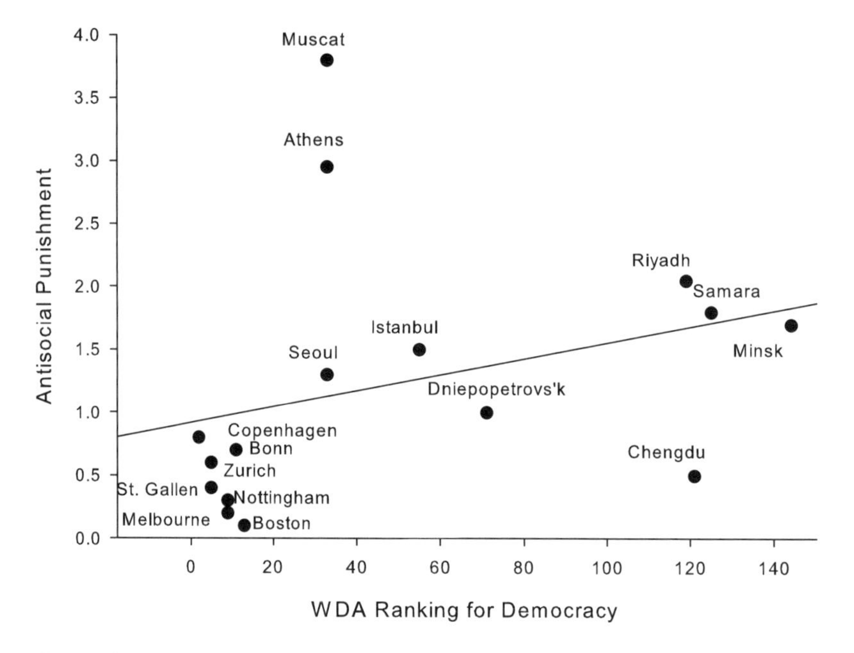

Figure 3.5. Countries judged highly democratic (political rights, civil liberties, press freedom, low corruption) by the World Democracy Audit engage in very little antisocial punishment, and conversely. (Statistics from Herrmann, Thöni, and Gächter, 2008.)

This behavior was later reported by Denant-Boemont, Masclet, and Noussair (2007) and Nikiforakis (2008), but because of its breadth, the Herrmann, Thöni, and Gächter study is distinctive for its implications for social theory. They found that in some countries, antisocial punishment was very rare, while in others it was quite common. As can be seen in figure 3.5, there is a strong negative correlation between the amount of anti-punishment exhibited and the World Development Audit's assessment of the level of democratic development of the society involved.

Figure 3.6 shows that a high level of antisocial punishment in a group translates into a low level of overall cooperation. The researchers first ran 10 rounds of the Public Goods Game without punishment (the N condition), and then another 10 rounds with punishment (the P condition). The figures show clearly that the more democratic countries enjoy a higher average payoff from payoffs in the Public Goods Game.

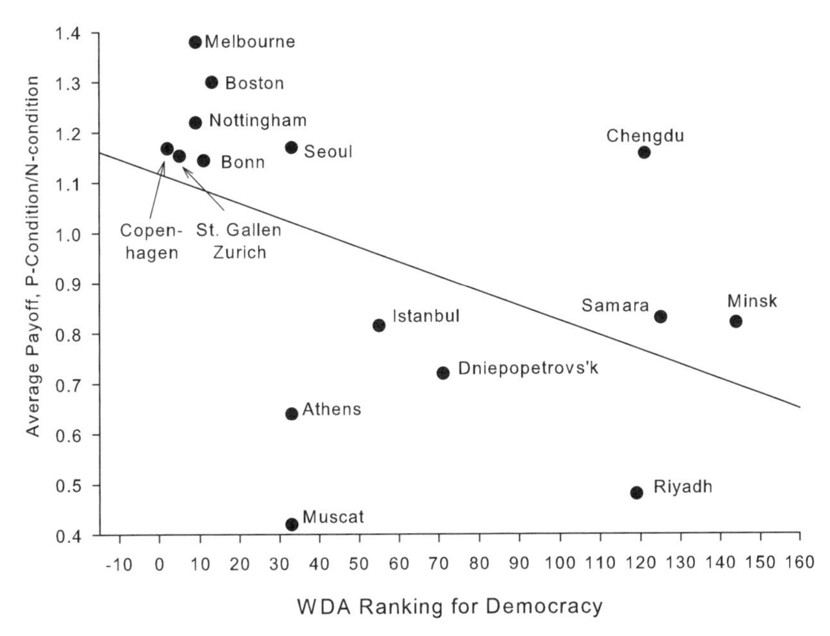

Figure 3.6. Antisocial punishment leads to low payoffs (Statistics from Herrmann, Thöni, and Gächter, Online Supplementary Material, 2008).

How might we explain this highly contrasting social behavior in university students in democratic societies with advanced market economies on the one hand, and more traditional societies based on authoritarian and parochial social institutions on the other? The success of democratic market societies may depend critically upon moral virtues as well as material interests, so the depiction of economic actors as "homo economicus" is as incorrect in real life as it is in the laboratory. These results indicate that individuals in modern democratic capitalist societies have a deep reservoir of public sentiment that can be exhibited even in the most impersonal interactions with unrelated others. This reservoir of moral predispositions is based upon an innate prosociality that is a product of our evolution as a species, as well as the uniquely human capacity to internalize norms of social behavior. Both forces predispose individuals to behave morally, even when this conflicts with their material interests, and to react to public disapprobation for free-riding with shame and penitence rather than antisocial self-aggrandizement.

More pertinent to the purposes of behavioral game theory, this experiment shows that laboratory games can be deployed to shed light on real-life social regularities that cannot be explained by participant observation or cross-country statistical analysis alone.

4

Rationalizability and Common Knowledge of Rationality

Men tracht un Got lacht
(Mortals scheme and God laughs)
Yiddish proverb

To determine what a rational player will do in a game, eliminate strategies that violate the cannons of rationality. Whatever is left we call *rationalizable*. We show that rationalizability in normal form games is equivalent to the iterated elimination of strongly dominated strategies, and the epistemological justification of rationalizability depends on the *common knowledge of rationality* (Tan and Werlang 1988).

If there is only one rationalizable strategy profile, it must be a Nash equilibrium, and it must be the choice of rational players, provided there is common knowledge of rationality.

There is no plausible set of epistemic conditions that imply the common knowledge of rationality. This perhaps explains the many non-obvious, indeed perplexing, arguments surrounding the iterated elimination of strongly dominated strategies, some of which are presented and analyzed below.

4.1 Epistemic Games

The Nash equilibrium criterion (§2.4) does not refer to the knowledge or beliefs of players. If players are Bayesian rational (§1.5), however, they then have beliefs concerning the behavior of the other players, and they maximize their expected utility by choosing best responses given these beliefs. Thus, to investigate the implications of Bayesian rationality, we must incorporate beliefs into the description of the game.

An *epistemic game* $\mathcal{G}$ consists of a normal form game with players $i = 1, \ldots, n$ and a finite pure-strategy set S_i for each player i, so $S = \prod_{i=1}^{n} S_i$ is the set of pure-strategy profiles for $\mathcal{G}$, with payoffs $\pi_i : S \rightarrow \mathbf{R}$. In addition, $\mathcal{G}$ includes a set of possible states Ω of the game, a knowledge partition $\mathcal{P}_i$

of Ω for each player i, and a subjective prior (§1.5) $p_i(\cdot;\omega)$ over Ω that is a function of the current state ω. A state ω specifies, possibly among other aspects of the game, the strategy profile s used in the game. We write this $s = \mathbf{s}(\omega)$. Similarly, we write $s_i = \mathbf{s}_i(\omega)$ and $s_{-i} = \mathbf{s}_{-i}(\omega)$.

The subjective prior $p_i(\cdot;\omega)$ represents the i's beliefs concerning the state of the game, including the choices of the other players, when the actual state is ω. Thus, $p_i(\omega';\omega)$ is the probability i places on the current state being ω' when the actual state is ω. Recall from §1.6 that a *partition* of a set X is a set of mutually disjoint subsets of X whose union is X. We write the cell of the partition $\mathcal{P}_i$ containing state ω as $\mathbf{P}_i\omega$, and we interpret $\mathbf{P}_i\omega \in \mathcal{P}_i$ as the set of states that i considers possible (i.e., among which i cannot distinguish) when the actual state is ω. Therefore, we require that $\mathbf{P}_i\omega = \{\omega' \in \Omega | p_i(\omega'|\omega) > 0\}$. Because i cannot distinguish among states in the cell $\mathbf{P}_i\omega$ of his knowledge partition $\mathcal{P}_i$, his subjective prior must satisfy $p_i(\omega'';\omega) = p_i(\omega'';\omega')$ for all $\omega'' \in \Omega$ and all $\omega' \in \mathbf{P}_i\omega$. Moreover, we assume a player believes the actual state is possible, so $p_i(\omega|\omega) > 0$ for all $\omega \in \Omega$.

If $\psi(\omega)$ is a proposition that is true or false at ω for each $\omega \in \Omega$, we write $[\psi] = \{\omega \in \Omega | \psi(\omega) = \text{true}\}$; i.e., $[\psi]$ is the set of states for which ψ is true.

The possibility operator $\mathbf{P}_i$ has the following two properties: for all $\omega, \omega' \in \Omega$,

$$\text{(P1)} \qquad \omega \in \mathbf{P}_i\omega$$
$$\text{(P2)} \qquad \omega' \in \mathbf{P}_i\omega \Rightarrow \mathbf{P}_i\omega' = \mathbf{P}_i\omega$$

P1 says that the current state is always possible (i.e., $p_i(\omega|\omega) > 0$), and P2 follows from the fact that $\mathcal{P}_i$ is a partition: if $\omega' \in \mathbf{P}_i\omega$, then $\mathbf{P}_i\omega'$ and $\mathbf{P}_i\omega$ have nonempty intersection, and hence must be identical.

We call a set $E \subseteq \Omega$ an *event*, and we say that player i *knows* the event E at state ω if $\mathbf{P}_i\omega \subseteq E$; i.e., $\omega' \in E$ for all states ω' that i considers possible at ω. We write $\mathbf{K}_iE$ for the event that i knows E.

Given a possibility operator $\mathbf{P}_i$, we define the *knowledge operator* $\mathbf{K}_i$ by

$$\mathbf{K}_iE = \{\omega | \mathbf{P}_i\omega \subseteq E\}.$$

The most important property of the knowledge operator is $\mathbf{K}_iE \subseteq E$; i.e., if an agent knows an event E in state ω (i.e., $\omega \in \mathbf{K}_iE$), then E is true in state ω (i.e., $\omega \in E$). This follows directly from P1.

We can recover the possibility operator $\mathbf{P}_i\omega$ for an individual from his knowledge operator $\mathbf{K}_i$, because

$$\mathbf{P}_i\omega = \bigcap\{E|\omega \in \mathbf{K}_i E\}. \tag{4.1}$$

To verify this equation, note that if $\omega \in \mathbf{K}_i E$, then $\mathbf{P}_i\omega \subseteq E$, so the left hand side of (4.1) is contained in the right hand side. Moreover, if ω' is not in the right hand side, then $\omega' \notin E$ for some E with $\omega \in \mathbf{K}_i E$, so $\mathbf{P}_i\omega \subseteq E$, so $\omega' \notin \mathbf{P}_i\omega$. Thus the right hand side of (4.1) is contained in the left.

To visualize a partition $\mathcal{P}$ of the universe into knowledge cells $\mathbf{P}_i\omega$, think of the universe Ω as a large cornfield consisting of a rectangular array of equally spaced stalks. A fence surrounds the whole cornfield, and fences running north/south and east/west between the rows of corn divide the field into plots, each completely fenced in. States ω are cornstalks. Each plot is a cell $\mathbf{P}_i\omega$ of the partition, and for any event (set of cornstalks) E, $\mathbf{K}_i E$ is the set of plots completely contained in E (Collins 1997).

For example, suppose $\Omega = S = \prod_{i=1}^n S_i$, where S_i is the set of pure strategies of player i in a game $\mathcal{G}$. Then, $P_{3t} = \{s = (s_1, \ldots, s_n) \in \Omega | s_3 = t \in S_3\}$ is the event that player 3 uses pure strategy t. More generally, if $\mathcal{P}_i$ is i's knowledge partition, and if i knows his own choice of pure strategy but not that of the other players, each $P \in \mathcal{P}_i$ has the form $P_{it} = \{s = (t, s_{-i}) \in S | t \in S_i, s_{-i} \in S_{-i}\}$. Note that if $t, t' \in S_i$, then $t \neq t' \Rightarrow P_{it} \cap P_{it'} = \emptyset$ and $\cup_{t \in S_i} P_{it} = \Omega$, so $\mathcal{P}_i$ is indeed a partition of Ω.

If $\mathbf{P}_i$ is a possibility operator for i, the sets $\{\mathbf{P}_i\omega | \omega \in \Omega\}$ form a partition $\mathcal{P}$ of Ω. Conversely, any partition $\mathcal{P}$ of Ω gives rise to a possibility operator $\mathbf{P}_i$, two states ω and ω' being in the same cell iff $\omega' \in \mathbf{P}_i\omega$. Thus, a knowledge structure can be characterized by its knowledge operator $\mathbf{K}_i$, by its possibility operator $\mathbf{P}_i$, by its partition structure $\mathcal{P}$, or even by the subjective priors $p_i(\cdot|\omega)$.

To interpret the knowledge structure, think of an event as a set of possible worlds in which some proposition is true. For instance, suppose E is the event "it is raining somewhere in Paris" and let ω be a state in which Alice is walking through the Jardin de Luxembourg where it is raining. Because the Jardin de Luxembourg is in Paris, $\omega \in E$. Indeed, in every state $\omega' \in \mathbf{P}_A\omega$ that Alice believes is possible, it is raining in Paris, so $\mathbf{P}_A\omega \subseteq E$; i.e., Alice knows that it is raining in Paris. Note that $\mathbf{P}_A\omega \neq E$, because, for instance, there is a possible world $\omega' \in E$ in which it is raining in Montmartre but not in the Jardin de Luxembourg. Then, $\omega' \notin \mathbf{P}_A\omega$, but $\omega \in E$.

Since each state ω in epistemic game $\mathcal{G}$ specifies the players' pure strategy choices $\mathbf{s}(\omega) = (\mathbf{s}_1(\omega), \ldots, \mathbf{s}_n(\omega)) \in S$, the players' subjective priors must specify their beliefs $\phi_1^\omega, \ldots, \phi_n^\omega$ concerning the choices of the other players. We have $\phi_i^\omega \in \Delta S_{-i}$, which allows i to assume other players' choices are correlated. This is because, while the other players choose independently, they may have communalities in beliefs that lead them independently to choose correlated strategies.

We call ϕ_i^ω player i's *conjecture* concerning the behavior of the other players at ω. Player i's conjecture is derived from i's subjective prior by noting that $[s_{-i}] =_{\text{def}} [\mathbf{s}_{-i}(\omega) = s_{-i}]$ is an event, so we define $\phi_i^\omega(s_{-i}) = p_i([s_{-i}]; \omega)$, where $[s_{-i}] \subset \Omega$ is the event that the other players choose strategy profile s_{-i}. Thus, at state ω, each player i takes the action $\mathbf{s}_i(\omega) \in S_i$ and has the subjective prior probability distribution ϕ_i^ω over S_{-i}. A player i is deemed *Bayesian rational* at ω if $\mathbf{s}_i(\omega)$ maximizes $\pi_i(s_i, \phi_i^\omega)$, where

$$\pi_i(s_i, \phi_i^\omega) =_{\text{def}} \sum_{s_{-i} \in S_{-i}} \phi_i^\omega(s_{-i})\pi_i(s_i, s_{-i}). \tag{4.2}$$

In other words, player i is Bayesian rational in epistemic game $\mathcal{G}$ if his pure-strategy choice $\mathbf{s}_i(\omega) \in S_i$ for every state $\omega \in \Omega$ satisfies

$$\pi_i(\mathbf{s}_i(\omega), \phi_i^\omega) \geq \pi_i(s_i, \phi_i^\omega) \qquad \text{for } s_i \in S_i. \tag{4.3}$$

We take the above to be the standard description of an epistemic game, so we assume without comment that if $\mathcal{G}$ is an epistemic game, then the players are $i = 1, \ldots, n$, the state space is Ω, the strategy profile at ω is $\mathbf{s}(\omega)$, the conjectures are ϕ_i^ω, i's subjective prior at ω is $p_i(\cdot|\omega)$, and so on.

4.2 A Simple Epistemic Game

Suppose Alice and Bob each choose heads (h) or tails (t), neither observing the other's choice. We can write the universe as $\Omega = \{\text{hh, ht, th, tt}\}$, where xy means Alice chooses x and Bob chooses y. Alice's knowledge partition is then $\mathcal{P}_A = \{\{\text{hh, ht}\}, \{\text{th, tt}\}\}$, and Bob's knowledge partition is $\mathcal{P}_B = \{\{\text{hh, th}\}, \{\text{ht, tt}\}\}$. Alice's possibility operator $\mathbf{P}_A$ satisfies $\mathbf{P}_A\text{hh} = \mathbf{P}_A\text{ht} = \{\text{hh, ht}\}$ and $\mathbf{P}_A\text{th} = \mathbf{P}_A\text{tt} = \{\text{th, tt}\}$, whereas Bob's possibility operator $\mathbf{P}_B$ satisfies $\mathbf{P}_B\text{hh} = \mathbf{P}_B\text{th} = \{\text{hh, th}\}$ and $\mathbf{P}_B\text{ht} = \mathbf{P}_B\text{tt} = \{\text{ht, tt}\}$.

In this case, the event "Alice chooses h" is $E_A^h = \{\text{hh, ht}\}$, and because $\mathbf{P}_A\text{hh}, \mathbf{P}_A\text{ht} \subset E$, Alice knows E_A^h whenever E_A^h occurs (i.e., $E_A^h = \mathbf{K}_i E_A^h$). The event E_B^h expressing "Bob chooses h" is $E_B^h = \{\text{hh, th}\}$, and Alice does not know E_B^h because at th Alice believes tt is possible, but tt $\notin E_B^h$.

4.3 An Epistemic Battle of the Sexes

Alfredo \ Violetta	g	o
g	2,1	0,0
o	0,0	1,2

Consider the Battle of the Sexes (§2.8), depicted to the right. Suppose there are four types of Violettas, V_1, V_2, V_3, V_4, and four types of Alfredos, A_1, A_2, A_3, A_4. Violetta V_1 plays $t_1 = o$ and conjectures that Alfredo chooses o. Violetta V_2 plays $t_2 = g$ and conjectures that Alfredo chooses g. Violetta V_3 plays $t_3 = g$ and conjectures that Alfredo plays his mixed-strategy best response. Finally, Violetta V_4 plays $t_4 = o$ and conjectures that Alfredo plays his mixed-strategy best response. Correspondingly, Alfredo A_1 plays $s_1 = o$ and conjectures that Violetta chooses o. Alfredo A_2 plays $s_2 = g$ and conjectures that Violetta plays g. Alfredo A_3 plays $s_3 = g$ and conjectures that Violetta plays her mixed-strategy best response. Finally, Alfredo A_4 plays $s_4 = o$ and conjectures that Violetta plays her mixed-strategy best response.

A state of the game is $\omega_{ij} = (A_i, V_j, s_i, t_j)$, where $i, j = 1, \ldots, 4$. We write $\omega_{ij}^A = A_i, \omega_{ij}^V = V_j, \omega_{ij}^s = s_i, \omega_{ij}^t = t_j$.

Define $E_i^A = \{\omega_{ij} \in \Omega | \omega_{ij}^A = A_i\}$ and $E_i^V = \{\omega_{ij} \in \Omega | \omega_{ij}^V = V_j\}$. Then, E_i^A is the event that Alfredo's type is A_i, and E_j^V is the event that Violetta's type is V_j. Since each type is associated with a given pure strategy, Alfredo's knowledge partition is $\{E_i^A, i = 1, \ldots, 4\}$ and Violetta's knowledge partition is $\{E_i^V, i = 1, \ldots, 4\}$.

Note that both players are Bayesian rational at each state of the game because each strategy choice is a best response to the player's conjecture. Also, a Nash equilibrium occurs at $\omega_{11}, \omega_{22}, \omega_{33}$ and ω_{44}, although at only the first two of these are the players' conjectures correct. Of course, there is no mixed-strategy Nash equilibrium, because each player chooses a pure strategy in each state. However, if we define a Nash equilibrium *in conjectures* at a state as a situation in which each player's conjecture is a best response to the other player's conjecture, then ω_{ii} is a Nash equilibrium in conjectures for $i = 1, \ldots, 4$, and ω_{34} and ω_{43} are also equilibria in conjectures. Note that in this case, if Alfredo and Violetta have common priors and mutual knowledge of rationality, their choices form a Nash equilibrium in conjectures. We will generalize this in theorem 8.2.

4.4 Dominated and Iteratedly Dominated Strategies

We say $s_i' \in S_i$ is *strongly dominated* by $s_i \in S_i$ if, for every $\sigma_{-i} \in \Delta^* S_{-i}$, $\pi_i(s_i, \sigma_{-i}) > \pi_i(s_i', \sigma_{-i})$. We say s_i' is *weakly dominated* by s_i if, for every $\sigma_{-i} \in \Delta^* S_{-i}$, $\pi_i(s_i, \sigma_{-i}) \geq \pi_i(s_i', \sigma_{-i})$ and for at least one choice of σ_{-i}, the inequality is strict. A strategy may fail to be strongly dominated by any pure strategy but may nevertheless be strongly dominated by a mixed strategy (§4.11).

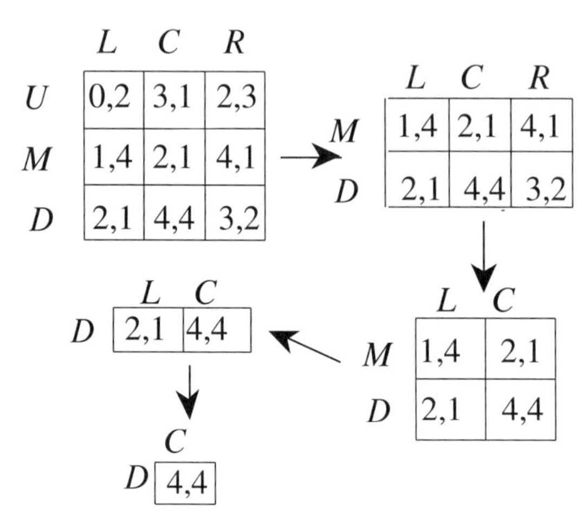

Figure 4.1. The iterated elimination of strongly dominated strategies

Having eliminated dominated strategies for each player, it often turns out that a pure strategy that was not dominated at the outset is now dominated. Thus, we can undertake a second round of eliminating dominated strategies. Indeed, this can be repeated until no remaining pure strategy can be eliminated in this manner. In a finite game, this occurs after a finite number of rounds and always leaves at least one pure strategy remaining for each player. If strongly (respectively, weakly) dominated strategies are eliminated, we call this the *iterated elimination of strongly (respectively, weakly) dominated strategies*. We call a pure strategy eliminated by this procedure an *iteratedly dominated strategy*.

Figure 4.1 illustrates the iterated elimination of strongly dominated strategies. First, U is strongly dominated by D for player 1. Second, R is strongly dominated by $0.5L+0.5C$ for player 2 (note that a pure strategy in this case is not dominated by any other pure strategy, but is strongly dominated by a mixed strategy). Third, M is strongly dominated by D, and finally, L is strongly dominated by C. Note that $\{D,C\}$ is indeed the unique Nash equilibrium of the game.

4.5 Eliminating Weakly Dominated Strategies

It seems completely plausible that a rational player will never use a weakly dominated strategy, since it cannot hurt, and can possibly help, to switch to a strategy that is not weakly dominated. However, this intuition is faulty for many reasons that will be explored in the sequel. We begin with an example, due to Rubinstein (1991), that starts with the Battle of the Sexes game $\mathcal{G}$ (§2.8), where if players choose gg, Alfredo gets 3 and Violetta gets 1, if they choose oo, Alfredo gets 1 and Violetta gets 3, and if they choose og or go, both get nothing. Now, suppose Alfredo says to Violetta before they make their choices, "I have the option of throwing away 1 before I choose, if I so desire." Now the new game $\mathcal{G}^+$ is shown in figure 4.2.

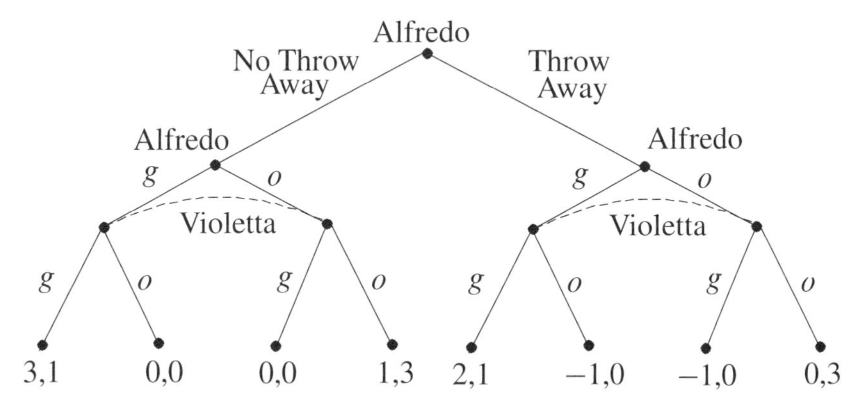

Figure 4.2. Battle of the Sexes with money burning

This game has many Nash equilibria. Suppose apply the iterated elimination of weakly dominated strategies to the normal form of this game. The normal form is shown in figure 4.3, where nx means "don't burn, choose x," bx means "burn money (throw away 1) and choose x," gg means "choose g," oo means "choose o," go means "choose g if Alfredo does not burn and choose o if Alfredo burns," and og means "choose o if Alfredo does not burn and choose g if Alfredo burns."

Let us assume rational players reject weakly dominated strategies, and assume it is common knowledge that Alfredo and Violetta are rational. Then,

	gg	go	og	oo
ng	3,1	3,1	0,0	0,0
no	0,0	0,0	1,3	1,3
bg	2,1	$-1,0$	2,1	$-1,0$
bo	$-1,0$	0,3	$-1,0$	0,3

Figure 4.3. Normal form of the Battle of the Sexes with money burning

bo is weakly dominated by ng, so Alfredo will not use bo. But then Violetta knows Alfredo is rational, so she eliminates bo from the running, after which oo is weakly dominated by og. Since Violetta is rational, she eliminates og, after which go is weakly dominated by gg, so Violetta eliminates these two strategies. Since Alfredo knows Violetta has eliminated these strategies, then no is weakly dominated by bg, so Alfredo eliminates no. But, Violetta then knows that Alfredo has made this elimination, so og is weakly dominated by gg, and she eliminates og. But, Alfredo knows this as well, so now bg is weakly dominated by ng, leaving only the Nash equilibrium (ng,gg). Thus, we have found that a purely hypothetical possibility that Alfredo might burn money, although he never does, allows him to enjoy the high-payoff Nash equilibrium in which he earns 3 and Violetta earns 1.

Of course, this result is not plausible. The fact that Alfredo has the capacity to do something bizarre, like burning money, should not lead rational players inexorably to choose an asymmetric equilibrium favoring Alfredo. The culprit here is the assumption of common knowledge of rationality, or the assumption that rational agents eliminate weakly dominated strategies, or both.

4.6 Rationalizable Strategies

Suppose $\mathcal{G}$ is an epistemic game. We denote the set of mixed strategies with support in S as $\Delta^* S = \prod_{i=1}^n \Delta S_i$, where ΔS_i is the set of mixed strategies for player i. We denote the mixed strategy profiles of all $j \neq i$ by $\Delta^* S_{-i}$.

In §1.5, we found that an agent whose choices satisfy the Savage axioms behaves as if maximizing a preference function subject to a subjective prior over the states of nature. We tailor this definition to epistemic game theory by saying that player i is rational at state ω if his pure strategy $\mathbf{s}_i(\omega)$ is a best response to his conjecture ϕ_i^ω of the other players' strategies at ω, as expressed in equation 4.3. Since a strongly dominated strategy can never be a best response, it follows that a rational player never uses a strongly dominated strategy. Moreover, if i knows that j is rational and hence never uses a strongly dominated strategy, then i can eliminate pure strategies in S_i that are best responses only to strategies in $\Delta^* S_{-i}$ that do not use strongly dominated pure strategies in S_{-i}. Moreover, if i knows that j knows that k is rational, then i knows that j will eliminate pure strategies that are best responses to k's strongly dominated strategies, and hence i can eliminate pure strategies that are best replies only to j's eliminated strategies. And so on. Pure strategies that survive this back-and-forth iterated elimination of pure strategies are call *rationalizable* (Bernheim 1984; Pearce 1984).

One elegant formal characterization of rationalizable strategies is in terms of *best response sets*. In epistemic game $\mathcal{G}$, we say a set $X = \prod_{i=1}^n X_i$, where each $X_i \subseteq S_i$, is a best response set if, for each i and each $x_i \in X_i$, i has a conjecture $\phi_{-i} \in \Delta X_{-i}$ such that x_i is a best response to ϕ_{-i}, as defined by (4.3). It is clear that the union of two best response sets is also a best response set, so the union of all best response sets is a maximal best response set. We define a strategy to be *rationalizable* if it is a member of this maximal best response set.

Note that the pure strategies for each player used with positive probability in a Nash equilibrium form a best response set in which each player conjectures the actual mixed-strategy choice of the other players. Therefore, any pure strategy used with positive probability in a Nash equilibrium is rationalizable. In a game with a completely mixed Nash equilibrium (§2.3), it follows that all strategies are rationalizable.

This definition of rationalizability is not constructive; i.e., knowing the definition does not tell us how to find the set that satisfies it. The following construction leads to the same set of rationalizable strategies. Let $S_i^0 = S_i$ for all i. Having defined S_i^k for all i and for $k = 0, \ldots, r-1$, we define S_i^r to be the set of pure strategies in S_i^{r-1} that are best responses to some conjecture $\phi_i \in \Delta S_{-i}^{r-1}$. Since $S_i^r \subseteq S_i^{r-1}$ for each i and there is only a finite number of pure strategies, there is some $r > 0$ such that $S_i^r = S_i^{r-1}$, and clearly for any $l > 0$, we then have $S_i^r = S_i^{r+l}$. We define i's rationalizable strategies as S_i^r.

These constructions refer only obliquely to the game's epistemic conditions, and in particular to the common knowledge of rationality (CKR) on which the rationalizability criterion depends. CKR obtains when each player is rational, each knows the others are rational, each knows the others know the others are rational, and so on. There is a third construction of rationalizability that makes its relationship to common knowledge of rationality more transparent.

Let $s_1, \ldots, s_n$ be the strategy profile chosen when $\phi_1, \ldots, \phi_n$ are the players' conjectures. The rationality of player i requires that s_i maximize i's expected payoff, given ϕ_i. Moreover, because i knows that j is rational, he knows that s_j is a best response, given some probability distribution over S_{-j}—namely, s_j is a best response to ϕ_j. We say ϕ_i is *first-order consistent* if ϕ_i places positive probability only on pure strategies of j that have the property of being best responses, given some probability distribution over S_{-j}. By the same reasoning, if i places positive probability on the pair s_j, s_k, because i knows that j knows that k is rational, i knows that j's conjecture is first-order consistent, and hence i places positive probability only on pairs s_j, s_k where j is first-order consistent and j places positive probability on s_k. When this is the case, we say that i's conjecture is *second-order consistent*. Clearly, we can define consistency of order r for all positive integers r, and a conjecture that is r-consistent for all r is simply called *consistent*. We say $s_1, \ldots, s_n$ is rationalizable if there is some consistent set of conjectures $\phi_1, \ldots, \phi_n$ that places positive probability on $s_1, \ldots, s_n$.

I leave it to the reader to prove that these three constructions define the same set of rationalizable strategies.

4.7 Eliminating Strongly Dominated Strategies

Consider the constructive approach to rationalizability developed in §4.6. It is clear that a strongly dominated strategy will be eliminated in the first round of the rationalizability construction if and only if it is eliminated in the first round of the iterated elimination of strongly dominated strategies. This observation can be extended to each successive stage in the construction of rationalizable strategies, which shows that all strategies that survive the iterated elimination of strongly dominated strategies are rationalizable. Are there other strategies that are rationalizable? The answer is that strongly dominated strategies exhaust the set of rationalizable strategies, given our assumption that players can have correlated conjectures. For details, see Bernheim (1984) or Pearce (1984).

4.8 Common Knowledge of Rationality

We will now define CKR formally. Let $\mathcal{G}$ be an epistemic game. For conjecture $\phi_i \in \Delta S_{-i}$, define $\text{argmax}_i(\phi_i) = \{s_i \in S_i | s_i \text{ maximizes } \pi_i(s_i', \phi_i)\}$; i.e., $\text{argmax}_i(\phi_i)$ is the set of i's best responses to the conjecture ϕ_i. Let $B_i(X_{-i})$ be the set of pure strategies of player i that are best responses to some mixed-strategy profile $\sigma_{-i} \in X_{-i} \subseteq S_{-i}$; i.e., $B_i(X_{-i}) = \{s_i \in S_i | (\exists \phi_i \in \Delta^* X_{-i})\ s_i \in \text{argmax}_i(\phi_i)\}$. We abbreviate $\phi([\mathbf{s}_j(\omega) = s_j]) > 0$ as $\phi(s_j) > 0$, and $\phi([\mathbf{s}_{-i}(\omega) = s_{-i}]) > 0$ as $\phi(s_{-i}) > 0$. We define

$$K_i^1 = [(\forall j \neq i)\phi_i^\omega(s_j) > 0 \Rightarrow s_j \in B_j(S_{-j})]. \tag{4.4}$$

K_i^1 is thus the event that i conjectures that a player j chooses s_j, only if s_j is a best response for j. In other words, K_i^1 is the event that i knows the other players are rational.

Suppose we have defined K_i^k for $k = 1, \ldots, r-1$. We define

$$K_i^r = K_i^{r-1} \cap [(\forall j \neq i)\phi_i^\omega(s_j) > 0 \Rightarrow s_j \in B_j(K_j^{r-1})].$$

Thus, K_i^2 is the event that i knows that every player knows that every player is rational. Similarly, K_i^r is the event that i knows that every chain of r recursive "j knows that k." We define $K^r = \cap_i K_i^r$, and if $\omega \in K^r$, we say there is *mutual knowledge of degree* r. Finally, we define the event CKR as

$$K^\infty = \bigcap_{r \geq 1}^{n} K^r.$$

Note that in an epistemic game, CKR cannot simply be *assumed* and is not a property of the players or of the informational structure of the game. This is because CKR generally holds only in certain states and fails in other states. For example, in chapter 5, we prove Aumann's famous theorem stating that in a generic extensive form game of perfect information, where distinct states are associated with distinct choice nodes, CKR holds only at nodes on the backward induction path (§5.11). The confusion surrounding CKR generally flows from attempting to abstract from the epistemic apparatus erected to define CKR and then to consider CKR to be some "higher form" of rationality that, when violated, impugns Bayesian rationality itself. There is no justification for such reasoning. There is nothing irrational about the failure of CKR. Nor is CKR some sort of "ideal" rationality that "boundedly rational" agents lamentably fail to attain. CKR is, unfortunately, just something that seemed might be plausible and useful, but turned out to have too many implausible and troublesome implications to be worth saving.

4.9 Rationalizability and Common Knowledge of Rationality

We will use the following characterization of rationalizability (§4.6). Let $S_i^0 = S_i$ for all i and define $S^0 = \prod_{i=1}^n S_i^0$ and $S_{-i}^0 = \prod_{j \neq i} S_j^0$. Having defined S^k and S_{-i}^k for all i and for $k = 0, \ldots, r-1$, we define $S_i^r = B_i(S_{-i}^{r-1})$. Then, $S^r = \prod_{i=1}^n S_i^r$ and $S_{-i}^r = \prod_{j \neq i} S_j^r$. We call S^r the set of pure strategies that survive r iterations of the elimination of unrationalizable strategies. Since $S_i^r \subseteq S_i^{r-1}$ for each i and there is only a finite number of pure strategies, there is some $r>0$ such that $S_i^r = S_i^{r-1}$, and for any $l > 0$, we then have $S_i^r = S_i^{r+l}$. We define i's rationalizable strategies as S_i^r.

THEOREM 4.1 *For all players i and* $r \geq 1$, *if* $\omega \in K_i^r$ *and* $\phi_i^\omega(s_{-i}) > 0$, *then* $s_{-i} \in S_{-i}^r$.

This implies that if there is mutual knowledge of degree r at ω, and i's conjecture at ω places strictly positive weight on s_{-i}, then s_{-i} survives r iterations of the elimination of unrationalizable strategies.

To prove this theorem, let $\omega \in K^1$ and suppose $\phi_i^\omega(s_j) > 0$. Then, $s_j \in B_j(S_{-j})$, and therefore $s_j \in S_j^1$, using the conjecture that maximizes s_j in $B_j(S_{-i})$. Since this is true for all $j \neq i$, $\phi^\omega(s_{-i}) > 0$ implies $s_{-i} \in S_{-i}^1$.

Now suppose we have proved the theorem for $k = 1, \ldots, r$ and let $\omega \in K_i^{r+1}$. Suppose $\phi_i^\omega(s_j) > 0$. We will show that $\omega \in S_j^{r+1}$. By the inductive hypothesis and the fact that $\omega \in K_i^{r+1} \subseteq K_i^r$, we have $s_j \in S_j^r$, so s_j is a best response to some $\phi_j \in S_{-j}^r$. But then $s_j \in S_j^{r+1}$ by construction. Since this is true for all $j \neq i$, if $\phi_i^\omega(s_{-i}) > 0$, then $s_{-i} \in S_{-i}^{r+1}$.

4.10 The Beauty Contest

In his overview of behavioral game theory Camerer (2003) summarizes a large body of evidence in the following way: "Nearly all people use one step of iterated dominance. ... However, at least 10% of players seem to use each of two to four levels of iterated dominance, and the median number of steps of iterated dominance is two." (p. 202) Camerer's observation would be unambiguous if the issue were decision theory, where a single agent faces a nonstrategic environment. But in strategic interaction, the situation is more complicated. In the games reported in Camerer (2003), players gain by using one more level of backward induction than the other players. Hence, players must assess not how many rounds of backward induction the others are capable of but rather how many the other players believe that other players will use. There is obviously an infinite recursion here, with little hope that considerations of Bayesian rationality will guide one to an answer. All we can say is that a Bayesian rational player maximizes expected payoff using a subjective prior over the expected number of rounds over which his opponents use backward induction. The Beauty Contest Game (Moulin 1986) is crafted to explore this issue.

In the Beauty Contest Game, each of $n > 2$ players chooses a whole number between 0 and 100. Suppose the average of these n numbers is k. Then, the players whose choices are closest to $2k/3$ share a prize equally. It is obviously strongly dominated to choose a number greater than $2/3 \times 100 \approx 67$ because such a strategy has payoff 0, whereas the mixed strategy playing 0 to 67 with equal probability has a strictly positive payoff. Thus, one round of eliminating strongly dominated strategies eliminates choices above 67. A second round of eliminating strongly dominated strategies eliminates choices above $(2/3)^2 \times 100 \approx 44$. Continuing in this manner, we see that the only rationalizable strategy is to choose 0. But this is a poor choice in real life. Nagel (1995) studied this game experimentally with various groups of size 14 to 16. The average number chosen was 35, which is between two and three rounds of iterated elimination of strongly dominated strategies. This again conforms to Camerer's generalization, but in this case, of course, people play the game *far* from the unique Nash equilibrium of the game.

4.11 The Traveler's Dilemma

Consider the following game G_n, known as the *Traveler's Dilemma* (Basu 1994). Two business executives pay bridge tolls while on a trip but do not have receipts. Their superior tells each of them to report independently an integral number of dollars between 2 and n on their expense sheets. If they report the same number, each will receive this much back. If they report different numbers, each will get the smaller amount, plus the low reporter will get an additional \$2 (for being honest) and the high reporter will lose \$2 (for trying to cheat).

	s_2	s_3	s_4	s_5
s_2	$2,2$	$4,0$	$4,0$	$4,0$
s_3	$0,4$	$3,3$	$5,1$	$5,1$
s_4	$0,4$	$1,5$	$4,4$	$6,2$
s_5	$0,4$	$1,5$	2.6	$5,5$

Figure 4.4. The Traveler's Dilemma

Let s_k be strategy report k. Figure 4.4 illustrates the game G_5. Note first that s_5 is only weakly dominated by s_4, but a mixed strategy $\epsilon s_2 + (1-\epsilon)s_4$ strongly dominates s_5 whenever $1/2 > \epsilon > 0$. When we eliminate s_5 for both players, s_3 only weakly dominates s_4, but a mixed strategy $\epsilon s_2 + (1-\epsilon)s_3$ strongly dominates s_4 for any $\epsilon > 0$. When we eliminate s_4 for both players, s_2 strongly dominates s_3 for both players. Hence (s_2, s_2) is the only strategy pair that survives the iterated elimination of strongly dominated strategies. It follows that s_2 is the only rationalizable strategy, and the only Nash equilibrium as well.

The following exercise asks you to show that for $n > 3$, s_n in the game G_n is strongly dominated by a mixed strategy of $s_2, \ldots, s_{n-1}$.

a. Show that for any $n > 4$, s_n is strongly dominated by a mixed strategy σ_{n-1} using only s_{n-1} and s_2.
b. Show that eliminating s_n in G_n gives rise to the game G_{n-1}.
c. Use the above reasoning to show that for any $n > 2$, the iterated elimination of strongly dominated strategies leaves only s_2, which is thus the only rationalizable strategy and hence also the only Nash equilibrium of G_n.

Suppose $n = 100$. It is not plausible to think that individuals would actually play 2,2 because by playing a number greater than, say, 92, they are assured of at least 90.

4.12 The Modified Traveler's Dilemma

One might think that the problem is that pure strategies are dominated by mixed strategies, and as we will argue in chapter 6, rational agents have no incentive to play mixed strategies in one-shot games.

However, we can change the game a bit so that 2,2 is the only strategy profile that survives the iterated elimination of pure strategies strictly dominated by pure strategies. In figure 4.5, I have added 1% of s_2 to s_4 and 2% of s_2 to s_3, for both players.

	s_2	s_3	s_4	s_5
s_2	2.00, 2.00	4.00, 0.04	4.00, 0.02	4.00, 0.00
s_3	0.04, 4.00	3.08, 3.08	5.08, 1.04	5.08, 1.00
s_4	0.02, 4.00	1.04, 5.08	4.04, 4.04	6.04, 2.00
s_5	0.00, 4.00	1.00, 5.08	2.00, 6.04	5.00, 5.00

Figure 4.5. The Modified Traveler's Dilemma

It is easy to check that now s_4 strictly dominates s_5 for both players, and when s_5 is eliminated, s_3 strictly dominates s_4 for both players. When s_4 is eliminated, s_2 strictly dominates s_3.

This method will extend to a Modified Traveler's Dilemma of any size. To implement this, let

$$f(m,q) = \begin{cases} q-2 & q < m \\ q & q = m \\ m+2 & q > m \end{cases},$$

and define

$$\begin{aligned} \pi(2,q) &= f(2,q) \qquad \text{for } q = 2,\ldots,n \\ \pi(m,q) &= \sum_{k=3,l=2,\ldots,n} f(m,q) + f(2,q)\frac{n-k}{4(n+1)} \end{aligned}$$

It is easy to show that this Modified Traveler's Dilemma is strictly dominance-solvable and that the only rationalizable strategy again has payoff 2,2. Yet, it is clear that for large n, rational players would likely choose a strategy with a payoff near n. This shows that there is something fundamentally wrong with the rationalizability criterion. The culprit is the CKR, which is the only questionable assumption we made in defining rationalizability. It is not irrational to choose a high number in the Modified Traveler's Dilemma, and indeed doing so is likely to lead to a high payoff compared to the game's only rationalizable strategy. However, doing so is not compatible with the common knowledge of rationality.

4.13 Global Games

Suppose Alice and Bob can cooperate (C) and earn 4, but by defecting (D) either can earn x, no matter what the other player does. However, if one player cooperates and the other does not, the cooperator earns 0. Clearly, if $x > 4$, D is a strictly dominant strategy, and if $x < 0$, C is a strictly dominant strategy. If $0 < x < 4$, the players have a Pareto-optimal strategy C in which they earn 4, but there is a second Nash equilibrium in which both players play D and earn $x < 4$.

Alice \ Bob	D	C
D	x,x	x,0
C	0,x	4,4

Suppose, however, that x is private information, each player receiving an imperfect signal $\xi_i = x + \hat{\epsilon}_i$ that is uniformly distributed on the interval $[x - \epsilon/2, x + \epsilon/2]$, where $\hat{\epsilon}_A$ is distributed independently of $\hat{\epsilon}_B$. We can then demonstrate the surprising result that, no matter how small the error ϵ is, the resulting game has a unique rationalizable strategy, which is to play C for $x < 2$ and D for $x > 2$. Note that this is very far from the Pareto-optimal strategy, no matter how small the error.

To see that this is the only Nash equilibrium, note that a player surely chooses C when $\xi < -\epsilon/2$, and D when $\xi > 4+\epsilon/2$, so there is a smallest cutoff x^* such that, at least in a small interval around x^*, the player chooses D when $\xi < x^*$ and C when $\xi > x^*$. For a discussion of this and other details of the model, see Carlsson and van Damme (1993), who invented and analyzed this game, which they term a *global game*. By the symmetry of the problem, x^* must be a cutoff for both players. If Alice is at the cutoff, then with equal probability Bob is above or below the cutoff, so he plays D and C with equal probability. This means that the payoff for Alice playing D is x^* and for playing C is 2. Because these must be equal if Alice is to have cutoff x^*, it follows that $x^* = 2$. Thus, there is a unique cutoff and hence a unique Nash equilibrium $x^* = 2$.

To prove that $x^* = 2$ is the unique rationalizable strategy, suppose Alice chooses cutoff x_A and Bob chooses x_B as a best response. Then when Bob receives the signal $\xi_B = x_B$, he knows Alice's signal is uniformly distributed on $[x_B - \epsilon, x_B + \epsilon]$. To see this, let $\hat{\epsilon}_i$ be player i's signal error, which is uniformly distributed on $[-\epsilon/2, \epsilon/2]$. Then

$$\xi_B = x + \hat{\epsilon}_B = \xi_A - \hat{\epsilon}_A + \hat{\epsilon}_B.$$

Because $-\hat{\epsilon}_A + \hat{\epsilon}_B$ is the sum of two random variables distributed uniformly on $[-\epsilon/2, \epsilon/2]$, ξ_B must be uniformly distributed on $[-\epsilon, \epsilon]$. It follows that the probability that Alice's signal is less than x_A is $q \equiv (x_A - x_B + \epsilon)/(2\epsilon)$, provided this is between zero and one. Then, x_B is determined by equating the payoff from D and C for Bob, which gives $4q = x_B$. Solving for x_B, we find that

$$x_B = \frac{2(x_A + \epsilon)}{2 + \epsilon} = x_A - \frac{(x_A - 2)\epsilon}{2 + \epsilon}. \tag{4.5}$$

The largest candidate for Alices cutoff is $x_A = 4$, in which case Bob will choose cutoff $f_1 \equiv 4 - 2\epsilon/(2+\epsilon)$. This means that no cutoff for Bob that is greater than f_1 is a best response for Bob, and therefore no such cutoff is rationalizable. But then the same is true for Alice, so the highest possible cutoff is f_1. Now, using (4.5) with $x_A = f_1$, we define $f_2 = 2(f_1+\epsilon)/(2+\epsilon)$, and we conclude that no cutoff greater than f_2 is rationalizable. We can repeat this process as often as we please, each iteration k defining $f_k = 2(f_{k-1}+\epsilon)/(2+\epsilon)$. Because the $\{f_k\}$ are decreasing and positive, they must have a limit, and this must satisfy the equation $f = 2(f+\epsilon)/(2+\epsilon)$, which has the solution $f = 2$. Another way to see this is to calculate f_k explicitly. We find that

$$f_k = 2 + 2\left(\frac{2}{2+\epsilon}\right)^k,$$

which converges to 2 as $k \to \infty$, no matter how small $\epsilon > 0$ may be. To deal with cutoffs below $x = 2$, note that (4.5) must hold in this case as well. The smallest possible cutoff is $x = 0$, so we define $g_1 = 2\epsilon/(2+\epsilon)$, and $g_k = 2(g_{k-1}+\epsilon)/(2+\epsilon)$ for $k > 1$. Then, similar reasoning shows that no cutoff below g_k is rationalizable for any $k \geq 1$. Moreover the $\{g_k\}$ are increasing and bounded above by 2. The limit is then given by solving $g = 2(g+\epsilon)/(2+\epsilon)$, which gives $g = 2$. Explicitly, we have

$$g_k = 2 - 2\left(\frac{2}{2+\epsilon}\right)^k,$$

which converges to 2 as $k \to \infty$. This proves that the only rationalizable cutoff is $x^* = 2$.

When the signal error is large, the Nash equilibrium of this game is plausible, and experiments show that subjects often settle on behavior close to that predicted by the model. However, the model predicts a cutoff of 2 for all $\epsilon > 0$ and a jump to cutoff 4 for $\epsilon = 0$. This prediction is not verified experimentally. In fact, subjects tend to treat public information and private information scenarios the same and tend to implement the payoff-dominant outcome rather than the less efficient Nash equilibrium outcome

(Heinemann, Nagel, and Ockenfels 2004; Cabrales, Nagel, and Armenter 2007).

4.14 CKR Is an Event, Not a Premise

Rational agents go through some process of eliminating unrationalizable strategies. CKR implies that players continue eliminating as long as there is anything to eliminate. By contrast, as we have seen, the median number of steps of iterated dominance found in experiments is 2, and few player use more than 4 (Camerer 2003). This evidence indicates that CKR does not hold in the games analyzed in this chapter. Yet, it is easy to construct games in which we would expect CKR to hold. For instance, consider the following Benign Centipede Game. Alice and Bob take turns for 100 rounds. In each round $r < 100$, the player choosing can cooperate, in which case we continue to the next round, or the player can quit, in which case each player has a payoff of $(1 - r/100)$ dollars and the game is over. If both players cooperate for all 100 rounds, each player gets \$10.

CKR for this game implies Alice and Bob will both choose 100, and they will each earn \$10. For, in the final round, because Bob is rational, he will choose to continue, to earn \$10 as opposed to $(1-100/100) = 0$ dollars by quitting. Since Alice knows that Bob is rational, she knows she will earn \$10 by continuing, as opposed to \$0.01 by quitting. Now, in round 98, Bob earns \$0.02 by quitting, which is more than he could earn by continuing and having Alice quit, in which case he would earn \$0.01. However, Bob knows that Alice knows that Bob is rational, and Bob knows that Alice is rational. Hence, Bob knows that Alice will continue, so he continues in round 98. The argument is valid back to round 1, so CKR implies cooperation on each round.

There is little doubt that real-life players will play the strategy dictated by CKR in this case, although they do not in the Beauty Contest Game, the Traveler's Dilemma, and many other such games. Yet, there are no epistemic differences in what the players know about each other in the Benign Centipede Game as opposed to the other games discussed above. Indeed, CKR holds in the Benign Centipede Game because players will continue to the final round in this game, and not vice-versa.

It follows from this line of reasoning that the notion that CKR is a *premise* concerning the knowledge agents have about one another is false. Rather, CKR is an *event* in which a strategy profile chosen by agents may or may not be included. Depending upon the particular game played, and under identical epistemic conditions, CKR may or may not hold.

I have stressed that a central weakness of epistemic game theory is the manner in which it represents the commonality of knowledge across individuals. Bayesian rationality itself supplies no analytical principles that are useful in deducing that two individuals have mutual, much less common, knowledge of particular events. We shall later suggest epistemic principles that do give rise to common knowledge (e.g., theorem 7.2), but these do not include common knowledge of rationality. To my knowledge, no one has ever proposed a set of epistemic conditions that jointly imply CKR. Pettit and Sugden (1989) conclude their critique of CKR by asserting that "the situation where the players are ascribed common knowledge of their rationality ought strictly to have no interest for game theory." (p. 182) Unless and until someone comes up with a epistemic derivation of CKR that explains why it is plausible in the Benign Centipede Game but not in the Beauty Contest game, this advice of Pettit and Sugden deserves to be heeded.

For additional analysis of CKR as a premise, see §5.13.

5

Extensive Form Rationalizability

> The heart has its reasons of which reason knows nothing.
>
> Blaise Pascal

The extensive form of a game is informationally richer than the normal form since players gather information that allows them to update their subjective priors as the game progresses. For this reason, the study of rationalizability in extensive form games is more complex than the corresponding study in normal form games. There are two ways to use the added information to eliminate strategies that would not be chosen by a rational agent: backward induction and forward induction. The latter is relatively exotic (although more defensible) and will be addressed in chapter 9. Backward induction, by far the most popular technique, employs the iterated elimination of weakly dominated strategies, arriving at the *subgame perfect* Nash equilibria—the equilibria that remain Nash equilibria in all subgames. We shall call an extensive form game *generic* if it has a unique subgame perfect Nash equilibrium.

In this chapter we develop the tools of modal logic and present Robert Aumann's famous proof (Aumann 1995) that CKR implies backward induction. This theorem has been widely criticized, as well as widely misinterpreted. I will try to sort out the issues, which are among the most important in contemporary game theory. I conclude that Aumann is perfectly correct, and the real culprit is CKR itself.

5.1 Backward Induction and Dominated Strategies

Backward induction in extensive form games with *perfect information* (i.e., where each information set is a single node) operates as follows. Choose any terminal node $\tau \in T$ and find the parent node of this terminal node, say node ν. Suppose player i chooses at ν and suppose i's highest payoff at ν is attained at terminal node $\tau' \in T$. Erase all the branches from ν so ν becomes a terminal node and attach the payoffs from τ' to the new terminal node ν. Also, record i's move at ν, so you can specify i's equilibrium strategy when you have finished the analysis. Repeat this procedure for all the terminal nodes of the original game. When you are done, you will have an extensive form game that is one level less deep than the original game. Now repeat the process as many times as possible. If the resulting game tree has just one possible move at each node, then when you reassemble the moves you have recorded for each player, you will have a Nash equilibrium.

We call this *backward induction* because we start at the terminal nodes of the game and move backward. Note that if players move at more than a single node, backward induction eliminates *weakly* dominated strategies, and hence can eliminate Nash equilibria that use weakly dominated strategies. Moreover, backward induction is prima facie much stronger than normal form rationalizability (§4.6), which is equivalent to the iterated elimination of strongly dominated strategies.

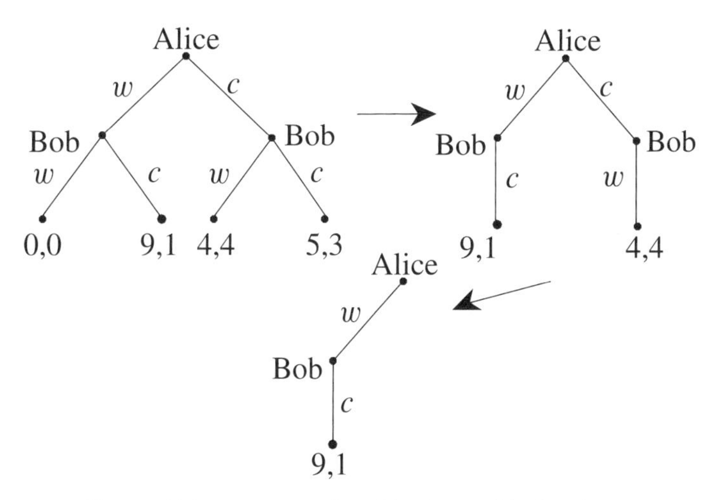

Figure 5.1. An example of backward induction

For an example of backward induction, consider figure 5.1. We start with the terminal node labeled (0,0) and follow it back to the Bob node on the left. At this node, w is dominated by c because $1 > 0$, so we erase the branch where Bob plays w and its associated payoff. We locate the next terminal node in the original game tree, (4,4), and follow back to the Bob node on the right. At this node, c is dominated by w, so we erase the dominated node and its payoff. Now we apply backward induction to this smaller game tree—this time, of course, it's trivial. We find the first terminal node, (9,1), that leads back to Alice's choice node. Here c

is dominated, so we erase that branch and its payoff. We now have our solution: Alice chooses w, Bob chooses cw, and the payoffs are (9,1).

It is clear from this example that by using backward induction and hence eliminating weakly dominated strategies, we have eliminated the Nash equilibrium c, ww. This is because when we assume Bob plays c in response to Alice's w, we eliminate the weakly dominated strategies ww and wc for Bob. We call c, ww an *incredible threat*. Backward induction eliminates incredible threats.

5.2 Subgame Perfection

Let ν be an information set of an extensive form game $\mathcal{G}$ that consists of a single node. Let $\mathcal{H}$ be the smallest collection of nodes including ν such that if h' is in $\mathcal{H}$, then all of the successor nodes of h' are in $\mathcal{H}$ and all nodes in the same information set as h' are in $\mathcal{H}$. We endow $\mathcal{H}$ with the information set structure, branches, and payoffs inherited from $\mathcal{G}$, the players in $\mathcal{H}$ being the subset of players of $\mathcal{G}$ who move at some information set of $\mathcal{H}$. It is clear that $\mathcal{H}$ is an extensive form game. We call $\mathcal{H}$ a *subgame* of $\mathcal{G}$.

If $\mathcal{H}$ is a subgame of $\mathcal{G}$ with root node ν, then every pure-strategy profile of $\mathcal{G}$ that reaches ν has a counterpart s_H in $\mathcal{H}$, specifying that players in $\mathcal{H}$ make the same choices with s_H at a node in $\mathcal{H}$ as they do with s_G at the same node in $\mathcal{G}$. We call s_H the *restriction* of s_G to the subgame $\mathcal{H}$. Suppose $\sigma_G = \alpha_1 s_1 + \cdots + \alpha_k s_k$ ($\sum_i \alpha_i = 1$) is a mixed strategy of $\mathcal{G}$ that reaches the root node ν of $\mathcal{H}$, and let $I \subseteq \{1, \ldots, k\}$ be the set of indices such that $i \in I$ iff s_i reaches h. Let $\alpha = \sum_{i \in I} \alpha_i$. Then, $\sigma_H = \sum_{i \in I} (\alpha_i/\alpha) s_i$ is a mixed strategy defined on $\mathcal{H}$, called the *restriction* of σ_G to $\mathcal{H}$. We have $\alpha > 0$ because σ_G reaches ν, and the coefficient α_i/α represents the probability of playing s_i, conditional on reaching h.

It is clear that if s_G is a pure-strategy Nash equilibrium for a game $\mathcal{G}$ and if $\mathcal{H}$ is a subgame of $\mathcal{G}$ whose root node is reached using s_G, then the restriction s_H of s_G to $\mathcal{H}$ must be a Nash equilibrium in $\mathcal{H}$. However, if the root node of $\mathcal{H}$ is not reached by s_G, then the restriction of s_G to $\mathcal{H}$ need *not* be a Nash equilibrium. This is because if a node is not reached by s_G, then the payoff to the player choosing at that node does not depend on his choice in $\mathcal{G}$, but it may depend on his choice in $\mathcal{H}$. We say a Nash equilibrium of an extensive form game is *subgame perfect* if its restriction to every subgame is a Nash equilibrium of the subgame.

It is easy to see that a simultaneous move game has no proper subgames (a game is always a subgame of itself; we call the whole game an *improper* subgame), because all the nodes are in the same information set for at least one player. Similarly, a game in which Nature makes the first move also has no proper subgames if there is at least one player who does not know Nature's choice.

At the other extreme, in a game of perfect information (i.e., for which all information sets are singletons), *every* nonterminal node is the root node of a subgame. This allows us to find the subgame perfect Nash equilibria of such games by backward induction, as described in §5.1. This line of reasoning shows that, in general, backward induction consists of the iterated elimination of weakly dominated strategies and eliminates all nonsubgame perfect Nash equilibria.

5.3 Subgame Perfection and Incredible Threats

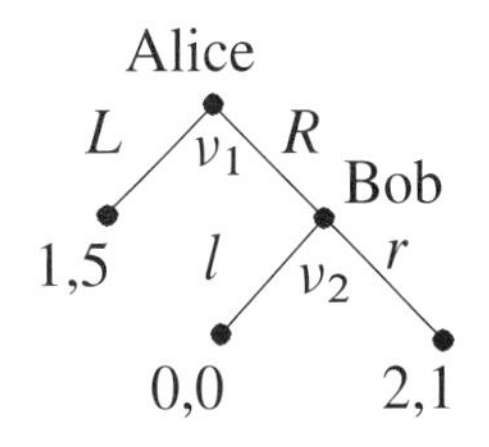

The game to the right has a Nash equilibrium in pure strategies Rr, in which Alice earns 2 and Bob earns 1. This equilibrium is subgame perfect because in the subgame starting with Bob's choice at ν_B, r is payoff-maximizing for Bob. This equilibrium is also the one chosen by backward induction. However, there is a second Nash equilibrium, Ll, in which Alice earns 1 and Bob earns 5. Bob much prefers this equilibrium, and if he can somehow induce Alice to believe that he will play l, her best response is L. However Bob communicates to Alice his intention to play l, if Alice believes Bob is rational, she knows he in fact will play r if the game actually reaches ν_B. Thus, Ll is thought to be an implausible Nash equilibrium, whereas the subgame perfect Nash equilibrium is held in high regard by game theorists.

5.4 The Surprise Examination

A group of game theorists once took an intensive Monday-through-Friday logic course. After several weeks, the professor announced that there would be a surprise examination one day the following week. Each student thought to himself, "The exam cannot be given next Friday because then it would not be a surprise." Each then concluded that, for similar reasons, the exam could not be given next Thursday, next Wednesday, next Tuesday, or next Monday. Each student thus concluded that the professor was mistaken.

The professor gave the exam the next Tuesday, and all of the students were surprised.

This is one version of a famous logic problem called the Surprise Examination or the Hanging Paradox. For an overview of the many proposed solutions to the problem, see Chow (1998). Interpretations vary widely, and there is no single accepted solution. There are a number of cogent analyses using standard logic and modal logic to show that the professor's statement is impermissively self-referential or self-contradictory, and because a false statement can validly imply anything, there is no paradox in the professor's prediction being correct.

Backward induction indicates that the exam cannot be given. But, if a student believes this, then it will be a surprise no matter what day it is given. Thus, the incoherence of backward induction should convince a rational student that the professor's prediction is indeed reasonable. But, what exactly is incoherent about backward induction? I present this paradox to indicate the danger of using the informal logic of backward induction. We develop a more analytically precise approach below.

5.5 The Common Knowledge of Logicality Paradox

Let us say an agent is *logical* in making inferences concerning a set of propositions if the agent rules out all statements that are inconsistent with this set. We then define *common knowledge of logicality* (CKL) for a set $i = 1, \ldots, n$ of agents in the usual way: for any set of integers $i_1, \ldots, i_k \in [1, \ldots, n]$, i_1 knows that i_2 knows that ... knows that i_{k-1} knows that i_k is logical.

A father has \$690,000 to leave to his children, Alice and Bob, who do not know the size of his estate. He decides to give one child \$340,000 and the other \$350,000, each with probability 1/2. However, he does not want one child to feel slighted by getting a smaller amount, at least during his lifetime. So, he tells his children: "I will randomly pick two numbers, without replacement, from a set $S \subseteq [1, \ldots, 100]$, assign to each of you randomly one of these numbers, and give you an inheritance equal to \$10,000 times the number you have been assigned. Knowing the number assigned to you will not allow you to conclude for sure whether you will inherit more or less than your sibling." The father, confident of the truth of his statement, which we take to be common knowledge for all three individuals, sets $S = \{34, 35\}$.

Alice ponders this situation, reasoning as follows, assuming common knowledge of logicality: "Father knows that if $1 \in S$ or $100 \in S$, then there is a positive probability one of these numbers will be chosen and assigned to me, in which case I would be certain of the relative position of my inheritance." Alice knows her father knows she is logical, so she knows that $1 \notin S$ and $100 \notin S$. But Alice reasons that her father knows that she knows that he knows that she is logical, so she concludes that her father knows that he cannot include 2 or 99 in S. But Alice know this as well, by CKL, so she reasons that her father cannot include 3 or 98 in S. Completing this recursive argument, Alice concludes that S must be empty.

However, the father gave one child the number 34 and the other 35, neither child knowing for sure which had the higher number. Thus, the father's original assertion was true, and Alice's reasoning was faulty. We conclude that *common knowledge of logicality is false* in this context. CKL fails when the father includes 35 in S, because this is precluded by CKL.

CKL appears prima facie to be an innocuous extension of logicality and indeed usually is not even mentioned in such problems, but, in fact, it leads to faulty reasoning and must be rejected. In this regard, CKL is much like CKR, which also appears to be an innocuous extension of rationality but in fact is often counterindicated.

5.6 The Repeated Prisoner's Dilemma

Suppose Alice and Bob play the Prisoner's Dilemma, one stage of which is shown to the right, 100 times. Common sense tells us that players will cooperate for at least 95 rounds, and this is indeed supported by experimental evidence (Andreoni and Miller 1993).

	C	*D*
C	3,3	0,4
D	4,0	1,1

However, a backward induction argument indicates that players will defect in the very first round. To see this, note that the players will surely defect in round 100. But then, nothing they do in round 99 can help prolong the game, so they will both defect in round 99. Repeating this argument 99 times, we see that they will both defect on round 1.

Although in general backward induction removes weakly iterated dominated strategies, in this case it removes only strongly iteratedly dominated strategies, so the only rationalizable strategy, according to the analysis of the previous chapter, is the universal defect Nash equilibrium. This presents a problem for the rationalizability concept that is at least as formidable as in the case of the normal form games presented in the previous chapter.

In this case, however, the extensive form provides an argument as to why the logic of backward induction is compromised. The backward induction reasoning depends on CKR in precisely the same manner as in the previous chapter. However, in the current case, the first time either player chooses C, both players know that CKR is false. At the terminal nodes of the Repeated Prisoner's Dilemma, players have chosen C many times. Therefore, we cannot assume CKR at the terminal nodes because these nodes could not be reached given CKR. This critique of backward induction has been made by Binmore (1987), Bicchieri (1989), Pettit and Sugden (1989), Basu (1990), and Reny (1993), among others.

The critique, however, is incorrect. The backward induction argument is simply a classic example of reductio ad adsurdum: assume a proposition and then show that the proposition is false. In this case, we assume CKR and we show by *reductio* that the 100th round will not be reached. There is no flaw in this argument. It is incoherent to base a critique of the proposition that CKR implies backward induction on what would happen if CKR were false.

The misleading attractiveness of this flawed critique of the proposition that CKR implies backward induction lies in the observation that the first time either player chooses C, both players know that CKR is false, and hence they are free to devise a modus operandi that serves their interests for the remainder of the game. For instance, both may employ the tit-for-tat strategy of playing C in one round and copying one's partner's previous move in each subsequent round, except for playing universal D as the game nears the 100th round termination point.

This argument is completely correct but is not a critique of the proposition that CKR implies backward induction. Indeed, assuming CKR, neither player will choose C in any period.

As I shall argue below, the problem with backward induction is that CKR is not generally a permissible assumption, and hence backward induction cannot be justified on rationality grounds.

5.7 The Centipede Game

In Rosenthal's Centipede Game, Alice and Bob start out with \$2 each and alternate rounds. In the first round, Alice can defect (D) by stealing \$2 from Bob, and the game is over. Otherwise, Alice cooperates (C) by not stealing, and Nature gives her \$1. Then Bob can defect (D) and steal \$2 from Alice, and the game is over, or he can cooperate (C), and Nature gives him \$1. This continues until one or the other defects or until 100 rounds have elapsed. The game tree is illustrated in figure 5.2.

Formally, the reduced normal form of the Centipede Game can be described as follows. Alice chooses an odd number k_a between 1 and 101, and Bob chooses an even number k_b between 2 and 100, plus either C or D if $k_b = 100$. The lower of the two choices, say k^*, determines the payoffs. If $k^* = 100$, the payoff from $(k^*, C) = (52, 52)$ and the payoff from $(k^*, D) = (50, 53)$. Otherwise, if k^* is an odd number, the payoffs are $(4 + (k^* - 1)/2, (k^* - 1)/2)$, and if k^* is even, the payoffs are $(k^*/2, 3 + k^*/2)$. You can check that these choices generate exactly the payoffs as described above.

To determine the strategies in this game, note that Alice and Bob each has 50 places to move, and in each place each can play D or C. We can thus describe a strategy for each as a sequence of 50 letters, each of which is a D or a C. This means there are $2^{50} = 1125899906842624$ pure strategies for each. Of course, the first time a player plays D, what he does after that does not affect the payoff of the game, so the only payoff-relevant question is at what round, if any, a player first plays D. This leaves 51 strategies for each player.

We can apply backward induction to the game, finding that in the unique subgame perfect Nash equilibrium of this game, both players defect the first time they get to choose. To see this, note that in the final round Bob will defect, and hence Alice will defect on her last move. But then Bob will defect on his next-to-last move, as will Alice on her next-to-last move. Similar reasoning holds for all rounds, proving that Bob and Alice will defect on their first move in the unique subgame perfect equilibrium.

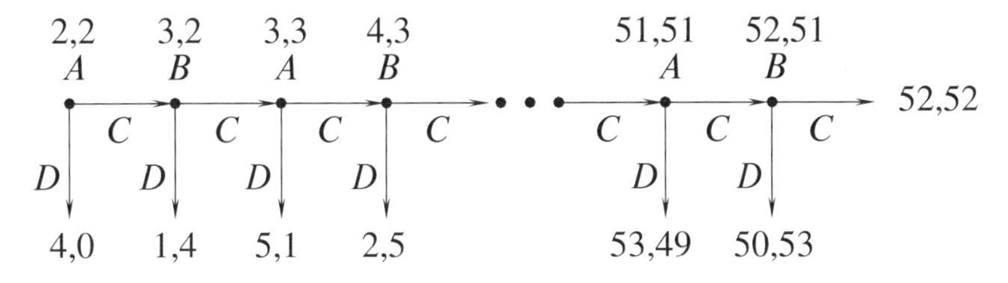

Figure 5.2. Rosenthal's Centipede Game

Now, of course, common sense tells you that this is not the way real players would act in this situation, and empirical evidence corroborates this intuition (McKelvey and Palfrey 1992). It may seem that the culprit is subgame perfection because backward induction finds only subgame perfect equilibria. This, however, is not the problem. The culprit is the Nash equilibrium criterion itself because in *any* Nash equilibrium Alice defects in round 1.

While backward induction does not capture how people really play the Centipede Game, normal form rationalizability does a better job because it suggests that cooperating until near the end of the game does not conflict with CKR. This is because all pure strategies for Bob are normal form rationalizable except $k_b = (100, \mathrm{C})$, as are all pure strategies for Alice except $k_a = 101$ (i.e., cooperate in every round). To see this, we can show that there is a mixed-strategy Nash equilibrium of the game where Bob uses $k_b = 2$ and any one of his other pure strategies except $(100,C)$, and Alice uses $k_a = 1$. This shows that all pure strategies for Bob except $(100,C)$ are rationalizable. Alice's $k_a = 101$ is strictly dominated by a mixed strategy using $k_a = 99$ and $k_a = 1$, but each of her other pure strategies is a best response to some rationalizable pure strategy of Bob. This shows that these pure strategies of Alice are themselves rationalizable.

This does not explain why real people cooperate until near the end of the game, but it does show that it does not conflict with CKR in the normal form game to do so. This is little consolation, however, since cooperating becomes compatible with CKR only by ignoring information that the players surely have—namely, that embodied in the extensive form structure of the game. In the context of this additional information, CKR certainly implies the validity of the backward induction argument, and hence of the assertion that CKR ensures defection on round 1.

5.8 CKR Fails Off the Backward Induction Path

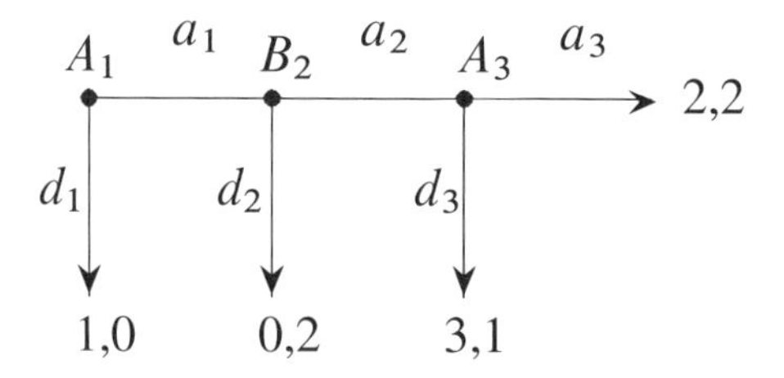

Figure 5.3. A short Centipede Game

This section presents a formal epistemic argument supporting the contention that CKR is violated off the subgame perfect game path in a generic extensive form game. This is the thrust of Aumann's (1995) general proof that CKR implies backward induction, but here we present the proof for a very simple game where the intuition is relatively clear. Figure 5.3 depicts a very short Centipede Game (§5.7) played by Alice (A) and Bob (B), where Alice moves at A_1 and A_3 and Bob moves at B_2. Let R_A and R_B stand for "Alice is rational," and "Bob is rational," respectively, let $\mathbf{K}_A$ and $\mathbf{K}_B$ be knowledge operators, and let π_A and π_B be the payoffs to Alice and Bob, respectively. We will have much more to say in later chapters concerning what it means to be rational and what it means to assert that an agent "knows something." For now, we simply assume that a rational agent always chooses a best response, we assume $\mathbf{K}p \Rightarrow p$ (i.e., if an agent knows something, then it is true), and we assume $\mathbf{K}p \wedge \mathbf{K}(p \Rightarrow q) \Rightarrow \mathbf{K}q$ (i.e., if an agent knows p and also knows that p implies q, then the agent knows q). We assume that both players know all the rules of the game, and since the game is one of perfect information, when the game is at a particular node, it is common knowledge that this is the case.

We have $A_3 \wedge R_A \Rightarrow d_3$, which we read, "at node A_3, if Alice is rational, then she will choose d_3." This is true because $a_3 \Rightarrow (\pi_A = 2)$, and $d_3 \Rightarrow (\pi_A = 3)$, and since these implications simply follow from the rules of the game, they are known to Alice, so $\mathbf{K}_A(a_3 \Rightarrow (\pi_A = 2)) \wedge \mathbf{K}_A(d_3 \Rightarrow (\pi_A = 3))$. This assertion implies $A_3 \wedge R_A \Rightarrow d_3$. Now, if Bob knows that Alice is rational, and if the rules of the game are common knowledge, then $a_2 \Rightarrow \mathbf{K}_B a_2 \Rightarrow K_B d_3 \Rightarrow \mathbf{K}_B(\pi_B = 1)$. Moreover, $d_2 \Rightarrow \mathbf{K}_B d_2 \Rightarrow \mathbf{K}_B(\pi_B = 3)$, so $B_2 \wedge R_B \wedge \mathbf{K}_B R_A \Rightarrow d_2$. Now, if Alice knows that Bob is rational at A_1 and that Bob knows that she is rational at B_2, then $\mathbf{K}_A(a_1 \Rightarrow d_2)$, so $\mathbf{K}_A(a_1 \Rightarrow (\pi_A = 0))$. However, $\mathbf{K}_A(d_1 \Rightarrow (\pi_A = 1))$. Hence, since Alice is rational at A_1, she will choose d_1. In short, we have

$$R_A \wedge \mathbf{K}_A(R_B \wedge \mathbf{K}_B R_A) \Rightarrow d_1. \tag{5.1}$$

We have thus shown that if there are two levels of mutual knowledge of rationality, then the backward induction solution holds. But this presupposes that the set of assumptions is *consistent*; i.e., it assumes that we cannot also prove from these assumptions that Alice will play a_1. Note that if Alice plays a_1, then the premise of (5.1) is false, and Bob knows this, which says that

$$\neg\mathbf{K}_B R_A \vee \neg\mathbf{K}_B\mathbf{K}_A R_B \vee \neg\mathbf{K}_B\mathbf{K}_A\mathbf{K}_B R_A. \tag{5.2}$$

In words, if Alice chooses a_1, then Bob does not know Alice is rational, or Bob does not know that Alice knows that Bob is rational, or Bob does not know that Alice knows that Bob knows that Alice is rational. One level of mutual knowledge, which imply $\mathbf{K}_B R_A$, eliminate the first alternative, and two levels, which implies $\mathbf{K}_B\mathbf{K}_A R_B$, eliminates the second alterative. Thus, if Alice chooses a_1, it must be the case that $\neg\mathbf{K}_B\mathbf{K}_A\mathbf{K}_B R_A$; i.e., Alice's choice violates third level mutual knowledge of rationality and hence violates common knowledge of rationality.

We conclude that no node after the first can be attained while conserving more than two levels of mutual knowledge of rationality. Nor is there anything special about this game. As we expressed in the previous section and will prove in §5.11, in all finite extensive form games of perfect information with unique subgame perfect equilibria, the only nodes in the game at which common knowledge of rationality can hold are along the backward induction path of play. In the current case, there are just two such nodes, the root node and the first terminal node.

5.9 How to Play the Repeated Prisoner's Dilemma

	C	D
C	R,R	S,T
D	T,S	P,P

In cases where a stage game is repeated a finite number of times, it is reasonable to assume Bayesian rationality (§1.5), avoid backward induction, and use decision theory to determine player behavior. Consider, for instance, the Prisoner's Dilemma (§2.10), the stage game of which is shown to the right with $T > R > P > S$, repeated until one player defects or 100 rounds have been played. Backward induction implies that both players will defect in the very first round, and indeed, this is the only Nash equilibrium of the game. However, player 1 may say to himself, "If my partner and I both play D, we will each earn only P. I am willing to cooperate for at least 95 rounds, and if my partner is smart, he will also be willing to cooperate for many rounds. I suspect my partner will reason similarly. Thus we stand to earn on the order of $95R$. If I am wrong about my partner, I will lose only $S - P$, so it's worth a try, because if I am right, I will go home with a tidy bundle."

More formally, suppose I conjecture that my partner will cooperate up to round k and then defect, with probability g_k. Then, I will choose a round m to defect in that maximizes the expression

$$\pi_m = \sum_{i=1}^{m-1}((i-1)R + S)g_i + ((m-1)R + P)g_m \qquad (5.3)$$
$$+ ((m-1)R + T)(1 - G_m),$$

where $G_m = g_1 + \cdots + g_m$. The first term in this expression represents the payoff if my partner defects first, the second term if we defect simultaneously, and the final term if I defect first. In many cases, maximizing this expression suggests cooperating for many rounds for all plausible probability distributions. For instance, suppose g_k is uniformly distributed in the rounds $m = 1, \ldots, 99$. Suppose, for concreteness, $(T, R, P, S) = (4, 2, 1, 0)$. Then, you can check by using equation (5.3) that it is a best response to cooperate up to round 98. Indeed, suppose you expect your opponent to defect in round 1 with probability 0.95 and otherwise defect with equal probability on any round from 2 to 99. Then it is still optimal to defect in round 98. Clearly, the backward induction assumption is not plausible unless you think your opponent is highly likely to be an obdurate backward inductor.

The reasoning dilemma begins if I then say to myself, "My partner is just as capable as I am of reasoning as above, so he will also cooperate at least up to round 98. Thus, I should set $m = 97$. But, of course, my partner also knows this, so he will surely defect in round 96, in which case I should surely defect in round 95." This sort of self-contradictory reasoning shows that there is something faulty in the way we have set up the problem. If the $\{g_k\}$ distribution is reasonable, then I should use it. It is self-contradictory to use this distribution to show that it is the wrong distribution to use. But my rational partner will know this as well, and I suspect he will revert to the first level of analysis, which says to cooperate at least up to round 95. Thus, we two rational folks will cooperate for many rounds in this game rather than play the Nash equilibrium.

Suppose, however, that it is common knowledge that both I and my partner have the *same Bayesian priors* (§1.5) concerning when the other will defect. This is sometimes called *Harsanyi consistency* (Harsanyi 1967). Then, it is obvious that we will both defect at our first opportunity because the backward induction conclusion now follows from a strictly Bayesian argument: the only prior that is compatible with common knowledge of common priors is defection in round 1. However, there is no plausible reason for us to assume Harsanyi consistency in this case.

This argument reinforces our conclusion that *there is nothing sacrosanct about CKR*. Classical game theorists commonly argue that rationality *requires* that agents use backward induction, but this is simply not the case. If two players are rational and they know both are rational, and if each knows the other's conjecture, then they will play the unique Nash equilibrium of the game (§8.4). But, as we have seen, we may each reasonably conclude that we do not know the other's conjecture, but we know enough to cooperate for many rounds.

5.10 The Modal Logic of Knowledge

The Savage model in decision theory is agnostic as to how an agent's subjective prior is acquired. How people play a game depends on their beliefs about the beliefs of the other players, including their beliefs about others' beliefs about the players, and so on. To deal analytically with this situation, we develop a formal model of what it means to say that an individual "knows" a fact about the world.

The states of nature consist of a finite universe Ω of possible worlds, subsets of which are called *events*. Event E occurs at state ω if $\omega \in E$. When Alice is in state ω, she knows only that she is in a subset $\mathbf{P}_A\omega \subseteq \Omega$ of states; i.e., $\mathbf{P}_A\omega$ is the set of states Alice considers possible when the actual state is ω. We say that Alice *knows* the event E at state ω if $\mathbf{P}_A\omega \subseteq E$ because for every state ω' that Alice knows is possible, $\omega' \in E$.

Given a possibility operator $\mathbf{P}$, we define a corresponding *knowledge operator* $\mathbf{K}$ by $\mathbf{K}E = \{\omega|\mathbf{P}\omega \subseteq E\}$. Note that $\mathbf{K}E$ is an event consisting of all states at which the individual knows E. It is easy to check that the knowledge satisfies the following properties:

(K1)	$\mathbf{K}\Omega = \Omega$	omniscience
(K2a)	$\mathbf{K}(E \cap F) = \mathbf{K}E \cap \mathbf{K}F$	
(K2b)	$E \subseteq F \Rightarrow \mathbf{K}E \subseteq \mathbf{K}F$	
(K3)	$\mathbf{K}E \subseteq E$	knowledge
(K4)	$\mathbf{K}E = \mathbf{K}\mathbf{K}E$	transparency
(K5)	$\neg\mathbf{K}\neg\mathbf{K}E \subseteq \mathbf{K}E$	negative introspection

where $\neg$, means "not"; i.e., logical negation. Note that K2a implies K2b. To see this, assume K2a and $E \subseteq F$. Then, $\mathbf{K}E = \mathbf{K}(E \cap F) = \mathbf{K}E \cap \mathbf{K}F$, so $\mathbf{K}E \subseteq \mathbf{K}F$, which proves K2b. Property K3, often called the *axiom of knowledge*, asserts that what is known must be true (if we drop this principle, we get a model of *belief* rather than *knowledge*) and follows directly from P1. Property K4, called the *axiom of transparency*, says that if you know something, then you know that you know it. Property K5 says that if you do not know something, then you know that you do not know it. This is not a very intuitive statement, but it allows us to specify the properties of the knowledge operator *syntactically* without regard to its *semantic* interpretation in terms of possible worlds and the possibility operator $\mathbf{P}\omega$. We show that K5 follows from P1 and P2 by extending the definition of the $\mathbf{P}$ operator from states to events by $\mathbf{P}E = \bigcup_{\omega \in E} \mathbf{P}\omega$, then the knowledge and possibilities operators are *dual* in the sense that $\neg\mathbf{K}\neg E = \mathbf{P}E$ and $\neg\mathbf{P}\neg E = \mathbf{K}E$ for any event $E \subseteq \Omega$. To see the first,

$$\begin{aligned}\neg\mathbf{K}\neg E &= \{\omega|\mathbf{P}\omega \subsetneq \neg E\} = \{\omega|\mathbf{P}\omega \cap E \neq \emptyset\} \\ &= \{\omega'|\omega' \in \cup_{\omega \in E}\mathbf{P}\omega\} = \mathbf{P}E.\end{aligned}$$

To see the second, suppose $\omega \in \neg\mathbf{P}\neg E$. We must show $\mathbf{P}\omega \subseteq E$. If this is false, then $\mathbf{P}\omega \cap \neg E \neq \emptyset$, which implies $\omega \in \mathbf{P}\neg E$, which is a contradiction. To prove K5, which can be written $\mathbf{PK}E \subseteq \mathbf{K}E$, suppose $\omega \in \mathbf{PK}E$. Then $\mathbf{P}\omega \subseteq \mathbf{K}E$, so $\omega \in \mathbf{K}E$. This argument can be reversed to prove equality.

As we saw in §4.1, we can recover the possibility operator $\mathbf{P}\omega$ for an individual from his knowledge operator $\mathbf{K}$ because

$$\mathbf{P}\omega = \bigcap\{E|\omega \in \mathbf{K}E\}. \tag{5.4}$$

To verify this equation, note that if $\omega \in \mathbf{K}E$, then $\mathbf{P}\omega \subseteq E$, so the left-hand side of (5.4) is contained in the right-hand side. Moreover, if ω' is not in the right-hand side, then $\omega' \notin E$ for some E with $\omega \in \mathbf{K}E$, so $\mathbf{P}\omega \subseteq E$, so $\omega' \notin \mathbf{P}\omega$. Thus, the right-hand side of (5.4) is contained in the left.

We say an event E is *self-evident* to an agent if he knows E at each state $\omega \in E$. Thus, E is self-evident exactly when $\mathbf{K}E = E$, which means $\mathbf{P}\omega \subseteq E$ for every $\omega \in E$. Clearly, Ω itself is self-evident, and if E and F are self-evident, then $E \cap F$ is self-evident. Thus, for each state ω, $\mathbf{P}\omega$ is the *minimal* self-evident event containing ω. Every self-evident event is the union of minimal self-evident events. The minimal self-evident events coincide with the cells of the partition $\mathcal{P}$.

5.11 Backward Induction and Extensive Form CKR

In this section we show how a little modal logic can clarify issues concerning rational behavior and choice. We take the case of backward induction, presenting Aumann's (1995) proof that common knowledge of rationality in generic extensive form games of perfect information is possible only at nodes of the game tree that lie along the backward induction path of play.

Consider a finite generic extensive form epistemic game of perfect information $\mathcal{G}$ (a game is generic if, for each player, no two payoffs at terminal nodes are equal). A pure-strategy profile s assigns an action s^v at each non-terminal node v. Indeed, if s_i is the pure-strategy profile of player i and if i moves at v, then $s^v = s_i^v$. We denote by b the unique backward induction strategy profile. Thus, if player i moves at node v, then

$$\pi_i^v(b) > \pi_i^v(b/a^v) \qquad \text{for } a^v \neq b^v, \tag{5.5}$$

where $\pi_i^v(s)$ is the payoff of strategy profile s to player i, starting from node v (even if, starting from the beginning of the game, v would not be reached), and s/t^v denotes the strategy profile s for the player who chooses at v, replacing his action s_i^v with action t at v.

To specify Bayesian rationality in this framework, suppose players choose pure strategy profile $\mathbf{s}(\omega)$ in state ω. We then say i is Bayesian rational if, for every node v at which i chooses and for every pure strategy $t_i \in S_i$, we have

$$R_i \subseteq \neg\mathbf{K}_i\{\omega \in \Omega | \pi_i^v(\mathbf{s}/t_i) > \pi_i^v(\mathbf{s})\}; \tag{5.6}$$

i.e., i does not know that there is a better strategy than $\mathbf{s}_i(\omega)$ at v. Common knowledge of rationality, which we write as CKR, means R_i is common knowledge for all players i. Note that this definition is somewhat weaker than Bayesian rationality, which requires that agents have subjective priors over events and maximize utility subject to these priors.

Let $I^v \subseteq \Omega$ be the event that b^v is chosen at node v. Thus

$$I^v = \{\omega \in \Omega | \mathbf{s}(\omega)^v = b^v\}, \tag{5.7}$$

so the event I that the backward induction path is chosen is simply

$$I = \cap_v I^v.$$

The assertion that common knowledge of rationality implies backward induction is then simply expressed as

THEOREM 5.1 *CKR* $\subseteq I$.

Proof: We first show that at every terminal node v, CKR $\subseteq I^v$. We have

$$\begin{aligned}
\text{CKR} \subseteq R_i &\subseteq \neg\mathbf{K}_i\{\omega \in \Omega | \pi_i^v(\mathbf{s}/b_i) > \pi_i^v(\mathbf{s})\} \\
&= \neg\mathbf{K}_i\{\omega \in \Omega | \pi_i^v(b) > \pi_i^v(b/\mathbf{s}_i)\} \\
&= \neg\mathbf{K}_i\{\omega \in \Omega | \mathbf{s}_i^v \neq b^v\} = \neg\mathbf{K}_i\neg I^v.
\end{aligned}$$

The first line follows from (5.6) with $t_i = b_i$. The second line follows from the fact that v is a terminal node at which i chooses, so $\pi_i^v(b) = \pi_i^v(\mathbf{s}/b_i)$ and $\pi_i^v(\mathbf{s}) = \pi_i^v(b/\mathbf{s}_i)$.

Because i chooses at v, I^v is a union of cells of i's knowledge partition, so I^v is self-evident to i, and hence $I^v = \mathbf{K}_i I^v$. Thus we have CKR $\subseteq \neg\mathbf{K}_i\neg\mathbf{K}_i I^v$. By negative introspection (K5), this implies CKR $\subseteq \neg\neg\mathbf{K}_i I^v = \mathbf{K}_i I^v = I^v$.

This argument proves that at every state compatible with CKR, players must make the backward induction move at each terminal node. Note that this argument does not commit what we called in chapter 5 the "fallacy of backward induction" because this argument does not assume that a terminal node is reached that could not be reached if players used backward induction. Indeed, this argument does not assume anything about which nodes are reached or not reached.

The rest of the proof proceeds by mathematical induction. Assume that CKR $\subseteq I^w$ for all nodes w that follow a node v, where player i chooses. We can then write CKR $\subseteq \mathbf{K}_i I^{>v} = \cap_{w>v}\mathbf{K}_i I^w$, where $w > v$ means that node w follows node v in the game tree. We then have

$$R_i \subseteq \neg\mathbf{K}_i\{\omega \in \Omega | \pi_i^v(\mathbf{s}/b_i) > \pi_i^v(\mathbf{s})\},$$

by (5.6) with $t_i = b_i$. Now $\pi_i^v(\mathbf{s})$ depends only on s^v and $s^{>v}$, the restriction of s to the nodes following v. Thus, we can write

$$R_i \cap I^{>v} \subseteq \neg\mathbf{K}_i\{\omega \in \Omega | \pi_i^v(b) > \pi_i^v(b/\mathbf{s}^v)\} \cap I^{>v} = \neg\mathbf{K}_i\neg I^v \cap I^{>v},$$

where the first inclusion follows from the fact that for $\omega \in I^{>v}$, $\pi_i^v(\mathbf{s}/b_i) = \pi_i^v(b)$ and $\pi_i^v(\mathbf{s}) = \pi_i^v(b/\mathbf{s}^v)$. Thus,

$$\text{CKR} \subseteq R_i \cap I^{>v} \subseteq \neg\mathbf{K}_i\neg I^v \cap I^{>v} \subseteq I^v \cap I^{>v},$$

where the argument for the final inclusion is as before. ■

This theorem does not claim that rational agents will always play the subgame perfect equilibrium. Rather, it claims that if a player makes a move to a node that is not along the backward induction path of play, then common knowledge of rationality cannot obtain at that node or at any subsequent node of the game tree. There is nothing irrational about a player making such a move, as he may have some notion as to how rational agents will play the game based on considerations other than CKR.

Another way of saying this is that CKR is an event, not a premise (§4.14). In some cases CKR holds, not because CKR implies the outcome, but rather because the outcome implies CKR.

5.12 Rationality and Extensive Form CKR

Between 1987 and 1993, several influential papers questioned the classical game-theoretic argument that in an extensive form game of perfect information with a single subgame perfect Nash equilibrium, rational agents must play this equilibrium. Aumann (1995) was seen by many game theorists as a futile and inadequate response to these critics in defense of the conventional wisdom. The central criticism of Aumann's analysis was stated as follows by Binmore (1996):

> What keeps a rational player on the equilibrium path is his evaluation of what would happen if he were to deviate. But, if he were to deviate, he would behave irrationally. Other players would then be foolish if they were not to take this evidence of irrationality into account in planning their responses to the deviation. ...Aumann ...is insistent that his conclusions say nothing whatever about what players would do if vertices of the game tree off the backward-induction path were to be reached. But, if nothing can be said about what would happen off the backward-induction path, then it seems obvious that nothing can be said about the rationality of remaining on the backward-induction path." (p. 135)

Similarly, Ben-Porath (1997) asserts that

> Aumann assumes that in every vertex x there is common knowledge that a player will play rationally in the subgame that starts at x. This is assumed even for vertices x that cannot be reached if there is CKR at the beginning. Thus, the assumption is that a player i will ignore the fact that another player j behaved in a way which is consistent with CKR. (p. 43)

One correction is clearly in order. If a rational player were to deviate from the equilibrium path, says Binmore, "he would behave irrationally." The correct statement is that if a player deviated from the equilibrium path, he would violate CKR, not rationality. That said, although there may be versions of CKR that are vulnerable to this critique, Aumann's version, presented in §5.11, is not.

This argument should not be seen, however, as a defense of CKR. Aumann himself, in all his writings, states clearly that CKR is not dictated by the norms of social interaction among rational agents. CKR is not a strengthening of Bayesian rationality. Rather, CKR is a powerful and often highly implausible assumption concerning the communality of mental representations across Bayesian rational agents.

A major attraction of epistemic game theory lies in its allowing us to replace arguments about where a proposition ψ is true and false in a game by an analysis of the set of states $\omega \in \Omega$ in which $\psi(\omega)$ holds. Thus, the conclusion of Aumann's argument, equation (5.1), must be read as "In every state ω in which CKR holds, the backward induction path is chosen by $\mathbf{s}(\omega)$." Similarly, "CKR fails off the backward induction path" should be read "In every state ω for which $\mathbf{s}(\omega)$ is not the backward induction path, CKR fails."

The bottom line is that the critics, from Binmore (1987) to Reny (1993) are correct in stating that rationality does not imply backward induction. But Aumann (1995) is also correct in stating that in every state where CKR holds, the backward induction path is followed.

5.13 On the Nonexistence of CKR

The proof that CKR implies backward induction in Aumann (1995) is followed by the proof of the following theorem.

THEOREM 5.2 *In every game of perfect information, there is a knowledge system such that CKR$\neq\emptyset$.*

The proof is trivial. Assume Ω has exactly one state, in which each agent's strategy is the backward induction strategy.

The more interesting question, however, is: What are the characteristics of knowledge systems for which CKR$\neq\emptyset$, and are there plausible knowledge systems for which $\Omega=$ CKR? The answers to these questions are, to my knowledge, unknown.

It is easy, however, to construct a realistic epistemic game in which CKR$=\emptyset$. For instance, consider the situation described in §5.9. The game is the 100-round Repeated Prisoner's Dilemma, and each player has a subjective prior that includes a probability distribution over the strategies of the potential partners and chooses a strategy that maximizes his expected payoff subject to this conjecture. Unless all players' conjectures lead to defecting in round 1, CKR$=\emptyset$ for this epistemic game. Nothing, of course, constrains rational agents to hold such a pattern of conjectures, so CKR$=\emptyset$ should be considered the default situation.

More generally, in any epistemic game that has a perfect information extensive form and a unique subgame perfect Nash equilibrium $\mathbf{s}^*$, the priors $p_i(\cdot|\omega)$ fully determine the probability each player places on the occurrence of $\mathbf{s}^*$ namely, $p_i([\mathbf{s} = \mathbf{s}^*]|\omega)$. This is surely zero unless $\mathbf{s}_i(\omega) = \mathbf{s}^*$. Moreover, we must have $p_i([\mathbf{s} = \mathbf{s}^*]|\omega) = 1$ if $\omega \in$ CKR, which is a restriction on subjective priors that has absolutely no justification in general, although it can be justified in certain cases (e.g., the one- or two-round Prisoner's Dilemma or the game in §5.3).

It might be suggested that a plausible strategy selection mechanism epistemically justified by some principle other than CKR might succeed in selecting out the subgame perfect equilibrium—for instance, extensive form rationality as proposed by Pearce (1984) and Battigalli (1997). However, this selection mechanism is not epistemically grounded at all. There are alternative, epistemically grounded selection mechanisms for extensive form games, such as Fudenberg, Kreps, and Levine (1988), Börgers (1994), and Ben-Porath (1997), but these mechanisms do not justify backward induction.

6

The Mixing Problem: Purification and Conjectures

God does not play dice with the universe.

Albert Einstein

Economic theory stresses that a proposed mechanism for solving a coordination problem assuming self-regarding agents is plausible only if it is *incentive compatible*: each agent should find it in his interest to behave as required by the mechanism. However, a strictly mixed-strategy Nash equilibrium $\sigma^* = (\sigma_1^*, \ldots, \sigma_n^*)$ fails to be incentive compatible, because a self-regarding agent i is indifferent to any mixed strategy in the support of σ_i^*. This chapter deals with the solution to this problem. We conclude that, while ingenious justifications of the incentive compatibility of mixed-strategy Nash equilibria have been offered, they fail except in a large majority of cases. We suggest that the solution lies in recognizing both the power of social norms, and of a human other-regarding psychological predisposition to conform to social norms even when it is costly to do so.

6.1 Why Play Mixed Strategies?

In Throwing Fingers (§2.7), there is a unique mixed-strategy Nash equilibrium in which both players choose each of their pure strategies with probability 1/2. However, if both pure strategies have equal payoffs against the mixed strategy of the other player, Why bother randomizing? Of course, this problem is perfectly general. By the fundamental theorem (§2.5), any mixed strategy best response consists of equal-payoff pure strategies, so why should a player bother randomizing? Moreover, this argument holds for all other players as well. Therefore, no player should expect any other player to randomize. This is the mixing problem.

We assume that the game is played only once (this is called a *one-shot* game, even though it could be an extensive form game with many moves by each player, as in chess), so there is no past history on which to base an inference of future play, and each decision node is visited at most once, so a statistical analysis during the course of the game cannot be carried out. If the stage game itself consists of a finite repetition of a smaller stage game, as in playing Throwing Fingers n times, one can make a good case for randomizing in the stage game. But only a small fraction of games have such form.

One suggestion for randomizing is that perhaps it is easier for an opponent to discover a pure-strategy choice on your part than a mixed-strategy choice (Reny and Robson 2004). Indeed, when Von Neumann and Morgenstern (1944) introduced the concept of a mixed strategy in zero-sum games, they argued that a player would use this strategy to "protect himself against having his intentions found out by his opponent." (p. 146) This defense is weak. For one thing, it does not hold at all for many games, such as Battle of the Sexes (§2.8), where a player *gains* when his partner discovers his pure-strategy move. More important, if there are informational processes, reputation effects, or other mechanisms whereby the agents' "types" can be known or discovered, this should be formally modeled in the specification of the game.

Finally, it is always *costly* to randomize because one must have some sophisticated mental algorithm modeling the spinning of a roulette wheel or other randomizing device, so mixed strategies are in fact strictly more costly to implement than pure strategies. In general, we do not discuss implementation costs, but in this case they play a critical role in evaluating the relative costs of playing a mixed-strategy best response or any of the pure strategies in its support. Hence the *mixing problem*: why bother randomizing?

There have been two major approaches to solving the mixing problem. The first approach, which we develop in §6.2, is due to Harsanyi (1973). Harsanyi treats mixed-strategy equilibria as limit cases of slightly perturbed "purified" games with pure-strategy equilibria. This remarkable approach handles many simple games very nicely but fails to extend to more complex environments. The more recent approach, which uses interactive epistemology to define knowledge structures representing subjective degrees of uncertainty, is due to Robert Aumann and his coworkers. This approach, which we develop in §6.5, does not predict how agents will actually play because it determines only the *conjectures* each player has of the *other* players' strategies. It follows that this approach does not solve the mixing problem. However, in §6.6, we show that a simple extension of the Aumann conjecture approach is valid under precisely the same conditions as

Harsanyi purification, with the added attractions of handling pure-strategy equilibria (§6.6) and applying to cases where payoffs are determinate.

Our general conclusion, which is fortified by two examples below, is that purification is possible in some simple games but not in the sorts of games that apply to complex social interaction, such as principal-agent models or repeated games. Yet, such complex models generally rely on mixed strategies. Thus, game theory alone is incapable of explaining such complex social interactions even approximately and even in principle. This provides one more nail in the coffin of methodological individualism (§8.8).

6.2 Harsanyi's Purification Theorem

	L	R
U	0,0	0,−1
D	−1,0	1,3

To understand Harsanyi's (1973) defense of the mixed-strategy Nash equilibrium in one-shot games, consider the game to the right, which has a mixed-strategy equilibrium $(\sigma_1^*, \sigma_2^*) = (3U/4 + D/4, L/2 + R/2)$. Because player 1, Alice, is indifferent between U and D, if she had some personal idiosyncratic reason for slightly preferring U to D, she would play pure strategy U rather than the mixed strategy σ_1^*. Similarly, some slight preference for R would lead player 2, Bob, to play pure strategy R. The mixed-strategy equilibrium would then disappear, thus solving the mixing problem.

	L	R
U	$\epsilon\theta_1$,$\epsilon\theta_2$	$\epsilon\theta_1$,−1
D	−1,$\epsilon\theta_2$	1,3

To formalize this, let θ be a random variable uniformly distributed on the interval $[-1/2, 1/2]$ (in fact, θ could be any bounded random variable with a continuous density) and suppose that the distribution of payoffs from U for the population of player 1's is $2 + \epsilon\theta_1$, and the distribution of payoffs from L for the population of player 2's is $1 + \epsilon\theta_2$, where θ_1 and θ_2 are independently distributed as θ. Suppose Alice and Bob are chosen randomly from their respective populations of player 1's and player 2's to play the game, and each knows only the distribution of payoffs to their partners. Suppose Alice uses U with probability β and Bob uses L with probability α. You can check that Bob infers that the payoff from U for Alice is distributed as $\pi_U = \epsilon\theta_1$, and that the payoff from D for Alice is $\pi_D = 1 - 2\alpha$. Similarly, Alice infers that the payoff from L for Bob is $\pi_L = \epsilon\theta_2$, and the payoff from R for Bob is $\pi_R = 3 - 4\beta$.

Now, β is the probability $\pi_U > \pi_D$, which is the probability that $\theta_1 > (1-2\alpha)/\epsilon$, which gives

$$\alpha = \mathrm{P}[\pi_L > \pi_R] = \mathrm{P}\left[\theta_B > \frac{3-4\beta}{\epsilon}\right] = \frac{8\beta - 6 + \epsilon}{2\epsilon} \tag{6.1}$$

$$\beta = \mathrm{P}[\pi_U > \pi_D] = \mathrm{P}\left[\theta_A > \frac{1-2\alpha}{\epsilon}\right] = \frac{4\alpha - 2 + \epsilon}{2\epsilon}. \tag{6.2}$$

Solving these simultaneously for α and β, we find

$$\alpha = \frac{1}{2} - \frac{\epsilon}{8-\epsilon^2} \qquad \beta = \frac{3}{4} - \frac{\epsilon^2}{4(8-\epsilon^2)}, \tag{6.3}$$

This is our desired equilibrium. Now, this *looks* like a mixed-strategy equilibrium, but it is not. The probability that an Alice randomly chosen from the population chooses pure strategy U is β, and the probability that a Bob randomly chosen from the population chooses pure strategy L is α. Thus, for instance, if an observer measured the frequency with which player 1's chose U, he would arrive at a number near α, despite the fact that no player 1 ever randomizes. Moreover, when ϵ is very small, $(\alpha, \beta) \approx (\sigma_1^*, \sigma_2^*)$. Thus, if we observed a large number of pairs of agents playing this game, the frequency of playing the various strategies would closely approximate their mixed-strategy equilibrium values.

To familiarize himself with this analysis, the reader should derive the equilibrium values of α and β, assuming that player i's idiosyncratic payoff is uniformly distributed on the interval $[a_i, b_i]$, $i = 1, 2$, and show that α and β tend to the mixed-strategy Nash equilibrium as $\epsilon \to 0$. Then, he should find the purified solution to Throwing Fingers (§2.7) assuming player 1 favors H with probability $\epsilon\theta_1$, where θ_1 is uniformly distributed on $[-0.5, 0.5]$ and player 2 favors H with probability $\epsilon\theta_2$, where θ_2 is uniformly distributed on [0,1], and then show that as $\epsilon \to 0$, the strategies in the perturbed game move to the mixed-strategy equilibrium.

Govindan, Reny, and Robson (2003) present a very general statement and elegant proof of Harsanyi's purification theorem. They also correct an error in Harsanyi's original proof (see also van Damme 1987, ch. 5). The notion of a *regular equilibrium* used in the theorem is the same as that of a *hyperbolic fixed point* in dynamical systems theory (Gintis 2009) and is satisfied in many simple games with isolated and strictly perfect Nash equilibria (meaning that if we add very small errors to each strategy, the equilibrium is displaced only a small amount).

The following theorem is weakened a bit from Govindan, Reny, and Robson (2003) to make it easier to follow. Let $\mathcal{G}$ be a finite normal form game with pure-strategy set S_i for player i, $i = 1, \ldots, n$, and payoffs $u_i : S \rightarrow \mathbf{R}$, where $S = \prod_{i=1}^{n} S_i$ is the set of pure-strategy profiles of the game.

A Nash equilibrium s is *strict* if there is a neighborhood of s (considered as a point in n-space) that contains no other Nash equilibrium of the game. The distance between equilibria is the Euclidean distance between the strategy profiles considered as points in $\mathbf{R}^{|S|}$ ($|S|$ is the number of elements in S). Another way of saying this is that a Nash equilibrium is strict if the connected component of Nash equilibria to which it belongs consists of a single point.

Suppose for each player i and each pure-strategy profile $s \in S$, there is a random perturbation $\nu_i(s)$ with probability distribution μ_i such that the actual payoff to player i from $s \in S$ is $u_i(s) + \epsilon\nu_i(s)$, where $\epsilon > 0$ is a small number. We assume the ν_i are independent, and each player knows only his own outcomes $\{\nu_i(s)|s \in S_i\}$. We have the following theorem.

THEOREM 6.1 *Suppose σ^* is a regular mixed-strategy Nash equilibrium of $\mathcal{G}$. Then, for every $\delta > 0$, there is an $\epsilon > 0$ such that the perturbed game with payoffs $\{u_i(s) + \epsilon\nu_i(s)|s \in S\}$ has a strict Nash equilibrium $\hat{\sigma}$ within ϵ of σ^*.*

6.3 A Reputational Model of Honesty and Corruption

Consider a society in which sometimes people are Needy, and sometimes others help the Needy. In the first period, a pair is selected randomly, one being designated Needy and the other Giver. Giver and Needy then play a stage game $\mathcal{G}$ in which if Giver helps, a benefit b is conferred on Needy at a cost c to Giver, where $0 < c < b$; or, if Giver defects, both players receive 0. In each succeeding period, Needy from the previous period becomes Giver in the current period. Giver is paired with a new, random Needy, and the game $\mathcal{G}$ is played by the new pair. If we assume that helping behavior is public information, there is a Nash equilibrium of the following form, provided the discount factor δ is sufficiently close to unity. At the start of the game, each player is labeled "in good standing." In every period Giver helps if and only if his partner Needy is in good standing. Failure to do so puts a player "in bad standing," where he remains for the rest of the game.

To see that this is a Nash equilibrium in which every Giver helps in every period for δ sufficiently close to 1, let v_c be the present value of the game to a Giver, and let v_b be the present value of the game for an individual who is not currently Giver or Needy. Then we have $v_c = -c + \delta v_b$ and $v_b = p(b+\delta v_c)+(1-p)\delta v_b$, where p is the probability of begin chosen as Needy. The first equation reflects the fact that a Giver must pay c now and becomes a candidate for Needy in the next period. The second equation expresses the fact that a candidate for Needy is chosen with probability p and then gets b, plus is Giver in the next period, and with probability $1-p$ remains a candidate for Needy in the next period. If we solve these two equations simultaneously, we find that $v_c > 0$ precisely when $\delta > c/(c+p(b-c))$. Because the right hand side of this expression is strictly less than 1, there is a range of discount factors for which it is a best response for a Giver to help, and thus remain in good standing.

Suppose, however, the informational assumption is that each new Giver knows only whether his partner Needy did or did not help his own partner in the previous period. If Alice is Giver and her partner Needy is Bob, and Bob did not help when he was Giver, it could be because when he was Giver, Carole, his Needy partner, had defected when she was Giver, or because Bob failed to help Carole even though she had helped Donald, her previous Needy partner when she was Giver. Because Alice cannot condition her action on Bob's previous action, Bob's best response is to defect on Carole, no matter what she did. Therefore, Carole will defect on Donald, no matter what he did. Thus, there can be no Nash equilibrium with the pure strategy of helping.

This argument extends to the richer informational structure where a Giver knows the previous k actions for any finite k. Here is the argument for $k=2$, which the reader is encouraged to generalize. Suppose the last five players are Alice, Bob, Carole, Donald, and Eloise, in that order. Alice can condition her choice on the actions taken by Bob, Carole, and Donald, but not on Eloise's action. Therefore, Bob's best response to Carole will not be conditioned on Eloise's action, and hence Carole's response to Donald will not be conditioned on Eloise's action. So, finally, Donald's response to Eloise will not be conditioned on her action, so her best response is to defect when she is Giver. Thus, there is no helping Nash equilibrium.

Suppose, however, back in the $k=1$ case, that instead of defecting unconditionally when facing a Needy who has defected improperly, a Giver helps with probability $p = 1 - c/b$ and defects with probability $1 - p$. The gain from helping unconditionally is then $b - c$, while the gain from following this new strategy is $p(b-c) + (1-p)pb$, where the first term is the probability p of helping times the reward b in the next period if one helps minus the cost c of helping in the current period, and the second term is the probability $1 - p$ of defecting times the probability p that you will be helped anyway when your are Needy, times the benefit b. Equating this expression with $b - c$, the cost of helping unconditionally, we get $p = 1 - c/b$, which is a number strictly between zero and one and hence a valid probability.

Consider the following strategy. In each round, Giver helps if his partner helped in the previous period, and otherwise helps with probability p and defects with probability $1 - p$. With this strategy each Giver i is indifferent to helping or defecting, because helping costs i the amount c when he is Giver but i gains b when he is Needy, for a net gain of $b - c$. However, defecting costs zero when Giver, but gives $bp = b - c$ when he is Needy. Because the two actions have the same payoff, it is incentive-compatible for each Giver to help when his partner Needy helped, and to defect with probability p otherwise. This strategy thus gives rise to a Nash equilibrium with helping in every period.

The bizarre nature of this equilibrium is clear from the fact that there is no reason for any player to follow this strategy as opposed to any other since all strategies have the same payoff. So, for instance, if you slightly favor some players (e.g., your friends or coreligionists) over others (e.g., your enemies and religious infidels), then you will help the former and defect on the latter. But then, if this is generally true, each player Bob knows that he will be helped or not by Alice independent of whether he helps Carole when she is Needy, so Bob has no incentive to help Carole. In short, if we add a small amount of "noise" to the payoffs in the form of a slight preference for some potential partners over others, there is no longer a Nash equilibrium with helping. Thus, this repeated game model with private signals cannot be purified (Bhaskar 1998b).

However, if players receive a subjective payoff from following the rules (we term this a *normative predisposition* in chapter 7) greater than the largest subjective gain from helping a friend or loss from defecting on an enemy, complete cooperation can be reestablished even with private signals. Indeed, even more sophisticated models can be constructed in which Alice can calculate the probability that the Giver with whom she is paired when she is Needy will condition his action on hers, a calculation that depends on the statistical distribution of her friends and enemies and on the statistical distribution of the strength of the predisposition to observe social norms. For certain parameter ranges of these variables, Alice will behave "meritocratically," and for others, she will act "corruptly" by favoring friends over enemies.

6.4 Purifying Honesty and Corruption

Suppose police are hired to apprehend criminals, but only the word of the police officer who witnessed the transgression is used to determine the punishment of the offender—there is no forensic evidence involved in the judgment, and the accused has no means of self-defense. Moreover, it costs the police officer a fixed amount f to file a criminal report. How can this society erect incentives to induce the police to act honestly?

Let us assume that the society's elders set up criminal penalties so that it is never profitable to commit a crime, provided the police are honest. The police, however, are self-regarding and so have no incentive to report a crime, which costs them f. If the elders offer the police an incentive w per criminal report, the police will file zero reports for $w < f$ and as many as possible for $w > f$. However, if the elders set $w = f$, the police will be indifferent between reporting and not reporting a crime, and there will be a Nash equilibrium at which the officers report all crimes they observe and none others.

This Nash equilibrium cannot be purified. If there are small differences in the cost of filing a report or if police officers derive small differences in utility from reporting crimes, depending on their relationship to the perpetrator, the Nash equilibrium will disappear. We can foresee how this model could be transformed into a full-fledged model of police honesty and corruption by adding effective monitoring devices, keeping tabs on the reporting rates of different officers, and the like. We could also add a in the form of a police culture favoring honesty or condemning corruption, and explore the interaction of moral and material incentives in controlling crime.

6.5 Epistemic Games: Mixed Strategies as Conjectures

Let $\mathcal{G}$ be an epistemic game where each player i has a subjective prior $p_i(\cdot;\omega)$. We say a probability distribution p over the state space Ω is a *common prior* for $\mathcal{G}$ if, for each player i and for each $P \in \mathcal{P}_i$, $p(P) > 0$ and i's subjective prior $p_i(\cdot|P)$ satisfies $p_i(\omega|P) = p(\omega)/p(P)$ for $\omega \in P$.

The sense in which conjectures solve the problem of why agents play mixed strategies is given by the following theorem due to Aumann and Brandenburger (1995), which we will prove in §8.7.

THEOREM 6.2 *Let $\mathcal{G}$ be an epistemic game with $n > 2$ players. Suppose the players have a common prior p and it is commonly known at $\omega \in \Omega$ that ϕ^{ω} is the set of conjectures for $\mathcal{G}$. Then, for each $j = 1, \ldots, n$, all $i \neq j$ induce the same conjecture $\sigma_j(\omega) = \phi_j^{\omega}$ about j's conjectured mixed strategy, and $\sigma(\omega) = (\sigma_1(\omega), \ldots, \sigma_n(\omega))$ form a Nash equilibrium of $\mathcal{G}$.*

Several game theorists have suggested that this theorem resolves the problem of mixed-strategy Nash equilibria. In their view, each player chooses a pure strategy, but there is a Nash equilibrium in player *conjectures* (see, for instance, §4.3). However, the fact that player conjectures are mutual best responses does not permit us to deduce anything concerning the relative frequency of player pure-strategy choices except that pure strategies not in the support of the equilibrium mixed strategy will have frequency zero. This suggested solution to the mixing problem is thus incorrect, assuming one cares about explaining behavior, not just conjectures in people's heads.

There are many stunning indications of contemporary game theorists' disregard for explaining behavior, but perhaps none more stunning than the complacency surrounding the acceptance of this argument. The methodological commitment behind this complacency was eloquently expressed by Ariel Rubinstein in his presidential address to the Econometric Society (Rubinstein 2006). "As in the case of fables, models in economic theory ... are not meant to be testable. ... a good model can have an enormous influence on the real world, not by providing advice or by predicting the future, but rather by influencing culture." It is hard not be sympathetic with Rubinstein's disarming frankness, despite his being dead wrong: the value of a model is its contribution to explaning reality, not its contribution to society's stock of pithy aphorisms.

6.6 Resurrecting the Conjecture Approach to Purification

Harsanyi purification is motivated by the notion that payoffs may have a statistical distribution rather than being the determinate values assumed in classical game theory. Suppose, however, that payoffs are indeed determinate, but the conjectures (§6.5) of individual players have a statistical distribution around the game's mixed-strategy equilibrium values. In this case, the epistemic solution to the mixing problem might be cogent. Indeed, as we shall see, the assumption of stochastic conjectures has advantages over the Harsanyi assumption of stochastic payoffs. This avenue of research has not been studied in the literature, to my knowledge, but it clearly deserves to be explored.

Consider the Battle of the Sexes (§2.8) and suppose that Alfredos in the population go to the opera with mean probability α, but the conjecture as to α by the population of Violettas is distributed as $\alpha+\epsilon\theta_V$. Similarly, suppose that Violettas in the population go to the opera with mean probability β, but the conjecture as to β by the population of Alfredos is distributed as $\beta+\epsilon\theta_A$.

Let π_o^A and π_g^A be the expected payoffs to a random Alfredo chosen from the population from going to the opera and from gambling, respectively, and let π_o^V and π_g^V be the expected payoff to a random Violetta chosen from the population from going to the opera and from gambling, respectively. An easy calculation shows that

$$\pi_o^A - \pi_g^A = 3\beta - 2 + 3\epsilon\theta_A,$$
$$\pi_o^V - \pi_g^V = 3\alpha - 1 + 3\epsilon\theta_V.$$

Therefore,

$$\alpha = \mathrm{P}[\pi_o^A > \pi_g^A] = \mathrm{P}\left[\theta_A > \frac{2-3\beta}{3\epsilon}\right] = \frac{6\beta - 4 + 3\epsilon}{6\epsilon}, \tag{6.4}$$

$$\beta = \mathrm{P}[\pi_o^V > \pi_g^V] = \mathrm{P}\left[\theta_V > \frac{1-3\alpha}{3\epsilon}\right] = \frac{6\alpha - 2 + 3\epsilon}{6\epsilon}. \tag{6.5}$$

If we assume that the beliefs of the agents reflect the actual state of the two populations, we may solve these equations simultaneously, finding

$$\alpha^* = \frac{1}{3} + \frac{\epsilon}{6(1+\epsilon)}, \qquad \beta^* = \frac{2}{3} - \frac{\epsilon}{6(1+\epsilon)} \tag{6.6}$$

Clearly, as $\epsilon \to 0$, this pure-strategy equilibrium tends to the mixed-strategy equilibrium of the stage game (2/3,1/3), as prescribed by the purification theorem.

However, note that our calculations in arriving at (6.6) assumed that $\alpha, \beta \in (0,1)$. This is true, however, only when

$$\frac{1}{3} - \frac{\epsilon}{2} < \alpha < \frac{1}{3} + \frac{\epsilon}{2}, \qquad \frac{2}{3} - \frac{\epsilon}{2} < \beta < \frac{2}{3} + \frac{\epsilon}{2}. \tag{6.7}$$

Suppose, however, that $\alpha < 1/3 - \epsilon/2$. Then, all Violettas choose gambling, to which gambling is any Alfredo's best response. In this case, the only equilibrium is $\alpha = \beta = 0$. Similarly, if $\alpha > 1/3 + \epsilon/6$, then all Violettas choose opera, so all Alfredos choose opera, and we have the pure-strategy Nash equilibrium $\alpha = \beta = 1$. This approach to solving the mixing problem has the added attraction that it yields an approximation not only to the mixed-strategy equilibrium for some statistical distribution of beliefs but also to one or another of the two pure-strategy equilibria with other distributions of beliefs.

7

Bayesian Rationality and Social Epistemology

> Social life comes from a double source, the likeness of consciences and the division of social labor.
>
> Emile Durkheim

> There is no such thing as society. There are individual men and women, and there are families.
>
> Margaret Thatcher

At least since Schelling (1960) and Lewis (1969), game theorists have interpreted social norms as Nash equilibria. More recent contributions based upon the idea of social norms as selecting among Nash equilibria include Sugden (1986), Elster (19891,b), Binmore (2005), and Bicchieri (2006). There are two problems with this approach. The first is that the conditions under which rational individuals play a Nash equilibrium are extremely demanding (theorem §8.4), and are not guaranteed to hold simply because there is a social norm specifying a particular Nash equilibrium. Second, the most important and obvious social norms do not specify Nash equilibria at all, but rather are devices that implement *correlated equilibria* (§2.11, §7.5).

Informally, a correlated equilibrium of an epistemic game $\mathcal{G}$ is a Nash equilibrium of a game $\mathcal{G}^+$, in which $\mathcal{G}$ is augmented by an initial move by a new player, whom we call the *choreographer*, who observes a random variable γ on a probability space (Γ, p), and issues a "directive" $f_i(\gamma) \in S_i$ to each player i as to which pure strategy to choose. Following the choreographer's directive is a best response for each player, if other players also follow the choreographer's directives.

This chapter uses epistemic game theory to expand on the notion of social norms as choreographer of a correlated equilibrium, and to elucidate the socio-psychological prerequisites for the notion that social norms implement correlated equilibria.

The correlated equilibrium is a much more natural equilibrium criterion than the Nash equilibrium, because of a famous theorem of Aumann (1987), who showed that Bayesian rational agents in an epistemic game $\mathcal{G}$ with a common subjective prior play a correlated equilibrium of $\mathcal{G}$ (§2.11–2.13).

Thus, while rationality and common priors do not imply Nash equilibrium, these assumptions do imply correlated equilibrium and as we shall see, social norms act not only as choreographer, but also supply the epistemic conditions for common priors.

In a correlated equilibrium, rational players have no incentive to deviate from the instructions of the choreographer, but if the correlated equilibrium involves multiple strategies with equal payoffs, they have no incentive to follow them either. If a correlated equilibrium can be purified (see chapter 6), each agent effectively has a strict preference to follow the directives of the choreographer. However, in most complex games purification fails (§6.3, §6.4), in which case, as we shall see, we must assume that agents have a *normative predisposition* towards following the choreographer's instructions unless they have alternatives with strictly higher payoffs.

The isomorphism between correlated equilibrium and Bayesian rationality with common priors assumes that the choreographer has at least as much information as any player. This means that all information is *public*, an assumption that is violated in many practical cases. For instance, each agent's payoff might consist of a *public component* that is known to the choreographer and a *private component* that reflects the idiosyncrasies of the agent and is unknown to the choreographer. Suppose the maximum size of the private component in any state for an agent is α, but the agent's inclination to follow the choreographer has strength greater than α. Then, the agent continues to follow the choreographer's directions whatever the state of his private information. Formally, we say an individual has an *α-normative predisposition* towards conforming to the social norm if he strictly prefers to play his assigned strategy so long as all his pure strategies have payoffs no more than α greater than when following the choreographer. We call an α-normative predisposition a *social preference* because it facilitates social coordination but violates self-regarding preferences for $\alpha > 0$. There are evolutionary reasons for believing that humans have evolved such social preferences for fairly high levels of α in a large fraction of the population through gene-culture coevolution (Gintis 2003a).

7.1 The Sexes: From Battle to Ballet

Suppose there is a social norm specifying that when choosing between opera and gambling, the male of the pair decides on Monday through Friday, and the female on the weekend. This norm choreographs a correlated equilibrium in which Alfredo and Violetta go to the opera if and only if it is a weekend. Assuming that their planned meeting occurs equally likely on each day of the week, Alfredo's payoff is $2(5/7) + 1(2/7) = 12/7$ and Violetta's is $1(5/7) + 2(2/7) = 9/7$. This correlated equilibrium is not a Nash equilibrium of the underlying game and, like the pure-strategy Nash equilibria of the game, it is Pareto-efficient.

7.2 The Choreographer Trumps Backward Induction

Suppose Alice and Bob play the 100-round repeated Prisoner's Dilemma under conditions of common knowledge of rationality. They thus defect on every round (§5.12). They then discover a choreographer who chooses a number k, with $1 \leq k \leq 99$, and with probability 1/2 advises Alice to cooperate up to the kth round and Bob to cooperate up to the $1+k^{\text{th}}$ round, and with probability 1/2, reverses the advice to Alice and Bob. Both players are advised to defect forever after the first defection.

Assuming that both Alice and Bob believe that each has probability $\theta(k) = 1/2$ of having the lower number when advised to defect on the kth round, we can show that this is a correlated equilibrium with cooperation up to round $k-1$. Suppose Bob takes the choreographer's advice, cooperating up to the suggested round, and then defecting thereafter. Then, Alice's payoff from cooperating, assuming the payoffs to the prisoner's dilemma stage game are $t > r > p > s$ (corresponding to $4 > 3 > 1 > 0$ in §5.6), is given by

$$\frac{1}{2}[r(k-1) + t + (n-k)p] + \frac{1}{2}[r(k-2) + s + (n-k+1)p]$$
$$= r(k-2) + \frac{s+t+p+r}{2} + (n-k)p.$$

If Alice disobeys the choreographer, she can only possibly gain by defecting either one or two rounds earlier. The payoff to defecting on round $k-1$ is

$$\frac{1}{2}[r(k-2) + t + (n-k+1)p] + \frac{1}{2}[r(k-2) + (n-k+2)p]$$
$$= r(k-2) + \frac{t}{2} + (n-k+1)p.$$

The payoff to obeying the choreographer rather than defecting one round earlier is thus $(r+s-p)/2 > 0$. If Alice defects two rounds earlier, her payoff is $r(k-3)+t+(n-k+2)p$, which is less than obeying, provided $r-p>(t-s)/4$. Thus, given this inequality (which clearly holds for the game in §5.6), we have a correlated equilibrium. If k is large, this correlated equilibrium has a high payoff, despite CKR.

7.3 Property Rights and Correlated Equilibrium

The Hawk-Dove-Game (§2.9) is an inefficient way to allocate property rights, especially if the cost of injury w is not much larger than the value v of the property. To see this, note that players choose hawk with probability v/w, and you can check that the ratio of the payoff to the efficient payoff $v/2$ is $1-v/w$. When w is near v, this is close to zero.

The Hawk-Dove-Game is thus a beautiful example of the Hobbesian state of nature, where life is nasty, brutish, and short (Hobbes 1968[1651]). However, suppose some members of the population institute a social norm respecting property rights, based on the fact that whenever two players have a property dispute, one of them must have gotten there first, and the other must have come later. We may call the former the "incumbent" and the latter the "contester."

Note that if all individuals obey the property rights social norm, then there can be no efficiency losses associated with the allocation of property. To show that we indeed have a correlated equilibrium, it is sufficient to show that if we add to the Hawk-Dove-Game a new strategy, P, called the property strategy, that always plays hawk when incumbent and dove when contester, then property is a best response to itself. When we add P to the normal form matrix of the game, we get the *Hawk-Dove-Property Game* depicted in figure 7.1. Note that the payoff to property against property, $v/2$, is greater than $3v/4-w/4$, which is the payoff to hawk against property, and is also greater than $v/4$, which is the payoff to dove against property. Therefore, property is a strict Nash equilibrium. It is also efficient, because there is never a hawk-hawk confrontation in the property correlated equilibrium, so there is never any injury.

The property strategy is not a Nash equilibrium of the Hawk-Dove-Game, but is a correlated equilibrium of the larger social system with the property norm. This example will be elaborated upon in chapter 11.

	H	D	B
H	$(v-w)/2$	v	$3v/4-w/4$
D	0	$v/2$	$v/4$
B	$(v-w)/4$	$3v/4$	$v/2$

Figure 7.1. The Hawk-Dove-Property Game

7.4 Convention as Correlated Equilibrium

The town of Pleasantville has one traffic intersection, one road going north-south and the other east-west. If an east-west car meets a north-south car at the intersection and both stop, one is randomly chosen to go first across the intersection and the second follows, with an average loss of time of one second each. If one car stops and the other goes, only the car that stopped will lose a second. If they both go, however, they may crash, so there is an expected loss of $c > 1$ for each.

There is clearly a unique symmetrical Nash equilibrium to this game, in which each car goes with probability $\alpha = 1/c$ and the expected payoff to each player is -1; that is, they do no better than both waiting. However, there is an obvious social norm in which one car, say the east-west car, always goes, and north-sound car always waits. This is now a correlated equilibrium that implements an asymmetric Nash equilibrium of the underlying game.

With these examples in mind, we can tackle the underlying theory.

7.5 Correlated Strategies and Correlated Equilibria

We will use epistemic game theory (§6.5) to show that if players are Bayesian rational in an epistemic game $\mathcal{G}$ and have a common prior over Ω, the strategy profiles $\mathbf{s}:\Omega \to S$ that they play form a correlated equilibrium (Aumann 1987). The converse also holds: for every correlated equilibrium of a game, there is an extension to an epistemic game $\mathcal{G}$ with a common prior $p \in \Omega$ such that in every state ω it is rational for all players to carry out the move indicated by the correlated equilibrium.

Informally, a correlated equilibrium of an epistemic game $\mathcal{G}$ is a Nash equilibrium of a game $\mathcal{G}^+$, which is $\mathcal{G}$ augmented by an initial move by

Nature, who observes a random variable γ on a probability space (Γ, p) and issues a directive $f_i(\gamma) \in S_i$ to each player i as to which pure strategy to choose. Following Nature's directive is a best response, if other players also follow Nature's directives, provided players have the common prior p.

The intuition behind the theorem is that in an epistemic game, the state space Ω includes all information concerning the players' actions, so common priors imply that all agents agree as to the probability distributions over the actions they will take. Hence, assuming each agent i has a single best response $\mathbf{s}_i(\omega)$ in every state ω (i.e., the equilibrium is a strict correlated equilibrium), the move of each player is known to the others, and because the agents are rational, each must then play a best response to the actions of the others.

Formally, a *correlated strategy* of epistemic game $\mathcal{G}$ consists of a finite probability space (Γ, p), where $p \in \Delta\Gamma$, and a function $f : \Gamma \to S$. If we think of a choreographer who observes $\gamma \in \Gamma$ and directs players to choose strategy profile $f(\gamma)$, then we can identify a correlated strategy with a probability distribution $\tilde{p} \in \Delta S$, where, for $s \in S$, $\tilde{p}(s) = p([f(\gamma) = s])$ is the probability that the choreographer chooses s. We call $\tilde{p}$ the *distribution* of the correlated strategy. Any probability distribution on S that is the distribution of some correlated strategy f is called a *correlated distribution.*

Suppose $f^1, \ldots, f^k$ are correlated strategies and let $\alpha = (\alpha_1, \ldots, \alpha_k)$ be a lottery (i.e., $\alpha_i \geq 0$ and $\sum_i \alpha_i = 1$). Then, $f = \sum_i \alpha_i f^i$ is also a correlated strategy defined on $\{1, \ldots, k\} \times \Gamma$. We call such an f a *convex sum* of $f^1, \ldots, f^k$. Any convex sum of correlated strategies is clearly a correlated strategy. It follows that any convex sum of correlated distributions is itself a correlated distribution.

Suppose $\sigma = (\sigma_1, \ldots, \sigma_n)$ is a Nash equilibrium of a game $\mathcal{G}$, where for each $i = 1, \ldots n$,

$$\sigma_i = \sum_{k=1}^{n_i} \alpha_{ki} s_{ki}$$

where n_i is the number of pure strategies in S_i and α_{ki} is the weight given by σ_i on the kth pure strategy $s_{ki} \in S_i$. Note that σ thus defines a probability distribution $\tilde{p}$ on S such that $\tilde{p}(s)$ is the probability that pure strategy profile $s \in S$ will be chosen when mixed strategy profile σ is played. Then, $\tilde{p}$ is a correlated distribution of an epistemic game associated with $\mathcal{G}$, which we will call $\mathcal{G}$ as well. To see this, define Γ_i as a set with n_i elements $\{\gamma_{1i}, \ldots, \gamma_{n_i i}\}$ and define $p_i \in \Delta S_i$ that places probability α_{ki} on γ_{ki}. Then, for $s = (s_1, \ldots, s_n) \in S$, define $p(s) = \prod_{i=1}^{n} p_i(s_i)$. Now, define $\Gamma = \prod_{i=1}^{n} \Gamma_i$ and let $f : \Gamma \to S$ be given by $f(\gamma_{k_1 1}, \ldots, \gamma_{k_n n}) = (s_{k_1 1}, \ldots, s_{k_n n})$. It is easy to check that f is a correlated strategy with correlated distribution $\tilde{p}$. In short, every Nash equilibrium is a correlated strategy, and hence any convex combination of Nash equilibria is a correlated strategy.

If f is a correlated strategy, then $\pi_i \circ f$ is a real-valued random variable on (Γ, p) with an expected value $\mathbf{E}_i[\pi_i \circ f]$, the expectation taken with respect to p. We say a function $g_i : \Gamma \to S_i$ is *measurable with respect to* f_i if $f_i(\gamma) = f_i(\gamma')$, then $g_i(\gamma) = g_i(\gamma')$. Clearly, player i can choose to follow $g_i(\gamma)$ when he knows $f_i(\gamma)$ iff g_i is measurable with respect to f_i. We say that a correlated strategy f is a *correlated equilibrium* if for each player i and any $g_i : \Gamma \to S_i$ that is measurable with respect to f_i, we have

$$\mathbf{E}_i[\pi_i \circ f] \geq \mathbf{E}_i[\pi_i \circ (f_{-i}, g_i)].$$

A correlated equilibrium induces a *correlated equilibrium probability distribution* on S, whose weight for any strategy profile $s \in S$ is the probability that s will be chosen by the choreographer. Note that a correlated equilibrium of $\mathcal{G}$ is a Nash equilibrium of the game generated from $\mathcal{G}$ by adding Nature, whose move at the beginning of the game is to observe the state of the world $\gamma \in \Gamma$, and to indicate a move $f_i(\gamma)$ for each player i such that no player has an incentive to do other than comply with Nature's recommendation, provided that the other players comply as well.

7.6 Correlated Equilibrium and Bayesian Rationality

THEOREM 7.1 *If the players in epistemic game $\mathcal{G}$ are Bayesian rational at ω, have a common prior p, and each player i chooses $\mathbf{s}_i(\omega) \in S_i$ in state ω, then the distribution of $\mathbf{s} = (\mathbf{s}_1, \ldots, \mathbf{s}_n)$ is a correlated equilibrium distribution given by correlating device f on probability space (Ω, p), where $f(\omega) = \mathbf{s}(\omega)$ for all $\omega \in \Omega$.*

To prove this theorem, we identify the state space for the correlated strategy with the state space Ω of $\mathcal{G}$, and the probability distribution on the state space with the common prior p. We then define the correlated strategy $f : \Omega \to S$ by setting $f(\omega) = (\mathbf{s}_1(\omega), \ldots, \mathbf{s}_n(\omega))$, where $\mathbf{s}_i(\omega)$ is i's choice in state ω (§6.5). Then, for any player i and any function $g_i : \Omega \to S_i$ that is $\mathcal{P}_i$-measurable (i.e., that is constant on cells of the partition $\mathcal{P}_i$), because i

is Bayesian rational, we have

$$\mathbf{E}[\pi_i(\mathbf{s}(\omega))|\omega] \geq \mathbf{E}[\pi_i(\mathbf{s}_{-i}(\omega), g_i(\omega))|\omega].$$

Now, multiply both sides of this inequality by $p(P)$ and add over the disjoint cells $P \in \mathcal{P}_i$, which gives, for any such g_i,

$$\mathbf{E}[\pi_i(\mathbf{s}(\omega))] \geq \mathbf{E}[\pi_i(\mathbf{s}_{-i}(\omega), g_i(\omega))].$$

This proves that $(\Omega, f(\omega))$ is a correlated equilibrium. Note that the converse clearly holds as well. ∎

7.7 The Social Epistemology of Common Priors

De gustibus, we are told, *non est disputandum*: the decision theorist does not question the origin or content of preferences. The Savage axioms, for instance, assume only a few highly general regularities of choice behavior (§1.5). However, we have seen that when we move from individual decision theory to epistemic game theory, this vaunted tolerance is no longer tenable. In its place we require common priors and sometimes, as we shall see, even common knowledge of conjectures (§8.1).

Common priors must be the result of a common process of belief formation. The subjectivist interpretation of probability (di Finetti 1974; Jaynes 2003) is persuasive as a model of human behavior but is a partial view because it cannot explain why individuals agree on certain probabilities.[1]

For standard explanations of common priors, we must turn to the frequency theories of von Mises (1981) and others or to the closely related propensity theory of Popper (1959), which interpret probability as the long-run frequency of an event or its propensity to occur at a certain rate. John Harsanyi (1967) has been perhaps the most eloquent proponent of this approach among game theorists, promoting the *Harsanyi doctrine*, which states that all differences in probability assessments among rational individuals must be due to differences in the information they have received. In fact, however, the Harsanyi doctrine applies only under a highly restricted set of circumstances.

[1]Savage's axiom A3 (see p. 15) suggests that there is something supraindividual about probabilities. This axiom says that the probability associated with an event must not depend on the desirability of the payoff contingent upon the event occurring. There is no reason why this should be the case unless there is some supraindividual standard for assessing probabilities.

Despite lack of agreement concerning the philosophical grounding of probabilistic statements (von Mises 1981; de Laplace 1996; Gillies 2000; Keynes 2004), there is little disagreement concerning the mathematical laws of probability (Kolmogorov 1950). Moreover, modern science is public and objective: except at the very cutting edge of research, there is broad agreement among scientists, however much they differ in creed, culture, or personal predilections.

This line of reasoning suggests that there is a basis for the formation of common priors to the extent that the event in question is what we may call a *natural occurrence*, such as "the ball is yellow," that can be inferred from first-order sense data. We say a natural occurrence is *mutually accessible* to a group of agents when this first-order sense data is accessible to all members of the group, so that if one member knows N, then he knows that all the other members know N. For instance, if i and j are both looking at the same yellow ball, if each sees the other looking at the ball, and if each knows the other has normal vision and is not delusional, then the ball's color is mutually accessible: i knows that j knows that the ball is yellow, and conversely. In short, we can assume that a social situation involving a set of individuals can share an *attentive state* concerning a natural occurrence such that, in a joint attentive state, the natural occurrence is mutually accessible (Tomasello 1999; Lorini, Tummolini, and Herzig 2005).

When we add to this sense data the possibility of joint attentive states of symmetric reasoners (§7.8), common knowledge of natural occurrences becomes plausible (theorem 7.2).

But higher-order epistemic constructs, such as beliefs concerning the intentions, beliefs, and prospective actions of other individuals, beliefs about the natural world that cannot be assessed through individual experience, as well as beliefs about suprasensory reality, do not fall into this category (Morris 1995; Gul 1998; Dekel and Gul 1997). How, then, do such higher-order constructs become commonly known?

The answer is that members of our species, *H. sapiens*, have the capacity to conceive that other members have minds and respond to experience in a manner parallel to themselves—a capacity that is extremely rare and may be possessed by humans alone (Premack and Woodruff 1978). Thus, if agent i believes something, and if i knows that he shares certain environmental experiences with agent j, then i knows that j probably believes this thing as well. In particular, humans have cultural systems that provide natural occurrences that serve as *symbolic cues* for higher-order beliefs and expectations. Common priors, then, are the product of common culture.

The neuropsychological literature on how minds know other minds deals with mirror neurons, the human prefrontal lobe, and other brain mechanisms that facilitate the sharing of knowledge and beliefs. From the viewpoint of modeling human behavior, these facilitating mechanisms must be translated into axiomatic principles of strategic interaction. This is a huge step to take for game theory, which has never provided a criterion *of any sort* for knowledge or beliefs.

7.8 The Social Epistemology of Common Knowledge

We have seen that we must add to Bayesian rationality the principle of normative predisposition to have a social epistemology sufficient to assert that rational agents with common priors, in the presence of the appropriate choreographer, will choose to play a correlated equilibrium. We now study the cognitive properties of agents sufficient to assert that social norms can foster common priors.

Many events are defined in part by the mental representations of the individuals involved. For instance, an individual may behave very differently if he construes an encounter as an impersonal exchange as opposed to a comradely encounter. Mental events fail to be mutually accessible because they are inherently private signals. Nevertheless, there are mutually accessible events N that reliably *indicate* social events E that include the states of mind of individuals in the sense that for any individual i, if i knows N, then i knows E (Lewis 1969; Cubitt and Sugden 2003).

For instance, if I wave my hand at a passing taxi in a large city, both I and the driver of the taxi will consider this an event of the form "hailing a taxi." When the driver stops to pick me up, I am expected to enter the taxi, give the driver an address, and pay the fare at the end of the trip. Any other behavior would be considered bizarre.

By an *indicator* we mean an event N that specifies a social event E to all individuals in a group; i.e., for any individual i, $\mathbf{K}_i N \Rightarrow \mathbf{K}_i E$. Indicators are generally learned by group members through acculturation processes. When one encounters a novel community, one undergoes a process of learning the various indicators of a social event specific to that community. In behavioral game theory an indicator is often called a *frame* of the social event it indicates, and then the *framing effect* includes the behavioral implications of expectations cued by the experimental protocols themselves.

We define individual i as a *symmetric reasoner* with respect to individual j for an indicator N of event E if, whenever i knows N, and i knows that j knows N, then i knows that j knows E; i.e., $\mathbf{K}_i N \wedge \mathbf{K}_i \mathbf{K}_j N \Rightarrow \mathbf{K}_i \mathbf{K}_j E$ (Vanderschraaf and Sillari 2007). We say the individuals in the group are symmetric reasoners if, for each i, j in the group, i is a symmetric reasoner with respect to j.

Like mutual accessibility, joint attentive states, and indicators, symmetric reasoning is an addition to Bayesian rationality that serves as a basis for the concordance of beliefs. Indeed, one may speculate that our capacity for symmetric reasoning is derived by analogy from our recognition of mutual accessibility. For instance, I may consider it just as clear that I am hailing a taxi as that the vehicle in question is colored yellow, and has a lighted sign saying "taxi" on the roof.

THEOREM 7.2 *Suppose individuals in a group are Bayesian rational symmetric reasoners with respect to the mutually accessible indicator N of E. If it is mutual knowledge that the current state $\omega \in N$, then E is common knowledge at ω.*

Proof: Suppose $\mathbf{P}_i\omega \subseteq N$ for all i. Then, for all i, $\mathbf{P}_i\omega \subseteq E$ because N indicates E. For any i, j, because N is mutually accessible, $\omega \in \mathbf{K}_i \mathbf{K}_j N$, and because i is a symmetric reasoner with respect to j, $\omega \in \mathbf{K}_i \mathbf{K}_j E$. Thus, we have $\mathbf{P}_i\omega \subseteq \mathbf{K}_j E$ for all i, j (the case $i = j$ holding trivially). Thus, N is an indicator of $\mathbf{K}_j E$ for all j. Applying the above reasoning to indicator $\mathbf{K}_k E$, we see that $\omega \in \mathbf{K}_i \mathbf{K}_j \mathbf{K}_k E$ for all i, j, and k. All higher levels of mutual knowledge are obtained similarly, proving common knowledge. ■

COROLLARY 7.2.1 *Suppose in state ω that N is a mutually accessible natural occurrence for a group of Bayesian rational symmetric reasoners. Then N is common knowledge in state ω.*

Proof: When $\omega \in N$ occurs, N is mutually known since N is a natural occurrence. Obviously, N indicates itself, so the assertion follows from theorem 7.2. ■

Note that we have adduced common knowledge of an event from simpler epistemic assumptions, thus affording us some confidence that the common knowledge condition has some chance of realization in the real world. This is in contrast to common knowledge of rationality, which is taken as primitive data and hence has little plausibility. Communality of knowledge should always be derived from more elementary psychological and social regularities.

7.9 Social Norms

We say an event E is *norm-governed* if there is a *social norm* $\mathcal{N}(E)$ that specifies *socially appropriate behavior* $\mathcal{N}(E) \subseteq S$, $\mathcal{N}(E) \neq \emptyset$, where S is the strategy profile set for an epistemic game. Note that we allow appropriate behavior to be correlated. How does common knowledge of a social situation E affect the play of a game $\mathcal{G}$? The answer is that each player i must associate a particular social norm $\mathcal{N}(E)$ with E that determines appropriate behavior in the game, i must be confident that other players also associate $\mathcal{N}(E)$ with E, i must expect that others will choose to behave appropriately according to $\mathcal{N}(E)$, and behaving appropriately must be a best response for i given all of the above.

Suppose E indicates $\mathcal{N}(E)$ for players because the players belong to a society in which common culture specifies that when a game $\mathcal{G}$ is played and event E occurs, then appropriate behavior is given by $\mathcal{N}(E)$. Suppose players are symmetric reasoners with respect to E. Then, reasoning similar to theorem 7.2 shows that $\mathcal{N}(E)$ is common knowledge. We then have the following theorem.

THEOREM 7.3 *Given epistemic game $\mathcal{G}$ with normatively predisposed players who are symmetric reasoners, suppose E is an indicator of social norm $\mathcal{N}(E)$. Then, if appropriate behavior according to $\mathcal{N}(E)$ is a correlated equilibrium for $\mathcal{G}$, the players will choose the corresponding correlated strategies.*

7.10 Game Theory and the Evolution of Norms

Social norms cannot be explained as a product of the interaction of Bayesian rational agents. Rather, as developed in chapters 10–12, social norms are explained by sociobiological models of gene-culture coevolution (Cavalli-Sforza and Feldman 1973; Boyd and Richerson 1985). Humans have evolved psychological predispositions that render social norms effective. Social evolution (Cavalli-Sforza and Feldman 1981; Dunbar 1993; Richerson and Boyd 2004) has favored the emergence both of social norms and human predispositions to follow social norms, to embrace common priors, and to recognize common knowledge of many events. Nor is this process limited to humans, as the study of territoriality in various nonhuman species makes clear (Gintis 2007b). The culmination of this process is a pattern of human attributes that can likely be subjected to axiomatic formulation much as we have done with the Savage axioms.

The notion of a social norm as a choreographer is only the first step in analyzing social norms—the step articulating the linkage between Bayesian rationality and game theory on the one hand, and macro-social institutions and their evolution on the other. We can add the dimension of coercion to the concept of a social norm by attaching rewards and punishments to behaviors based on their relationship to socially approved behavior. We can also treat social norms as strongly prosocial under some conditions, meaning that individuals prefer to follow these norms even when it is not in their personal interest to do so, provided others do the same (α-normative predisposition). Finally, we may be able to use a model of the social norm linkage to develop a theory of the evolution of norms, a task initiated by Binmore (1993, 1998, 2005).

7.11 The Merchants' Wares

Consider the coordination game $\mathcal{G}$ with the normal form matrix shown to the right. There are two pure-strategy equilibria: (2,2) and (1,1). There is also a mixed-strategy equilibrium with payoffs (1/3,1/3) in which players choose s_1 with probability 1/3. There is no principle of Bayesian rationality that would lead the players to coordinate on the higher-payoff, or any other, Nash equilibrium.

	s_1	s_2
s_1	2,2	0,0
s_2	0,0	1,1

There are two obvious social norms in this case: "when participating in a pure coordination game, choose the strategy that gives players the maximum (respectively, minimum) common payoff." The following is a plausible social norm that leads to a convex combination of the two coordinated payoffs.

There are two neighboring tribes whose members produce and trade apples and nuts. The members of one tribe wear long gowns, while the members of the other tribe wear short gowns. Individuals indicate a willingness to trade by visually presenting their wares, the quality of which is either 1 or 2, known prior to exchange to the seller but not to the buyer. After exchanging goods, both parties must be satisfied or the goods are restored to their original owners and no trade is consummated. The social norm $\mathcal{N}$

that governs exchange is "never try to cheat a member of your own tribe, and always try to cheat a member of the other tribe."

We assume that, when two individuals meet, the visibility of their wares, F, represents a mutually accessible natural occurrence that is an indicator that the appropriate social norm is $\mathcal{N}$, conforming to which is a best response for both parties. If both have long robes or short robes, $\mathcal{N}$ indicates that they each give full value 2, while if their robe styles differ, each attempts to cheat the other by giving partial value 1. Because all individuals have a normative predisposition and this is common knowledge, each follows the norm, as either trade is better than no trade. A trade is thus consummated in either case. The expected payoff to each player is $2p+(1-p)$, where p is the probability of meeting a trader from one's own tribe.

This analysis illustrates several key points. First, as we will see in §8.1, there are very straightforward sufficient conditions for two rational agents to play a Nash equilibrium (theorem 8.2): mutual knowledge of rationality, mutual knowledge of the game and its payoffs, and mutual knowledge of conjectures (what the other player will choose).

In The Merchants' Wares Problem it is clear that the problem is: Where does mutual knowledge of conjectures come from? As I have stressed, there is nothing in the theory of rational choice that permits us to conclude that two rational agents share beliefs concerning each other's beliefs about each other's beliefs. Rather, each player forms conjectures concerning the other's likely behavior, and the other's likely conjectures, from his *social knowledge*—in this case, from the common *mores* of the two tribes and from mutual knowledge that each knows the *mores* of the two tribes.

The conclusion is that there is no reason to posit that rational agents will choose the Pareto-superior equilibrium because we have seen that sometimes they do not. It is not *reason* but *humanity* that leads us to believe that the Pareto-superior equilibrium is obvious. We humans, by virtue of our gene-culture coevolutionary history and our civilized culture, harbor a default frame that says, "in a coordination game, unless you have some special information that suggests otherwise, conjecture that the other player also considers the frame to be a default frame and reasons as you do, and choose the action that assumes your partner is trying to do well by you." We will return to this point in chapter 12, where we locate it as part of an evolutionary epistemology.

8

Common Knowledge and Nash Equilibrium

> Where every man is Enemy to every man ... the life of man is solitary, poore, nasty, brutish, and short.
>
> Thomas Hobbes

> In the case of any person whose judgment is really deserving of confidence, how has it become so? Because he has kept his mind open to criticism of his opinions ....
>
> John Stuart Mill

This chapter applies the modal logic of knowledge developed in §4.1 and §5.10 to explore sufficient conditions for a Nash equilibrium in two-player games (§8.1). We then expand the modal logic of knowledge to multiple agents and prove a remarkable theorem, due to Aumann (1976), that asserts that an event that is self-evident for each member of a group is common knowledge (§8.3).

This theorem is surprising because it appears to prove that individuals know the content of the minds of others with no explicit epistemological assumptions. We show in §8.4 that this theorem is the result of implicit epistemological assumptions involved in construction of the standard semantic model of common knowledge, and when more plausible assumptions are employed, the theorem is no longer true.

Aumann's famous *agreement theorem* is the subject of §8.7, where we show that the Aumann and Brandenburger (1995) theorem, which supplies sufficient conditions for rational agents to play a Nash equilibrium in multi-player games, is essentially an agreement theorem. Because there is no principle of Bayesian rationality that gives us the commonality of beliefs on which agreement depends, our analysis entails the demise of methodological individualism, a theme explored in §8.8.

8.1 Conditions for a Nash Equilibrium in Two-Player Games

Suppose that rational agents know one another's conjectures (§4.1) in state ω, so that for all i and $j \neq i$,if $\phi_i^\omega(s_{-i}) > 0$ and $s_j \in S_j$ is player j's pure strategy in s_{-i}, then s_j is a best response to his conjecture ϕ_j^ω. We then have a genuine "equilibrium in conjectures," as now no agent has an incentive to change his pure strategy choice s_i, given the conjectures of the other players. We have the following theorem.

THEOREM 8.1 *Let $\mathcal{G}$ be an epistemic game players who are Bayesian rational at ω, and suppose that in state ω each player i knows the others' actions $\mathbf{s}_{-i}(\omega)$. Then $\mathbf{s}(\omega)$ is a Nash equilibrium.*

PROOF: To prove this theorem, which is due to Aumann and Brandenburger (1995), note that for each i, i knows the other players' actions at ω, so $\phi_i^\omega(s_{-i}) = 1$, which implies $\mathbf{s}_{-i}(\omega) = s_{-i}$ by K3, and i's Bayesian rationality at ω then implies $\mathbf{s}_i(\omega)$ is a best response to s_{-i}. ■

We say a Nash equilibrium in conjectures $(\phi_1^\omega, \ldots, \phi_n^\omega)$ occurs at ω if for each player i, $\mathbf{s}_i(\omega)$ is a best response to ϕ_i^ω, and for each i, $\phi_i^\omega \in \Delta^* S_{-i}$. We then have the following theorem.

THEOREM 8.2 *Suppose $\mathcal{G}$ is a two-player game, and at $\omega \in \Omega$, for $i = 1, 2$, $j \neq i$.*

a. Each player knows the other is rational: i.e., $\forall \omega' \in \mathbf{P}_i\omega$, $\mathbf{s}_j(\omega')$ is a best response to $\phi_j^{\omega'}$;

b. Each player knows the other's beliefs; i.e., $\mathbf{P}_i\omega \subseteq \{\omega' \in \Omega | \phi_j^{\omega'} = \phi_j^\omega\}$.

Then, the mixed-strategy profile $(\sigma_1, \sigma_2) = (\phi_2^\omega, \phi_1^\omega)$ is a Nash equilibrium in conjectures.

PROOF: To prove the theorem, which is due to Aumann and Brandenburger (1995) and Osborne and Rubinstein (1994), suppose s_1 has positive weight in $\sigma_1 = \phi_2^\omega$. Because $\phi_2^\omega(s_1) > 0$, there is some ω' such that $\omega' \in \mathbf{P}_2\omega$ and $s_1(\omega') = s_1$. By (a) s_1 is a best reply to $\phi_1^{\omega'}$, which is equal to ϕ_1^ω by (b). Thus s_1 is a best reply to $\sigma_2 = \phi_1^\omega$, and a parallel argument shows that s_2 is a best reply to σ_1, so (σ_1, σ_2) is a Nash equilibrium. ■

8.2 A Three-Player Counterexample

Unfortunately theorem 8.2, which says that Bayesian rationality and mutual knowledge of conjectures imply Nash equilibrium, does not extend to three or more players. For example figure 8.1 shows a game where Alice chooses the row (U, D), Bob chooses the column (L, R), and Carole chooses the matrix (E, W) (this example is due to Osborne and Rubinstein,

	L	R
U	2,3,0	2,0,0
D	0,3,0	0,0,0
	W	

	L	R
U	0,0,0	0,2,0
D	3,0,0	3,2,0
	E	

Figure 8.1. Alice, Bob and Carole

1994, p. 79). Note that every strategy of Carole's is a best response because her payoff is identically zero. We assume there are seven states, so $\Omega = \{\omega_1, \ldots, \omega_7\}$, as depicted in figure 8.2. States ω_1 and ω_7 represent Nash equilibria. There are also two sets of mixed-strategy Nash equilibria. In the first, Alice plays D, Carole plays $2/5W + 3/5E$, and Bob plays anything (Carole's strategy is indeed specified by the condition that it gives Bob equal payoffs for all strategies), while in the second, Bob plays L, Carole plays $3/5W + 2/5E$, and Alice plays anything (this time, Carole's strategy is specified by the condition that it equalizes all Alice's payoffs).

	ω_1	ω_2	ω_3	ω_4	ω_5	ω_6	ω_7
P	32/95	16/95	8/95	4/95	2/95	1/95	32/95
s_1	U	D	D	D	D	D	D
s_2	L	L	L	L	L	L	R
s_3	W	E	W	E	W	E	E
$\mathcal{P}_A$	$\{\omega_1\}$	$\{\omega_2$	$\omega_3\}$	$\{\omega_4$	$\omega_5\}$	$\{\omega_6\}$	$\{\omega_7\}$
$\mathcal{P}_B$	$\{\omega_1$	$\omega_2\}$	$\{\omega_3$	$\omega_4\}$	$\{\omega_5$	$\omega_6\}$	$\{\omega_7\}$
$\mathcal{P}_C$	$\{\omega_1\}$	$\{\omega_2\}$	$\{\omega_3\}$	$\{\omega_4\}$	$\{\omega_5\}$	$\{\omega_6\}$	$\{\omega_7\}$

Figure 8.2. Information structure for the Alice, Bob, and Carole Game. Note that P is the probability of the state, s_i is i's choice in the corresponding state, and $\mathcal{P}_i$ is the knowledge partition for individual i.

Because there is a common prior (the P row in figure 8.2) and every state is in the corresponding cell of partition for each player (the last three rows in the figure), these are true knowledge partitions. Moreover, the posterior probabilities for the players are compatible with the knowledge operators for each player. For instance, in state ω_4, $\mathbf{P}_A\omega_4 = \{\omega_4, \omega_5\}$, and the conditional probability of ω_4, given $\mathbf{P}_A\omega_4$, is 2/3, and that of ω_5 is 1/3. Therefore, Alice's conjecture for Bob is $\phi_{AB}^{\omega_4} = L$, and for Carole it is $\phi_{AC}^{\omega_4} = 2/3E + 1/3W$. Alice's move at ω_4, which is D, is therefore a best response, with a payoff of 2 as opposed to the payoff of 2/3 earned from playing U against L and $2/3E + 1/3W$. Moreover, Alice knows that Carole is rational at ω_4 (trivially, because her payoff does not depend on her move). Alice knows Bob's beliefs at ω_4 because Bob could be in $\mathcal{P}_B$ partition cell $\{\omega_3, \omega_4\}$ or $\{\omega_5, \omega_6\}$, in both of which he believes Alice plays D and Carole plays $2/3W + 1/3E$. She also knows that Bob plays L in both cells, and Bob is rational because L pays off 2 against D and $2/3W + 1/3E$, as opposed to a payoff of 2/3 from playing R. Similarly, at ω_4, $\mathbf{P}_B\omega_4 = \{\omega_3, \omega_4\}$, so Bob knows that Alice is in $\mathcal{P}_A$ partition cell $\{\omega_2, \omega_3\}$ or $\{\omega_4, \omega_5\}$, in both of which Alice knows that Bob plays L and Carole plays $2/3E+1/3W$. Thus, Bob knows Alice's beliefs and that Alice is rational in playing D. Similar reasoning shows that Carole knows Alice and Bob's beliefs and that they are rational at ω_4. Thus, all the conditions of the previous theorem are satisfied at ω_4 but, of course, the conjectures at ω_4 do not form a Nash equilibrium because $\phi_{AB}^{\omega_4} = L$ and $\phi_{BA}^{\omega_4} = D$ are not part of any Nash equilibrium of the game.

The reason theorem 8.2 does not extend to this three-player game is that Alice and Bob have different conjectures as to Carole's behavior, which is possible because Carole has more than one best response to Alice and Bob. They both know Carole is rational, and they both know Carole believes $\phi_C^{\omega} = \{D, L\}$ for $\omega \in \{\omega_2, \ldots, \omega_5\}$. However, these conditions do not determine Carole's mixed strategy. Thus, mutual knowledge of rationality and beliefs is not sufficient to ensure that a Nash equilibrium will be played.

8.3 The Modal Logic of Common Knowledge

Suppose we have a set of n agents, each of whom has a knowledge operator $\mathbf{K}_i$, $i = 1, \ldots, n$. We say $E \subseteq \Omega$ is a *public event* if E is self-evident for all $i = 1, \ldots, n$. By K1, Ω is a public event, and if E and F are public events, so is $E \cap F$, by K2a. Hence, for any $\omega \in \Omega$, there is a minimal public event $\mathbf{P}_*\omega$ containing ω; namely the intersection of all public events containing ω.

We can construct $\mathbf{P}_*\omega$ as follows. First, let

$$\mathbf{P}_*^1\omega = \bigcup_{j \in N} \mathbf{P}_j\omega, \tag{8.1}$$

which is the set of states that are possible for at least one agent at ω. Now, ω is possible for all players i from every state $\omega' \in \mathbf{P}_*^1\omega$, but an arbitrary $\omega' \in \mathbf{P}_*^1\omega$ is possible for some player i at ω, although not necessarily for all. So, $\mathbf{P}_*^1\omega$ may not be a public event. Thus, we define

$$\mathbf{P}_*^2\omega = \bigcup\{\mathbf{P}_*^1\omega' | \omega' \in \mathbf{P}_*^1\omega\}, \tag{8.2}$$

which is the set of states that are possible for some agent at some state in $\mathbf{P}_*^1\omega$; i.e., this is the set of states that are possible for some agent from some state ω' that is possible for some (possibly other) agent at ω. Using similar reasoning, we see that any state in $\mathbf{P}_*^1$ is possible for any player i and any state $\omega' \in \mathbf{P}_*^2$, but there may be states in $\mathbf{P}_*^2\omega$ that are possible for one or more agents but not for all agents. In general, having defined $\mathbf{P}_*^i\omega$ for $i = 1, \ldots, k-1$, we define

$$\mathbf{P}_*^k\omega = \bigcup\{\mathbf{P}_*^1\omega' | \omega' \in \mathbf{P}_*^{k-1}\omega\}. \tag{8.3}$$

Finally, we define

$$\mathbf{P}_*\omega = \bigcup_{k=1}^{\infty} \mathbf{P}_*^k\omega. \tag{8.4}$$

This is the set of states ω' such that there is a sequence of states $\omega = \omega_1, \omega_2, \ldots, \omega_{k-1}, \omega_k = \omega'$ such that ω_{r+1} is possible for some agent at ω_r for $r = 0, \ldots, r-1$. Of course, this is really a finite union because Ω is a finite set. Therefore, for some k, $\mathbf{P}_*^k\omega = \mathbf{P}_*^{k+i}\omega$ for all $i \geq 1$.

We can show that $\mathbf{P}_*\omega$ is the minimal public event containing ω. First, $\mathbf{P}_*\omega$ is self-evident for each $i = 1, \ldots, n$ because for every $\omega' \in \mathbf{P}_*\omega$, $\omega' \in \mathbf{P}_*^k\omega$ for some integer $k \geq 1$, so $\mathbf{P}_i\omega' \subseteq \mathbf{P}_*^{k+1}\omega \subseteq \mathbf{P}_*\omega$. Hence $\mathbf{P}_*\omega$ is a public event containing ω. Now let E be any public event containing ω. Then, E must contain $\mathbf{P}_i\omega$ for all $i = 1, \ldots, n$, so $\mathbf{P}_*^1\omega \subseteq E$. Assume we have proven $\mathbf{P}_*^j\omega \subseteq E$ for $j = 1, \ldots, k$. Because $\mathbf{P}_*^k\omega \subseteq E$ and E is a public event, then $\mathbf{P}_*^{k+1}\omega = \mathbf{P}_*^1(\mathbf{P}_*^k\omega) \subseteq E$. Thus, $\mathbf{P}_*\omega \subseteq E$.

The concept of a public event can be defined directly in terms of the agents' partitions $\mathcal{P}_1, \ldots, \mathcal{P}_n$. We say partition $\mathcal{P}$ is *coarser* than partition $\mathcal{Q}$ if every cell of $\mathcal{Q}$ lies in some cell of $\mathcal{P}$, and we say $\mathcal{P}$ is *finer* than $\mathcal{Q}$ if $\mathcal{Q}$ is coarser than $\mathcal{P}$. The public event partition $\mathcal{P}_*$ corresponding to $\mathbf{P}_*$ is then the finest common coarsening of the partitions $\mathcal{P}_1, \ldots, \mathcal{P}_n$ of the individual players.

To visualize these concepts, we return to the cornfield analogy (§4.1). To coarsen a partition, simply remove one or more fence segments, and then to be tidy, repeatedly remove any fence segments that have either end unconnected to another segment. To refine (i.e., make finer) a partition, simply partition one or more of its cells. If the field has two partitions, visualize one with fence segments colored red and the other with fence segments colored blue. Where the fence segments intersect, let them share a common fence pole. Where a red and a blue fence segment separate the same cornstalks, including the fence segments surrounding the whole cornfield, merge them into red and blue striped fence segments. The finest common coarsening of the two partitions is then the partition formed by removing all fence segments that are of only one color.

This visualization extends directly to the public event partition corresponding to the knowledge partitions in an n-player game. We give each player's fence partition a distinctive color, and we allow two or more agents to share fence segments by applying multiple colors to shared segments. We allow fence segments of different agents to pass through one another by placing a common fence pole at a point of intersection. Now, remove all fence segments that have fewer than n colors. What remains is the public event partition. Alternatively, the minimal public event $\mathbf{P}_*\omega$ containing state ω consists of the states that can be attained by walking from ω to any state in the field, provided one never climbs over a fence shared by all the players.

Clearly the operator $\mathbf{P}_*$ satisfies P1. To show that it also satisfies P2, suppose $\omega' \in \mathbf{P}_*\omega$. Then, by construction, $\mathbf{P}_*\omega' \subseteq \mathbf{P}_*\omega$. To show that $\mathbf{P}_*\omega' = \mathbf{P}_*\omega$, note that $\omega' \in \mathbf{P}_*^k\omega$ for some k. Therefore, by construction, there is a sequence $\omega = \omega_1, \omega_2, \ldots, \omega_{k-1}, \omega_k = \omega'$ such that $\omega_{j+1} \in \mathbf{P}_{i_j}\omega_j$ for some $i_j \in n$ for $j = 1, \ldots, k-1$. However, reversing the order of the sequence shows that $\omega \in \mathbf{P}_*\omega'$. Therefore, $\mathbf{P}_*\omega = \mathbf{P}_*\omega'$. This proves that P2 holds, so $\mathbf{P}_*$ has all the properties of a possibility operator.

It follows that $\mathbf{P}_*$ is a possibility operator. We define a *public event* operator $\mathbf{K}_*$ as the knowledge operator corresponding to the possibility operator $\mathbf{P}_*$, so $\mathbf{K}_*E = \{\omega|\mathbf{P}_*\omega \subseteq E\}$. We can then define an event E as a *public event* at $\omega \in \Omega$ if $\mathbf{P}_*\omega \subseteq E$. Thus, E is a public event if and only if E is self-evident to all players at each $\omega \in E$. Also, E is a public event if and only if E is the union of minimal public events of the form $\mathbf{P}_*\omega$. Moreover, K5 shows that if E is a public event, then at every $\omega \in E$ everyone knows that E is a public event at ω.

In the standard treatment of common knowledge (Lewis 1969; Aumann 1976), an event is common knowledge if everyone knows E, everyone knows that everyone knows E, and so on. A public event is always common knowledge, and conversely. To see this, suppose E is a public event. Then, for any $i, j, k = 1, \ldots, n$, $\mathbf{K}_i E = E$, $\mathbf{K}_j \mathbf{K}_i E = \mathbf{K}_j E = E$, $\mathbf{K}_k \mathbf{K}_j \mathbf{K}_i E = \mathbf{K}_k E = E$, and so on. Thus, all events of the form $\mathbf{K}_k \mathbf{K}_j \ldots \mathbf{K}_i E$ are self-evident for k, so E is common knowledge. Conversely, suppose that for any sequence $i, j, \ldots, k = 1, \ldots, n$, $\mathbf{K}_i \mathbf{K}_j \ldots \mathbf{K}_k E \subseteq E$. Then, for any $\omega \in E$, because $\mathbf{P}_i \omega \subseteq E$, we have $\mathbf{P}_*^1 \omega \subseteq E$, where $\mathbf{P}_*^1$ is defined in (8.1). We also have $\mathbf{K}_i \mathbf{P}_*^1 \omega \subseteq E$ because $\mathbf{K}_i \mathbf{K}_j E \subseteq E$ for $i, j = 1, \ldots, n$, so $\mathbf{P}_*^2 \omega \subseteq E$ from (8.2). From (8.3), we now see that $\mathbf{P}_*^k \omega \subseteq E$ for all k, so $\mathbf{P}_* \omega \subseteq E$. Therefore E is the union of public events and hence is a public event.

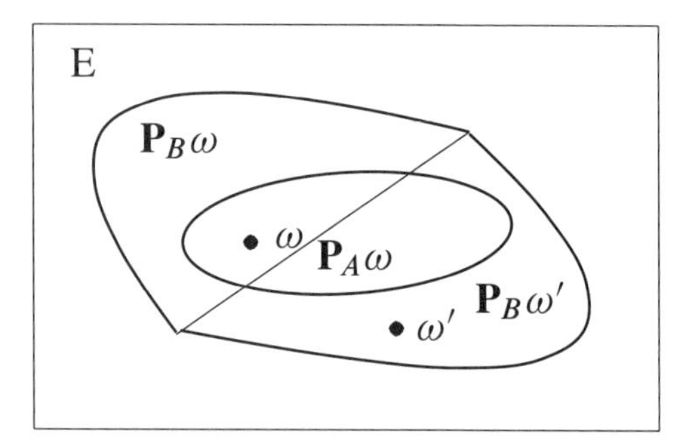

Figure 8.3. The case where, at ω, Bob knows that Alice knows E.

Figure 8.3 shows the situation where Alice knows E at ω, because her minimal self-evident event $\mathbf{P}_A \omega$ at ω lies within E. Moreover $\mathbf{P}_A \omega$ intersects two of Bob's minimal self-evident events, $\mathbf{P}_B \omega$ and $\mathbf{P}_B \omega'$. Because both of $\mathbf{P}_B \omega$ and $\mathbf{P}_B \omega'$ lie within E, Bob knows that Alice knows that E at ω (and at every other state in $\mathbf{P}_A \omega$).

8.4 The Commonality of Knowledge

We have defined a public event as an event that is self-evident to all players. We then showed that an event E is public if and only if it is common knowledge. It appears, then, that at a public event there is a perfect *commonality of knowledge*: players know a great deal about what other players know. Where does this knowledge come from? The answer is that we have tacitly assumed that the way each individual partitions Ω is known to all, not in the formal sense of a knowledge operator but rather in the sense that an expression of the form $\mathbf{K}_i \mathbf{K}_j E$ makes sense and means "i knows that j knows that E." Formally, to say that i knows that j knows E at ω means that at every state $\omega' \in \mathbf{P}_j \omega$, $\mathbf{P}_i \omega' \subseteq E$. But i knows that this is the case only if he knows $\mathbf{P}_j \omega$, which allows him to test $\mathbf{K}_i \omega' \subseteq E$ for each $\omega' \in \mathbf{P}_j \omega$.

For example, suppose Alice, Bob, and Carole meet yearly on a certain date at a certain time to play a game $\mathcal{G}$. Suppose, by chance, all three happen to be in Dallas, Texas, the day before, and although they do not see each other, each witnesses the same highly unusual event x. We define the universe $\Omega = \{\omega, \omega'\}$, where the unusual even occurs in ω but not in ω'. Then, $\mathbf{P}_A\omega = \mathbf{P}_B\omega = \mathbf{P}_C\omega = \{\omega\}$, and hence $\mathbf{K}_A\omega = \mathbf{K}_B\omega = \mathbf{K}_C\omega = \{\omega\}$. Thus ω is self-evident to all three individuals, and hence ω is a public event. Therefore at ω, Alice knows that Bob knows that Carole knows ω, and so on. But, of course, this is not the case. Indeed, none of the three individuals is aware that the others know the event x.

The problem is that we have misspecified the universe. Suppose an event ω is a four-vector, the first entry of which is either x or $\neg x$ (meaning "not x") and the other three are "true" or "false," depending on whether Alice, Bob, and Carole, respectively, knows or does not know whether x occurred. The universe Ω now has 16 distinct states, and the state ω that actually occurred is $\omega = [x, \text{true}, \text{true}, \text{true}]$. However, now $\mathbf{P}_A\omega = \{\omega' \in \Omega | \omega'[1] = x \wedge \omega'[2] = \text{true}\}$. Therefore, the state ω is now *not* self-evident for Alice. Indeed, the smallest self-evident event $\mathbf{P}_A\omega$ for Alice at ω in this case is Ω itself!

This line of reasoning reveals a central lacuna in epistemic game theory: its semantic model of common knowledge assumes too much. Economists have been misled by the elegant theorem that says mutual self-evidence implies common knowledge into believing that the axioms of rational choice imply something substantive concerning the commonality of knowledge across agents. They do not. Indeed, there is no formal principle specifying conditions under which distinct individuals attribute the same truth value to a proposition p with empirical content (we can assume rational agents all agree on mathematical and logical tautologies) or have a mental representation of the fact that others attribute truth value to p. We address this below by sketching the attributes of what we have termed *mutually accessible* events (§7.8).

8.5 The Tactful Ladies

While walking in a garden, Alice, Bonnie, and Carole encountered a violent thunderstorm and were obliged to duck hastily into a restaurant for tea. Carole notices that Alice and Bonnie have dirty foreheads, although each is unaware of this fact. Carole is too tactful to mention this embarrassing situation, which would surely lead them to blush, but she observes that, like her, each of the two ladies knows that someone has a dirty forehead but is also too tactful to mention this fact. The thought occurs to Carole that she also might have a dirty forehead, but there are no mirrors or other detection devices handy that might help resolve her uncertainty.

At this point, a little boy walks by the three young ladies' table and exclaims, "I see a dirty forehead!" After a few moments of awkward silence, Carole realizes that she has a dirty forehead and blushes.

How is this feat of logical deduction possible? Certainly, it is mutually known among the ladies that at least one of them has a dirty forehead, so the little boy did not inform any of them of this fact. Moreover, each lady can see that each of the other ladies sees at least one dirty forehead, so it is mutually known that each lady knew the content of the little boy's message before he delivered it. However, the little boy's remark does inform each lady that they all know that they all know that one of them has a dirty forehead. This is something that none of the ladies knew before the little boy's announcement. For instance, Alice and Bonnie each knows she might not have a dirty forehead, so Alice knows that Bonnie might believe that Carole sees two clean foreheads, in which case Alice and Bonnie know that Carole might not know that there is at least one dirty forehead. Following the little boy's announcement, however, and assuming the other ladies are logical thinkers (which they must be if they are Bayesian decision makers), Carole's inference concerning the state of her forehead is unavoidable.

To see why, suppose Carole does not have a dirty forehead. Carole then knows that Alice sees one dirty forehead (Bonnie's), so Alice has learned nothing from the little boy's remark. But Carole knows that Bonnie sees that Carole's forehead is not dirty, so if Bonnie's forehead is not dirty, then Alice would see two clean foreheads, and the little boy's remark would have implied that Alice knows that she is the unfortunate possessor of a dirty forehead. Because Alice did not blush, Carole knows that Bonnie would have concluded that she herself must have a dirty forehead and would have blushed. Because Bonnie did no such thing, Carole knows that her assumption that she has a clean forehead is false.

To analyze this problem formally, suppose Ω consists of eight states of the form $\omega = xyz$, where $x, y, z \in \{d, c\}$ are the states of Alice, Bonnie, and Carole, respectively, and where d and c stand for "dirty forehead" and "clean forehead," respectively. Thus, for instance, $\omega = ccd$ is the state of the world where Carole has a dirty forehead but Alice and Bonnie both have clean foreheads. When Carole sits down to tea, she knows $E_C = \{ddc, ddd\}$, meaning she sees that Alice and Bonnie have dirty foreheads, but her own forehead could be either clean or dirty. Similarly, Alice knows $E_A = \{cdd, ddd\}$ and Bonnie knows $E_B = \{dcd, ddd\}$. Clearly, no lady knows her own state. What does Bonnie know about Alice's knowledge? Because Bonnie does not know the state of her own forehead, she knows that Alice knows the event "Carole has a dirty forehead," which is $E_{BA} = \{cdd, ddd, ccd, dcd\}$. Similarly, Carole knows that Bonnie knows that Alice knows $E_{CBA} = \{cdd, ddd, ccd, dcd, cdc, ddc, ccc, dcc\} = \Omega$. Assuming Carole has a clean forehead, she knows that Bonnie knows that Alice knows $E'_{CBA} = \{cdc, ddc, dcc, ccc\}$. After the little boy's announcement, Carole then knows that Bonnie knows that Alice knows $E''_{CBA} = \{cdc, ddc, dcc\}$, so if Bonnie did not have a dirty forehead, she would know that Alice knows $E''_{BA} = \{dcc\}$, so Bonnie would conclude that Alice would blush. Thus, Bonnie's assumption that she herself has a clean forehead would be incorrect, and she would blush. Because Bonnie does not blush, Carole knows that her assumption that she herself has a clean forehead is incorrect.

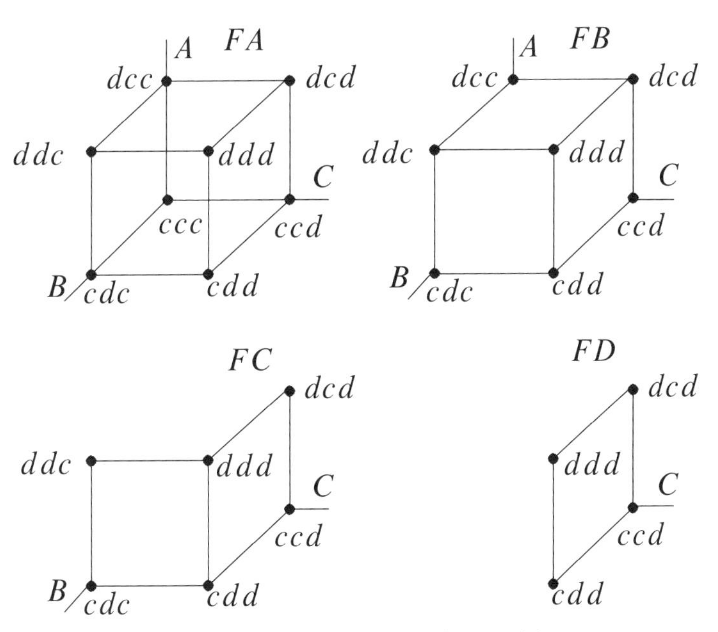

Figure 8.4. The Tactful Ladies Problem

There is an instructive visual way to approach the problem of the Tactful Ladies Problem, due to Fagin et al. (1995) and illustrated in figure 8.4.

Think of each of the ladies as owning one of the three axes in this figure, each corner of the cube representing one of the eight states of the world. The endpoints of lines parallel to an axis represent minimal self-evident events for the lady owning that axis; i.e., the lady in question cannot determine whether her own forehead is dirty.

Because the endpoints of every line segment represent a minimal self-evident event for one of the ladies, a node is reachable from another provided there is some path along the lines of the graph, connecting the first to the second. What, for instance, does it mean that ccc is reachable from ddd along the arrows in panel FA of the figure? First, at ddd, Alice believes cdd is possible, at cdd, B believes ccd is possible, and at ccd, C believes that ccc is possible. In other words, at ddd, Alice believes that it is possible that B believes that it is possible that C believes that ccc might be the true state. Indeed, it is easy to see that any sequence of moves around the cube corresponds to some statement of the form "x believes it is possible that y believes it is possible that $\ldots$, and so on. We define an event $E \subseteq \Omega$ as a public event, or common knowledge, if every state $\omega \in E$ is reachable from every other in this manner. Clearly, the only public event is Ω itself.

When the little boy announces b (that someone has a dirty forehead), assuming this statement is taken as truthful, then the three ladies all know that ccc cannot occur, so we can delete all the paths from some state to ccc. The result is shown in pane FB of the figure. Now, if dcc were the state, Alice would know she has a dirty forehead, and because she apparently does not know this, we can delete the lines terminating in dcc, leading to pane FC in the figure. Now, at ddc or cdc, Bonnie would know she has a dirty forehead, so we can delete the lines connecting to these two nodes. This leaves the nodes depicted in pane FD. Clearly, Carole knows at this event that she has a dirty forehead, but Alice and Bonnie do not.

8.6 The Tactful Ladies and the Commonality of Knowledge

The Tactful Ladies Problem involves many unstated epistemological assertions going far beyond the common knowledge of rationality involved in the conclusion that Carole knows the state of her forehead. Let us see exactly what they are.

Let x_i be the condition that i has a dirty forehead and let k_i be the knowledge operator for i, where $i = A, B, C$, standing for Alice, Bonnie, and

Carole, respectively. When we write i, we mean any $i = A, B, C$, and when we write i, j, we mean any $i, j = A, B, C$, with $j \neq i$, and when we write i, j, m we mean $i, j, m = A, B, C$ and $i \neq j \neq m \neq i$. Let y_i be the condition that i blushes. The six symbols x_i and y_i represent the possible states of affairs in a state space Ω. Let E be the event prior to the little boy's exclamation $b = x_A \vee x_B \vee x_C$.

The statement of the problem tells us that $x_i \in E$ and $k_i x_j \in E$; i.e., each lady sees the forehead of the other two ladies, but not her own. The problem also asserts that $k_i x_i \Rightarrow y_i \in E$ (a lady who knows she has a dirty forehead will blush), and $y_i \Rightarrow k_j y_i \in E$. It is easy to check that these conditions are compatible with $\neg k_i x_i \in E$; i.e., no lady knows the state of her own forehead at event E. These conditions also imply that $k_i b \in E$ (each lady knows the little boy's statement is true).

While the problem intends that $k_i x_j \Rightarrow k_i k_m x_j \in E$ (i.e., if i knows that j has a dirty forehead, she then knows that m knows this as well), this implication does not follow from any principle of rationality, so we must include it as a new principle. The concept needed is that of a mutually accessible natural occurrence. The mutual accessibility of x_i to j and m may appear to be a weak assumption, but in fact it is the *first time* we have made a substantive assertion that one agent knows that another agent knows something. With this assumption, $k_i k_j b \in E$ follows—each lady knows the others know b holds in E (recall that b is the little boy's statement that ccc is false). To see this, note that $k_i x_j \Rightarrow k_i k_m x_j \Rightarrow k_i k_m b$, which is true for all i and $m \neq i$.

Let E' be the state of knowledge following the exclamation $b = x_A \vee x_B \vee x_C$, which we assume is common knowledge. To prove that in E' one of the ladies (e.g., Carole) blushes, we will assume that y_i is mutually accessible to j, m, and j is a symmetric reasoner with respect to m concerning event y_i.

The reasoning following the little boy's statement can be summarized as follows. We will show that if Carole assumes $\neg x_C$ at any state in E', she will arrive at a contradiction. Assuming $\neg x_C$ is true and b is common knowledge, we have $k_C k_B(\neg x_B \Rightarrow k_A \neg x_B \Rightarrow k_A(\neg x_B \wedge \neg x_C \wedge b) \Rightarrow k_A x_A \Rightarrow y_A) \Rightarrow k_C k_B y_A \Rightarrow k_C y_A$, which is false in E'. Thus in E', $k_C k_B x_B \Rightarrow k_C y_B$, which is not true at any state in E'. Hence x_C is true in E', and since Carole knows the current state is in E', $k_C x_C$, she blushes.

8.7 Agreeing to Disagree

In a four-page paper buried in the *Annals of Statistics*, Robert Aumann (1976) proved a remarkable theorem. He showed that if two agents have the same priors concerning an event, if they update their priors using their private knowledge of the current state ω, and if their posterior probabilities are common knowledge, then these posterior probabilities must be equal. In short, two rational agents with common priors cannot "agree to disagree" even though the information upon which each bases his updating can be quite different. I will call any theorem with this conclusion an *agreement theorem*. Aumann commented "We publish this observation with some diffidence, because once one has the appropriate framework, it is mathematically trivial." (p. 1236) It is valuable to understand this theorem and its generalizations because, as it turns out, the common knowledge conditions for a Nash equilibrium are such as to entail an agreement theorem among the agents as to how they will play the game.

Suppose Alice and Bob have a common prior p over Ω, where $p(\omega) > 0$ for all $\omega \in \Omega$. Suppose the actual state is ω_α, leading Alice to update the probability of an event E from $p(E)$ to $p_A(E) = p(E|\mathbf{P}_A\omega_\alpha) = a$ and leads Bob to update $p(E)$ to $p_B(E) = p(E|\mathbf{P}_B\omega_\alpha) = b$. Then, if $p_A(E) = a$ and $p_B(E) = b$ are common knowledge, we must have $a = b$. Thus, despite the fact that Alice and Bob may have different information ($\mathbf{P}_A\omega_\alpha \neq \mathbf{P}_B\omega_\alpha$), their posterior probabilities cannot disagree if they are common knowledge.

To see this, suppose the minimal public event containing ω_α is $\mathbf{K}_*^{\omega_\alpha} = \mathbf{P}_A\omega_1 \cup \cdots \cup \mathbf{P}_A\omega_k$, where each of the $\mathbf{P}_A\omega_i$ is a minimal self-evident event for Alice. Because the event $p_A(E) = a$ is common knowledge, it is constant on $\mathbf{K}_*^{\omega_\alpha}$, so for any j, $a = p_A(E) = p(E|\mathbf{P}_A\omega_j) = p(E \cap \mathbf{P}_A\omega_j)/p(\mathbf{P}_A\omega_j)$, so $p(E \cap \mathbf{P}_A\omega_j) = ap(\mathbf{P}_A\omega_j)$. Thus,

$$
\begin{aligned}
p_A(E \cap \mathbf{K}_*^{\omega_\alpha}) &= p(E \cap \cup_i \mathbf{P}_A\omega_i) = p(\cup_i E \cap \mathbf{P}_A\omega_i) \\
&= \sum_i p(E \cap \mathbf{P}_A\omega_i) = a \sum_i p(\mathbf{P}_A\omega_i) = ap(\mathbf{K}_*^{\omega_\alpha}).
\end{aligned}
$$

However, by similar reasoning, $p_A(E \cap \mathbf{K}_*^{\omega_\alpha}) = bp(\mathbf{K}_*^{\omega_\alpha})$. Hence, $a = b$.

It may seem that this theorem would have limited applicability because when people disagree, their posterior probabilities are usually private information. But suppose Alice and Bob are risk-neutral, each has certain financial assets, they agree to trade these assets, and there is a small cost to trading. Let E be the event that the expected value of Alice's assets is greater than the expected value of Bob's assets. If they agree to trade, then Alice believes E with probability 1 and Bob believes E with probability 0, and this is indeed common knowledge, because their agreement indicates that they desire to trade. This is a contradictory situation, which proves that Alice and Bob cannot agree to trade.

Because in the real world people trade financial assets every day in huge quantities, this proves that either common knowledge of rationality or common priors must be false. In fact, both are probably false. As I argued in §5.11, a rational agent violates CKR whenever such a violation will increase his expected payoff, a situation that is often the case where the subgame perfect equilibrium has relatively low payoffs for the players (§5.9, 5.7). Moreover, there is little reason to believe that the Harsanyi doctrine (§7.7) holds with respect to stock market prices (Kurz 1997).

We can generalize Aumann's argument considerably. Let $f(P)$ be a real number for every $P \subseteq \Omega$. We say f satisfies the *sure-thing principle* on Ω if, for all $P, Q \subseteq \Omega$ with $P \cap Q = \emptyset$, if $f(P) = f(Q) = a$, then $f(P \cup Q) = a$. For instance, if p is a probability distribution on Ω and E is an event, then the posterior probability $f(X) = p(E|X)$ satisfies the sure-thing principle, as does the expected value $f(X) = \mathbf{E}[x|X]$ of a random variable x given $X \subseteq \Omega$. We then have the following *agreement theorem* (Collins 1997):

THEOREM 8.3 *Suppose that for each agent $i = 1, \ldots, n$, f_i satisfies the sure-thing principle on Ω and suppose it is common knowledge at ω that $f_i = s_i$. Then $f_i(\mathbf{K}_*^\omega) = s_i$ for all i, where $\mathbf{K}_*^\omega$ is the cell of the common knowledge partition that contains ω.*

PROOF: To prove this theorem, note that $\mathbf{K}_*^\omega$ is the disjoint union of i's possibility sets $\mathbf{P}_i\omega'$ and that $f_i = s_i$ on each of these sets. Hence, by the sure-thing principle, $f_i = s_i$ on $\mathbf{K}_*^\omega$. ■

COROLLARY 8.3.1 *Suppose agents $i = 1, \ldots, n$ have a common prior on Ω, indicating an event E has probability $p(E)$. Suppose each agent i now receives private information that the actual state ω is in $\mathbf{P}_i\omega$. Then, if the posterior probabilities $s_i = p(E|\mathbf{P}_i\omega)$ are common knowledge, $s_1 = \cdots = s_n$.*

COROLLARY 8.3.2 *Suppose rational, risk-neutral agents $i = 1, \ldots, n$ have the same subjective prior p on Ω, and each has a portfolio of assets*

X_i, all of which have equal expected value $\mathbf{E}_p(X_1) = \cdots = \mathbf{E}_p(X_n)$, and there is a small trading cost $\epsilon > 0$, so no pair of agents desires to trade. In state ω, where agents have posterior expected values $\mathbf{E}_p(X_i|\mathbf{P}_i\omega)$, it cannot be common knowledge that an agent desires to trade.

Finally, we come to our sought-after relationship between common knowledge and Nash equilibrium:

THEOREM 8.4 *Let $\mathcal{G}$ be an epistemic game with $n > 2$ players, and let $\phi^\omega = \phi_1^\omega, \ldots, \phi_n^\omega$ be a set of conjectures. Suppose the players have a common prior p, all players are rational at $\omega \in \Omega$, and it is commonly known at ω that ϕ is the set of conjectures for the game. Then, for each $j = 1, \ldots, n$, all $i = j$ induce the same conjecture $\sigma_j(\omega)$ about j's action, and $(\sigma_1(\omega), \ldots, \sigma_n(\omega))$ form a Nash equilibrium of $\mathcal{G}$.*

The surprising aspect of this theorem is that if conjectures are common knowledge, they must be independently distributed. This is true, essentially, because it is assumed that a player's prior ϕ_i^ω is independent of his own action $s_i(\omega)$. Thus, when strategies are common knowledge, they can be correlated, but their conditional probabilities given ω must be independently distributed.

PROOF: To prove this theorem, we note that, by (8.3), $\phi = \phi_1^\omega, \ldots, \phi_n^\omega$ are common knowledge at ω, and hence $(\sigma_1(\omega), \ldots, \sigma_n(\omega))$ are uniquely defined. Because all agents are rational at ω, each $s_i(\omega)$ maximizes $\mathbf{E}[\pi_i(s_i, \phi_i^\omega)]$. It remains only to show that the conjectures imply that agent strategies are uncorrelated. Let $F = \{\omega' | \phi^{\omega'} |\text{is common knowledge}\}$. Because $\omega \in F$ and $p(\mathbf{P}\omega) > 0$, we have $p(F) > 0$. Now, let $Q(a) = P([s]|F)$ and $Q(s_i) = P([s_i]|F)$, where in general we define $[x] = \{\omega \in \Omega | x(\omega) = x\}$ for some variable function $x : \Omega \to \mathbf{R}$ (thus, $[s] = \{\omega \in \Omega | s(\omega) = s\}$). Note that $Q(a) = P([s] \cap F)/P(F)$. Now, let $H_i = [s_i] \cap F$. Because F is commonly known and $[s_i]$ is known to i, H_i is known to i. Hence H_i is the union of minimal i-known events of the form $\mathbf{P}_i\omega'$, and $p([s_i] \cap \mathbf{P}_i\omega') = \phi_i^\omega(s_-i)p(\mathbf{P}_i\omega')$. Adding up over all the $\mathbf{P}_i^{\omega'}$ comprising H_i (a disjoint union), we conclude $P([s] \cap F) = P([s_{-i}] \cap H) = \phi_i^\omega(s_{-i})P(H_i) = \phi_i^\omega(s_{-i})Q(s_i)P(F)$. Dividing by $P(F)$, we get $Q(a) = \phi_i^\omega(s_{-i})Q(s_i) = Q(s_{-i})Q(s_i)$.

It remains to prove that if $Q(a) = Q(s_{-i})Q(s_i)$ for all $i = 1, \ldots, n$, then $Q(a) = Q(s_1) \cdots Q(s_n)$. This is clearly true for $n = 1, 2$. Suppose it is true for $n = 1, 2, \ldots, n-1$. Starting with $Q(a) = Q(s_1)Q(s_{-1})$, where $a = (s_1, \ldots, s_n)$, we sum over s_i, getting $Q(s_{-n}) = Q(s_1)Q(s_2, \ldots, s_{n-1})$. Similarly, $Q(s_{-i}) = Q(s_i)Q(s_2, \ldots, s_{i-1}, s_{i+1}, \ldots, s_{n-1})$ for any $i = 1, \ldots, n-1$. By the induction hypothesis, $Q(s_{-n}) = Q(s_1)Q(s_2)\cdots Q(s_{n-1})$, so $Q(a) = Q(s_1)\cdots Q(s_n)$. ■

Theorem 8.4 indicates that common priors and common knowledge of conjectures are the epistemic conditions we need to conclude that rational agents will implement a Nash equilibrium. The question, then, is under what conditions are common priors and common knowledge of conjectures likely to be instantiated in real-world strategic interactions?

8.8 The Demise of Methodological Individualism

There is a tacit understanding among classical game theorists that no information other than the rationality of the agents should be relevant to analyzing how they play a game. This understanding is a form of *methodological individualism*, a doctrine that holds that social behavior consists of the interaction of individuals, so nothing beyond the characteristics of individuals is needed, or even permitted, in modeling social behavior.

The most prominent proponent of methodological individualism was Austrian school economist and philosopher Ludwig von Mises, in his book *Human Action*, first published in 1949. While most of Austrian school economic theory has not stood the test of time, methodological individualism has, if anything, grown in stature among economists, especially since the "rational expectations" revolution in macroeconomic theory (Lucas 1981). "Nobody ventures to deny," writes von Mises, "that nations, states, municipalities, parties, religious communities, are real factors determining the course of human events." He continues: "Methodological individualism, far from contesting the significance of such collective wholes, considers it as one of its main tasks to describe and to analyze their becoming and their disappearing, their changing structures, and their operation." Von Mises' arguments in favor of this principle involve an appeal neither to social theory nor social fact. Rather, he asserts, "a social collective has no existence and reality outside of the individual members' actions. ...the way to a cognition of collective wholes is through an analysis of the individuals' actions." (p. 42).

This defense, of course, is merely a restatement of the principle. A passing familiarity with levels of explanation in natural science shows that it is not prima facie plausible. A computer, for instance, is composed of a myriad of solid-state and other electrical and mechanical devices, but stating that one can successfully model the operation of a computer using only models of the behavior of these underlying parts is just false, even in principle. Similarly, eukaryotic cells are composed of a myriad of organic chemicals, yet organic chemistry does not supply all the tools for modeling cell dynamics.

We learn from modern complexity theory that there are many levels of physical existence on earth, from elementary particles to human beings, each level solidly grounded in the interaction of entities at a lower level, yet having emergent properties that are ineluctably associated with the dynamic interaction of its lower-level constituents, yet are incapable of being explained on a lower level. The panoramic history of life synthesis of biologists Maynard Smith and Szathmáry (1997) elaborates this theme that every major transition in evolution has taken the form of a higher level of biological organization exhibiting properties that cannot be deduced from its constituent parts. Morowitz (2002) extends the analysis to emergence in physical systems. Indeed, the point should not be mystifying because there is nothing preventing the most economical model of a phenomenon from being the model itself (Chaitin 2004). Adding emergent properties as fundamental entities in the higher-level model thus may permit the otherwise impossible: the explanation of complex phenomena.

Epistemic game theory suggests that the conditions ensuring that individuals play a Nash equilibrium are not limited to their *personal* characteristics but rather include their *common* characteristics, in the form of common priors and common knowledge. We saw (theorem 7.2) that both individual characteristics and collective understandings, the latter being irreducible to individual characteristics, are needed to explain common knowledge. It is for this reason that methodological individualism is incorrect when applied to the analysis of social life.

Game theory has progressed by accepting no conceptual constructs above the level of the individual actor, as counseled by methodological individualism. Social theory operating at a higher level of aggregation, such as much sociological theory, has produced important insights but has not developed an analytical core on which solid cumulative explanatory progress can be based. The material presented here suggests the fruitfulness of dropping methodological individualist ideology but carefully articulating the analytical linkages between individually rational behavior and the social institutions that align the beliefs and expectations of individuals, making possible effective social intercourse.

Methodological individualism is inadequate, ultimately, because human nature in general, and human rationality in particular, are products of biological evolution. The evolutionary dynamic of human groups has produced *social norms* that coordinate the strategic interaction of rational individuals and regulate kinship, family life, the division of labor, property rights, cultural norms, and social conventions. It is a mistake (the error of methodological individualism) to think that social norms can be brought within the purview of game theory by reducing a social institution to the interaction of rational agents.

9

Reflective Reason and Equilibrium Refinements

> If we allow that human life can be governed by reason, the possibility of life is annihilated.
>
> Leo Tolstoy

> If one weight is twice another, it will take half as long to fall over a given distance.
>
> Aristotle, On the Heavens

In previous chapters, we have stressed the need for a social epistemology to account for the behavior of rational agents in complex social interactions. However, there are many relatively simple interactions in which we can use some form of reflective reason to infer how individuals will play. Since reflective reason is open to the players as well as to us, in such cases we expect Nash equilibria to result from play. However, in many cases there are a plethora of Nash equilibria, only some of which will be played by reasonable agents.

A *Nash equilibrium refinement* of an extensive form game is a criterion that applies to all Nash equilibria that are deemed reasonable but fails to apply to other Nash equilibria that are deemed unreasonable, based on our informal understanding of how rational individuals might play the game. A voluminous literature has developed in search of an adequate equilibrium refinement criterion. A number of criteria have been proposed, including *subgame perfect*, *perfect*, *perfect Bayesian*, *sequential*, and *proper* equilibrium (Harsanyi 1967; Myerson 1978; Selten 1980; Kreps and Wilson 1982; Kohlberg and Mertens 1986), which introduce player error, model beliefs off the path of play, and investigate the limiting behavior of perturbed systems as deviations from equilibrium play go to zero.[1]

I present a new refinement criterion that better captures our intuitions and elucidates the criteria we use implicitly to judge a Nash equilibrium as reasonable or unreasonable. The criterion does not depend on counterfactual

[1]Distinct categories of equilibrium refinement for normal-form games, not addressed in this chapter, are focal point (Schelling 1960; Binmore and Samuelson 2006), and risk dominance (Harsanyi and Selten 1988) criteria. The perfection and sequential criteria are virtually coextensive (Blume and Zame 1994) and extend the subgame perfection criterion.

or disequilibrium beliefs, trembles, or limits of nearby games. I call this the *local best response* (LBR) criterion. The LBR criterion appears to render the traditional refinement criteria superfluous.

The traditional refinement criteria are all variants of subgame perfection and hence suffer from the fact that there is generally no good reason for rational agents to choose subgame perfect strategies in situations where CKR cannot be assumed. The LBR criterion, by contrast, is a variant of *forward induction*, in which agents infer from the fact that a certain node in the game tree has been attained that certain future behaviors can be inferred from the fact that the other players are rational.

We assume a finite extensive form game $\mathcal{G}$ of perfect recall, with players $i = 1, \ldots, n$ and a finite pure-strategy set S_i for each player i, so $S = S_1 \times \cdots \times S_n$ is the set of pure-strategy profiles for $\mathcal{G}$, with payoffs $\pi_i : S \rightarrow \mathbf{R}$. Let S_{-i} be the set of pure-strategy profiles of players other than i and let $\Delta^* S_{-i} = \prod_{j \neq i} \Delta S_j$ be the set of mixed strategies over S_{-i}. Let $\mathcal{N}$ be the set of information sets of $\mathcal{G}$ and let $\mathcal{N}_i$ be the information sets where player i chooses.

A *behavioral strategy* p at an information set ν is a probability distribution over the actions A_ν available at ν. We say p is *part of* a strategy profile σ if p is the probability distribution over A_ν induced by σ. We say a mixed-strategy profile σ *reaches* an information set ν if a path through the game tree that occurs with strictly positive probability, given σ, passes through a node of ν.

For player i and $\nu \in \mathcal{N}_i$, we call $\phi^\nu \in \Delta^* S_{-i}$ a *conjecture* of i at ν. If ϕ^ν is a conjecture at $\nu \in \mathcal{N}_i$ and $j \neq i$, we write ϕ^ν_μ for the marginal distribution of ϕ^ν on $\mu \in \mathcal{N}_j$, so ϕ^ν_μ is i's conjecture at ν of j's behavioral strategy at μ.

Let N_ν be the set of Nash equilibrium strategy profiles that reach information set ν and let $\mathcal{N}^\sigma$ be the set of information sets reached when strategy profile σ is played. For $\tau \in N_\nu$, we write p^τ_μ for the behavioral strategy at μ (i.e., the probability distribution over the choices A_μ at μ) induced by τ. We say a set of conjectures $\{\phi^\nu | \nu \in \mathcal{N}\}$ *supports* a Nash equilibrium σ if, for any i and any $\nu \in \mathcal{N}^\sigma \cap \mathcal{N}_i$, σ_i is a best response to ϕ^ν.

We say a Nash equilibrium σ is an LBR equilibrium if there is a set of conjectures $\{\phi^\nu | \nu \in \mathcal{N}\}$ supporting σ with the following properties: (a) For each i, each $j \neq i$, each $\nu \in \mathcal{N}_i$, and each $\mu \in \mathcal{N}_j$, if $N_\mu \cap N_\nu \neq \emptyset$, then $\phi^\nu_\mu = p^\tau_\mu$ for some $\tau \in N_\mu \cap N_\nu$; and (b) If player i choosing at ν has several choices that lead to different information sets of the other players (we call such choices *decisive*), i chooses among those with the highest payoff.

We can state the first condition verbally as follows. An LBR equilibrium is a Nash equilibrium σ supported by a set of conjectures of players at each information set reached, given σ, where players are constrained to conjecture only behaviors on the part of other players that are part of a Nash equilibrium.

9.1 Perfect, Perfect Bayesian, and Sequential Equilibria

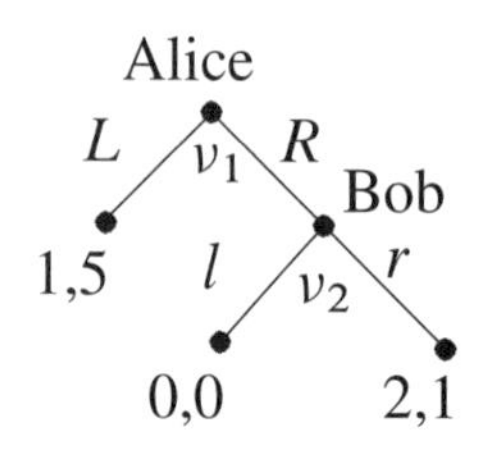

The traditional refinement criteria can be understood as reactions to a problem appearing in the game to the right. This game has two Nash equilibria, Ll and Rr. The former, however, includes an *incredible threat* because if Bob is rational, when he is faced with a choice at ν_B, he will surely choose r rather than l. If Alice believes Bob is rational, she will not find Ll plausible and will play R, to which r is Bob's best response.

Selten (1975) treated this as a problem with subgame perfection (§5.2). He noted that if we assume that there is always a small positive probability that a player will make a wrong move and choose a random action at each information set (called a tremble) then all nodes of the game tree must be visited with positive probability, and only subgame-perfect equilibria will remain. Selten defines an equilibrium σ as *perfect* if, for any $\delta > 0$, there is an $\epsilon > 0$ such that every instance of the game plus trembles of size less than ϵ has a Nash equilibrium within δ of σ. Clearly, in the game above, Rr is the only perfect equilibrium.

One weakness of this solution is that the Ll equilibrium is unreasonable even if there is zero probability of a tremble. A more pertinent equilibrium refinement, a *perfect Bayesian* equilibrium (Fudenberg and Tirole 1991), directly incorporates beliefs in the refinement. Let $\mathcal{N}$ be the set of information sets of the game. A Nash equilibrium σ determines a behavioral strategy p_ν for all $\nu \in \mathcal{N}$ (i.e., a probability distribution over the actions A_ν at ν). An *assessment* μ is defined to be a probability distribution over the nodes at each information set ν. An assessment must be consistent with the behavior strategy $\{p_\nu|\nu \in \mathcal{N}\}$. Consistency means that if σ reaches $\nu \in \mathcal{N}$ and x is a node in ν, then $\mu(x)$ must equal the probability of reaching x, given σ. On an information set not reached, given σ, μ can be defined arbitrarily. We say σ is a perfect Bayesian equilibrium if there is a consistent assessment μ such that, for every player i and every information set ν

where i chooses, p_ν maximizes i's payoff when played against $p_{-\nu}$, using the probability weights given by μ at ν.[2]

This is a rather complicated definition, but the intuition is clear. A Nash equilibrium is perfect Bayesian if there are consistent beliefs rendering each player's choice at every information set payoff-maximizing. Note that for the game above, the equilibrium Ll is not perfect Bayesian because at ν_B Bob's payoff from r is greater than his payoff from l.

Perhaps the most influential refinement criterion other than subgame perfect is the *sequential equilibrium* (Kreps and Wilson 1982). This criterion is a hybrid of perfect and perfect Bayesian. Rather than allowing arbitrary assessments off the path of play (i.e., where the Nash equilibrium does not reach with positive probability), the sequential equilibrium approach perturbs the strategy choices using some pattern of errors and requires that the assessment at an information set off the path of play be the limit of assessment in the perturbed game as the error rate goes to zero. If the pattern of errors is chosen appropriately, Bayes' rule plus σ uniquely determine a limit assessment μ at all nodes, which is the limit of consistent assessments as the error rate goes to zero. We say σ is a sequential equilibrium if there is a limit assessment μ such that, for every player i and every information set ν where i chooses, p_ν maximizes i's payoff when played against $p_{-\nu}$, using the probability weights given by μ at ν.

9.2 Incredible Threats

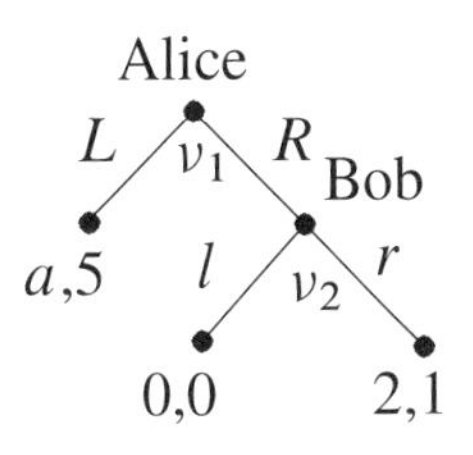

In the game to the right, first suppose $a = 3$. All Nash equilibria have the form (L, σ_B) for an arbitrary mixed strategy σ_B for Bob. At ν_A, any conjecture for Alice supports all Nash equilibria. Since no Nash equilibrium reaches ν_B, there are no constraints on Alice's conjecture. At ν_B, Bob's conjecture must put probability 1 on L, and any σ_B for Bob is a best response to this conjecture. Thus, all (L, σ_B) equilibria satisfy the LBR criterion. While only one of these equilibria is subgame perfect, none involves an incredible threat, and hence there is no reason a rational Bob would choose one strategy profile over another. This is why all satisfy the LBR criterion.

[2]We define $p_{-\nu}$ as the behavioral strategies at all information sets other than ν given by σ.

Now, suppose $a = 1$ in the figure. The Nash equilibria are now (R, r) and (L, σ_B), where $\sigma_B(r) \leq 1/2$. Alice conjectures r for Bob because this is the only strategy at ν_B that is part of a Nash equilibrium. Because R is the only best response to r, the (L, σ_B) are not LBR equilibria. Bob must conjecture R for Alice because this is her only choice in a Nash equilibrium that reaches ν_B. Bob's best response is r. Thus (R, r), the subgame perfect equilibrium, is the unique LBR equilibrium.

Note that this argument does not require any out-of-equilibrium belief or error analysis. Subgame perfection is assured by epistemic considerations alone; i.e., a Nash equilibrium in which Bob plays l with positive probability is an incredible threat.

One might argue that subgame perfection can be defended because there is, in fact, always a small probability that Alice will make a mistake and play R in the $a = 3$ case. However, why single out this possibility? There are many possible imperfections that are ignored in passing from a real-world strategic interaction to the game depicted in the above figure, and they may work in different directions. Singling out the possibility of an Alice error is thus arbitrary. For instance, suppose l is the default choice for Bob, in the sense that it costs him a small amount ϵ_d to decide to choose r over l, and suppose it costs Bob ϵ_B to observe Alice's behavior. The new decision tree is depicted in figure 9.1.

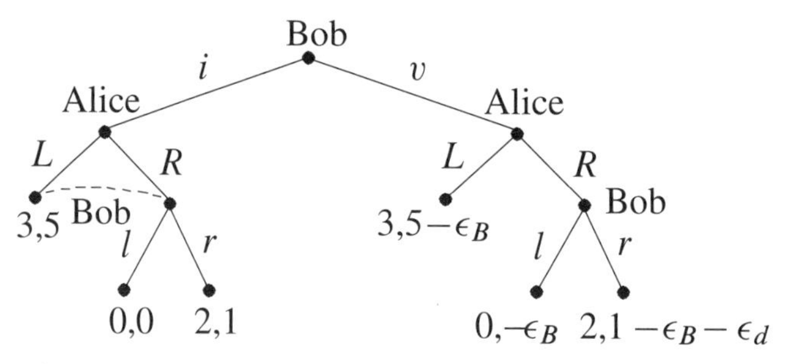

Figure 9.1. Adding an infinitesimal decision cost for Bob

In this new situation, Bob may choose not to observe Alice's choice (i), with payoffs as before and with Bob choosing l by default. But if Bob chooses to view (v), he pays inspection cost ϵ_B, observes Alice's choice, and shifts to the nondefault r when she accidentally plays R, at cost ϵ_d. If Alice plays R with probability ϵ_A, it is easy to show that Bob will choose to inspect only if $\epsilon_A \geq \epsilon_B/(1 - \epsilon_d)$.

The LBR criterion is thus the correct refinement criterion for this game. Standard refinements fail by accepting only subgame perfect equilibria whether or not there is any rational reason to do so (e.g., that the equilibrium involves an incredible threat). The LBR criterion gets to the heart of the matter, which is expressed by the argument that when there is an incredible threat, if Bob gets to choose, he will choose the strategy that gives him a higher payoff, and Alice knows this. Thus Alice maximizes by choosing R, not L. If there is no incredible threat, Bob can choose as he pleases.

In the remainder of this chapter, I compare the LBR criterion with traditional refinement criteria in a variety of typical game contexts. I address where and how the LBR criterion differs from traditional refinements, and which criterion better conforms to our intuition of rational play. For illustrative purposes, I include some cases where both perform equally well. Mainly, however, I treat cases where the traditional criteria perform poorly and the LBR criterion performs well. I am aware of no cases where the traditional criteria perform better than the LBR criterion. Indeed, I know of no cases where the LBR criterion, possibly strengthened by other epistemic criteria, does not perform well, assuming our intuition is that a Nash equilibrium will be played. My choice of examples follows Vega-Redondo (2003). I have tested the LBR criterion for all of Vega-Redondo's examples, and many more, but present only a few of the more informative examples here.

The LBR criterion shares with the traditional refinement criteria the presumption that a Nash equilibrium will be played, and indeed, in every example in this chapter, I would expect rational players to choose an LBR equilibrium (although this expectation is not backed by empirical evidence). In many games, however, such as Rosenthal's Centipede Game (Rosenthal 1981), Basu's Traveler's Dilemma (Basu 1994), and Carlsson and van Damme's Global Games (Carlsson and van Damme 1993), both our intuition and the behavioral game-theoretic evidence violate the presumption that rational agents play Nash equilibria. The LBR criterion does not apply to these games.

Many games have multiple LBR equilibria, only a strict subset of which would be played by rational players. Often, epistemic criteria supplementary to the LBR criterion single out this subset. In this chapter, I use the principle of insufficient reason and what I call the principle of honest communication to this end.

9.3 Unreasonable Perfect Bayesian Equilibria

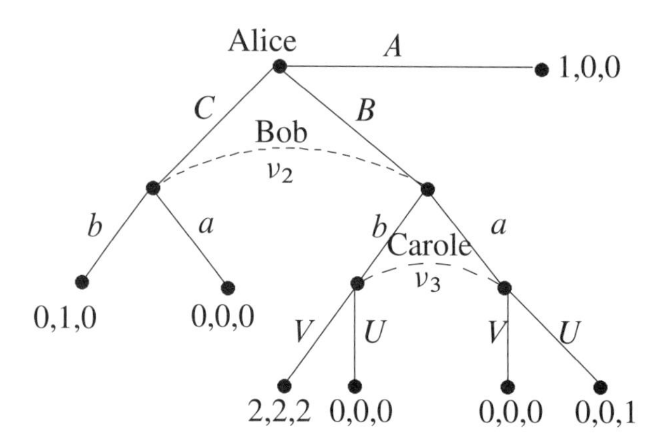

Figure 9.2. Only the equilibrium BbV is reasonable, but there is a connected set of Nash equilibria, including the pure strategy AbU, all members of which are perfect Bayesian.

Figure 9.2 depicts a game in which all Nash equilibria are subgame perfect and perfect Bayesian, but only one is reasonable, and this is the only equilibrium that satisfies the LBR criterion. The game has two sets of equilibria. The first, $\mathcal{A}$, chooses A with probability 1 and $\sigma_B(b)\sigma_C(V) \leq 1/2$, which includes the pure-strategy equilibrium AbU, where σ_A, σ_B, and σ_C are mixed strategies of Alice, Bob, and Carole, respectively. The second is the strict Nash equilibrium BbV. Only the latter is a reasonable equilibrium in this case. Indeed, while all equilibria are subgame perfect because there are no proper subgames, and AbU is perfect Bayesian if Carole believes Bob chose a with probability at least 2/3, it is not sequential because if Bob actually gets to move, he chooses b with probability 1 because Carole chooses V with positive probability in the perturbed game.

The forward induction argument for the unreasonability of the $\mathcal{A}$ equilibria is as follows. Alice can ensure a payoff of 1 by playing A. The only way she can secure a higher payoff is by playing B and having Bob play b and Carole play V. Carole knows that if she gets to move, Alice must have chosen B, and because choosing b is the only way Bob can possibly secure a positive payoff, Bob must have chosen b, to which V is the unique best response. Thus, Alice deduces that if she chooses B, she will indeed secure the payoff 2. This leads to the equilibrium BbV.

To apply the LBR criterion, note that the only moves Bob and Carole use in a Nash equilibrium where they get to choose (i.e., that reaches one of their information sets) are b and V, respectively. Thus, Alice must conjecture this, to which her best response is B. Bob conjectures V, so choose b, and Carole conjectures b, so chooses V. Therefore, only BbV is an LBR equilibrium.

9.4 The LBR criterion picks out the sequential equilibrium

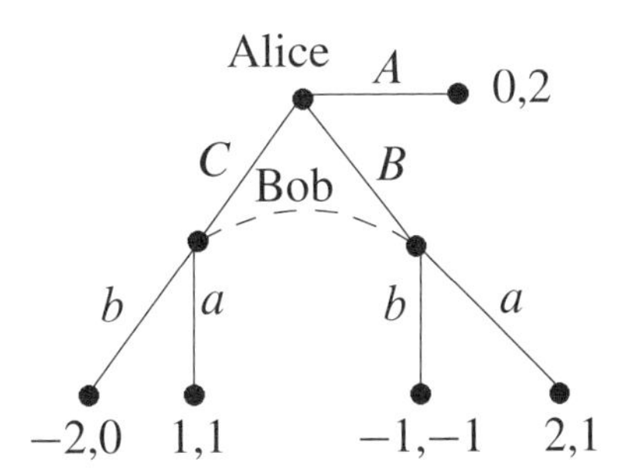

The figure to the right depicts another example where the LBR criterion rules out unreasonable equilibria that pass the subgame perfection and perfect Bayesian criteria, but sequentiality and the LBR criterion are equally successful in this case. In addition to the Nash equilibrium Ba, there is a set $\mathcal{A}$ of equilibria in which Alice plays A with probability 1 and Bob plays b with probability at least 2/3. Equilibria in the set $\mathcal{A}$ are not sequential, but Ba is sequential. The LBR criterion requires that Alice conjecture that Bob plays a if he gets to choose because this is Bob's only move in a Nash equilibrium that reaches his information set. Alice's only best response to this conjecture is B. Bob must conjecture B because this is the only choice by Alice that is part of a Nash equilibrium and reaches his information set, and a is a best response to this conjecture. Thus, Ba is an LBR equilibrium, and the others are not.

9.5 Selten's Horse: Sequentiality vs. the LBR criterion

Selten's Horse is depicted in figure 9.3. This game shows that sequentiality is neither strictly stronger than nor strictly weaker than the LBR criterion since the two criteria pick out distinct equilibria in this case.

There is a connected component $\mathcal{M}$ of Nash equilibria given by

$$\mathcal{M} = \{(A, a, p_\lambda \lambda + (1 - p_\lambda)\rho) | 0 \le p_\lambda \le 1/3\},$$

where p_λ is the probability that Carole chooses λ, all of which, of course, have the same payoff (3,3,0). There is also a connected component $\mathcal{N}$ of Nash equilibria given by

$$\mathcal{N} = \{(D, p_a a + (1 - p_a)d, \lambda) | 1/2 \le p_a \le 1\},$$

where p_a is the probability that Bob chooses a, all of which have the same payoff (4,4,4).

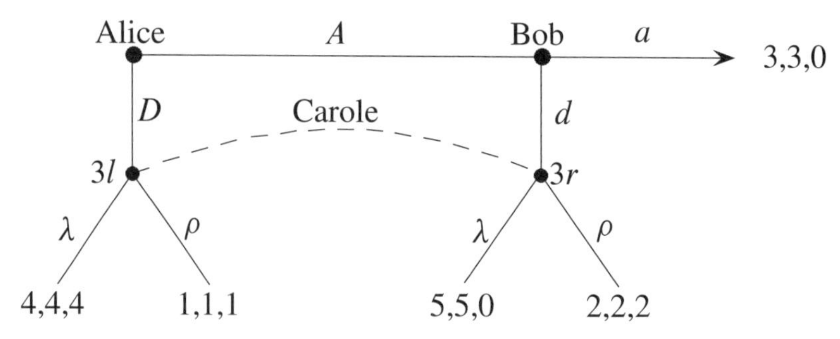

Figure 9.3. Selten's Horse

The $\mathcal{M}$ equilibria are sequential, but the $\mathcal{N}$ equilibria are not even perfect Bayesian since if Bob were given a choice, his best response would be d, not a. Thus, the standard refinement criteria select the $\mathcal{M}$ equilibria as reasonable.

The only Nash equilibrium in which Carole gets to choose is in the set $\mathcal{N}$, where she plays λ. Hence, for the LBR criterion, Alice and Bob must conjecture that Carole chooses λ. Also, a is the only choice by Bob that is part of a Nash equilibrium that reaches his information set. Thus, Alice must conjecture that Bob plays a and Carole plays λ, so her best response is D. This generates the equilibrium $Da\lambda$. At Bob's information set, he must conjecture that Carole plays λ, so a is a best response. Thus, only the pure strategy $Da\lambda$ in the $\mathcal{N}$ component satisfies the LBR criterion.

Selten's Horse is thus a case where the LBR criterion chooses an equilibrium that is reasonable even though it is not even perfect Bayesian, while the standard refinement criteria choose an unreasonable equilibrium in $\mathcal{M}$. The $\mathcal{M}$ equilibria are unreasonable because if Bob did get to choose, he would conjecture that Carole plays λ, because that is her only move in a Nash equilibrium where she gets to move, and hence violates the LBR condition of choosing an action that is part of a Nash equilibrium that reaches his choice node. However, if he is rational, he will violate the LBR criterion and play d, leading to the payoff (5,5,0). If Alice conjectures that Bob will play this way, she will play a, and the outcome will be the non-Nash equilibrium $Aa\lambda$. Of course, Carole is capable of following this train of thought, and she might conjecture that Alice and Bob will play non-Nash strategies, in which case, she could be better off playing the non-Nash ρ herself. But, of course, both Alice and Bob might realize that Carole might reason in this manner. And so on. In short, we have here a case where the sequential equilibria are all unreasonable, but there are non-Nash choices that are as reasonable as the Nash equilibrium singled out by the LBR criterion.

9.6 The Spence Signaling Model

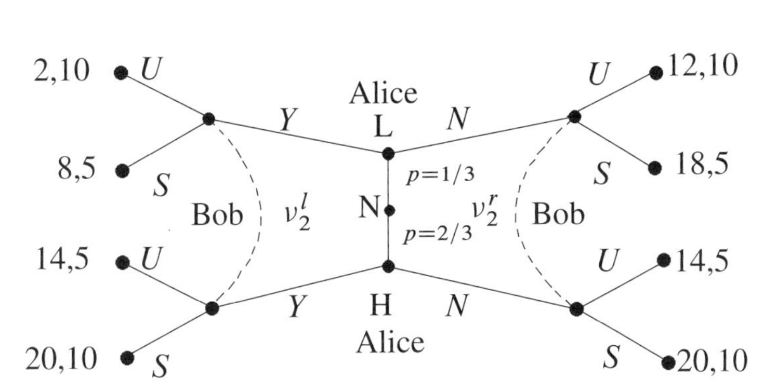

Figure 9.4. The LBR criterion rejects an unreasonable pooling equilibrium

figure 9.4 represents the famous Spence signaling model (Spence 1973). Alice is either a low quality worker (L) with probability $p = 1/3$ or a high quality worker (H) with probability $p = 2/3$. Only Alice knows her own quality. Bob is an employer who has two types of jobs to offer, one for an unskilled worker (U) and the other for a skilled worker (S). If Bob matches the quality of a hire with the skill of the job, his profit is 10; otherwise, his profit is 5. Alice can invest in education (Y) or not (N). Education does not enhance Alice's skill, but if Alice is low quality, it will cost her 10 to be educated, while if she is high quality, it will cost her nothing. Education is thus purely a signal, possibly indicating Alice's type. Finally, the skilled job pays 6 more than the unskilled job, the uneducated high-quality worker earns 2 more than the uneducated low-quality worker in the unskilled job, and the base pay for a low quality, uneducated worker in an unskilled job is 12. This gives the payoffs listed in figure 9.4.

This model has a separating equilibrium in which Alice gets an education only if she is high quality and Bob assigns educated workers to skilled jobs and uneducated workers to unskilled jobs. In this equilibrium, Bob's payoff is 10 and Alice's payoff is 17.33 prior to finding out whether she is of low or high quality. Low-quality workers earn 12, and high-quality workers earn 20. There is also a pooling equilibrium in which Alice never gets an education and Bob assigns all workers to skilled jobs. Indeed, any combination of strategies SS (assign all workers to skilled jobs) and SU (assign uneducated workers to skilled jobs and educated workers to unskilled jobs) is a best response for Bob in the pooling equilibrium. In this equilibrium, Bob earns 8.33 and Alice earns 19.33. However, uneducated workers earn 18 in this equilibrium and skilled workers earn 20.

Both sets of equilibria are sequential. For the pooling equilibrium, consider a completely mixed-strategy profile σ_n in which Bob chooses SS with probability $1-1/n$. For large n, Alice's best response is not to be educated, so the approximate probability of being at the top node a_t of Bob's right-hand information set ν_r is approximately 1/3. In the limit, as $n \to \infty$, the probability distribution over ν_t computed by Bayesian updating approaches (1/3,2/3). Whatever the limiting distribution over the left-hand information set ν_l (note that we can always ensure that such a limiting distribution exists), we get a consistent assessment in which SS is Bob's best response. Hence the pooling equilibrium is sequential.

For the separating equilibrium, suppose Bob chooses a completely mixed strategy σ_n with the probability of US (allocate uneducated workers to unskilled jobs and educated workers to skilled jobs) equal to $1-1/n$. Alice's best response is NY (only a high-quality Alice gets and education), so Bayesian updating calculates probabilities at ν_r as placing almost all the weight on the top node, and at Bob's left-hand information set ν_l, almost all weight is placed on the bottom node. In this limit, we have a consistent assessment in which Bob believes that only high-quality workers get an education, and the separating equilibrium is Bob's best response given this belief.

Both Nash equilibria specify that Bob choose S at ν_B^l, so Alice must conjecture this, and that Alice choose N at L, so Bob must conjecture this. It is easy to check that (NN, SS) and (NY, US) thus both satisfy the LBR criterion.

9.7 Irrelevant Node Additions

Kohlberg and Mertens (1986) use figure 9.5 with $1 < x \leq 2$ to show that an irrelevant change in the game tree can alter the set of sequential equilibria. We use this game to show that the LBR criterion chooses the reasonable equilibrium in both panels, while the sequential criterion does so only if we add an "irrelevant" node, as in the right panel of figure 9.5. The reasonable equilibrium in this case is ML, which is sequential. However, TR is also sequential in the left panel. To see this, let $\{\sigma_A(T), \sigma_A(B), \sigma_A(M)\} = \{1-10\epsilon, \epsilon, 9\epsilon\}$ and $\{\sigma_B(L), \sigma_B(R)\} = \{\epsilon, 1-\epsilon\}$. These converge to T

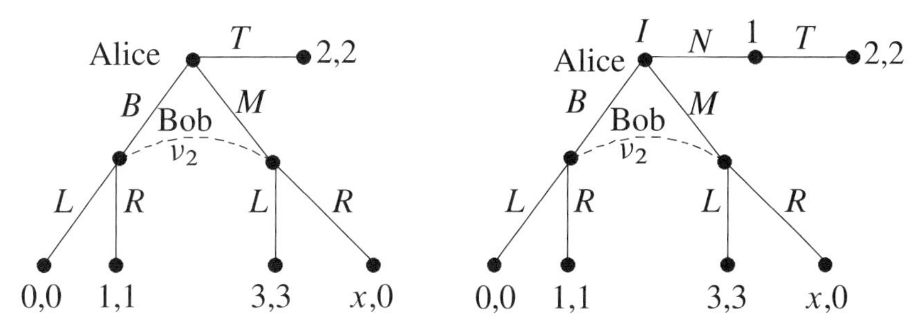

Figure 9.5. The only reasonable equilibrium for $1 < x \leq 2$ is ML, which is sequential and satisfies the LBR criterion. However, the unreasonable equilibrium TR is sequential in the left panel, but not the right, when an "irrelevant" node is added.

and R, respectively, and the conditional probability of being at the left node of ν_2 is 0.9, so Bob's mixed strategy is a distance ϵ from a best response. In the right panel, however, M strictly dominates B for Alice, so TR is no longer sequential.

To apply the LBR criterion, note that the only Nash equilibrium allowing Bob to choose is ML, which gives Alice a payoff of 3, as opposed to a payoff of 2 from choosing T. Therefore, conjecturing this, Alice maximizes her payoff by allowing Bob to choose; i.e., ML is the only LBR equilibrium.

9.8 Improper Sequential Equilibria

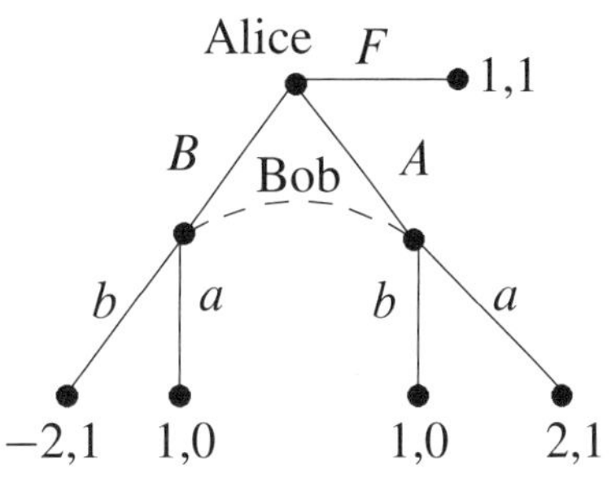

Consider the game to the right. There are two Nash equilibria, Fb and Aa. These are also sequential equilibria since if Alice intends F, but if the probability that B is chosen by mistake is much greater than the probability that A is chosen, then b is a best response. Conversely, if the probability of choosing A by mistake is much greater than the probability that B is chosen by mistake, then a is a best response. Since B is the more costly mistake for Alice, the proper equilibrium concept assumes it occurs very infrequently compared to the A mistake, Bob will play a when he gets to move, so Alice should chose A. Therefore, Aa is the only proper equilibrium according to the Myerson (1978) criterion.

To see that Aa is the only LBR equilibrium of the game, note that the only Nash equilibrium that reaches Bob's information set is Aa. The LBR criterion therefore stipulates that Alice conjecture that Bob chooses a, to which her best response is A.

The reader will note how simple and clear this justification is of Aa by comparison with the properness criterion, which requires an order-of-magnitude assumption concerning the rate at which trembles go to zero.

9.9 Second-Order Forward Induction

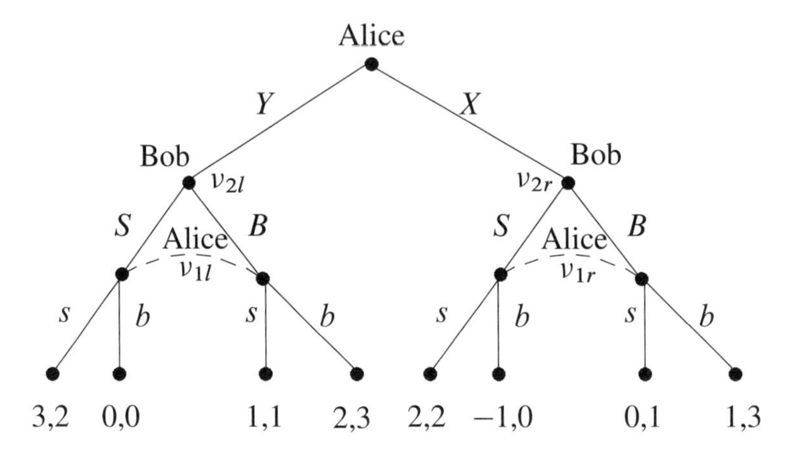

Figure 9.6. The Money Burning Game

Figure 9.6 depicts the famous Money Burning Game analyzed by Ben-Porath and Dekel (1992), illustrating second-order forward induction. By "burning money" (to the amount of 1), Alice can ensure a payoff of 2, so if she does not burn money, she must expect a payoff greater than 2. This induces Alice's Battle of the Sexes partner to act favorably toward her, giving her a payoff of 3.

The set of Nash equilibria can be described as follows. First, there is $\mathcal{YB}$, in which Alice plays Yb (do not burn money and choose favoring Bob) and Bob chooses any mixture of BB (play B no matter what Alice did) and SB (play S if Alice played X and play B if Alice played Y). This set represents the Pareto-optimal payoffs favoring Bob. The second is $\mathcal{YS}$, in which Alice plays Ys (don't burn money and choose favoring Alice), and Bob chooses any mixture of BS (play B against X and S against Y) and SS (play S no matter what). This set represents the Pareto-optimal payoffs favoring Alice. The third is $\mathcal{XS}$, in which Alice plays Xs and

Bob chooses any mixed-strategy combination of SB and SS in which the former is played with probability $\geq 1/2$. This is the money-burning Pareto-inferior Battle-of-the-Sexes equilibrium favoring Bob. The fourth set is the set $\mathcal{M}$ in which Alice plays Y and then $(1/4)b + (3/4)s$ and Bob chooses any mixed strategy that leads to the behavioral strategy $(3/4)B + (1/4)S$. Second-order forward induction selects out $\mathcal{YS}$.

All equilibria are sequential and result from two different orders in eliminating weakly dominated strategies. The only Nash equilibrium where Bob chooses at ν_{2l} involves choosing S there. Thus, Alice conjectures this, and knows that her best response compatible with ν_{2l} is Xs, which gives her a payoff of 2. There are three sets of Nash equilibria where Alice chooses Y. In one, she chooses Yb and Bob chooses B, giving her a payoff of 2. In the second, she chooses Ys and Bob chooses S, giving her a payoff of 3. In the third, Alice and Bob play the Battle-of-the-Sexes mixed-strategy equilibrium, with a payoff of 3/2 for Alice. Each of these is compatible with a conjecture of by that Bob plays a Nash strategy. Hence, her highest payoff is with Ys. Because Y is decisive and includes a Nash equilibrium, where Alice plays Ys, with a higher payoff for Alice than any Nash equilibrium using X, when Bob moves at ν_{2r}, the LBR criterion stipulates that he conjectures this, and hence his best response is S. Thus, Alice must conjecture that Bob plays S when she plays Y, to which Ys is the only best response. Thus, YSs is the only LBR equilibrium.

9.10 Beer and Quiche Without the Intuitive Criterion

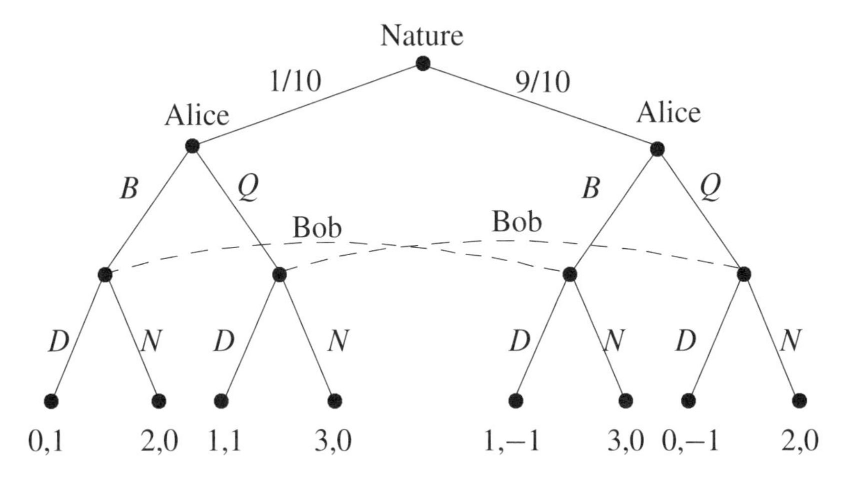

Figure 9.7. The Real Men Don't Eat Quiche Game

Figure 9.7 depicts perhaps the most famous example arguing the importance of beliefs in equilibrium refinement, the Cho and Kreps (1987) Real Men Don't Eat Quiche Game. It also illustrates the well-known *intuitive criterion*, which is complementary to sequentiality. However, the LBR criterion singles out the reasonable equilibrium without recourse to an additional criterion.

This game has two sets of (pooling) Nash equilibria. The first is $\mathcal{Q}$, in which Alice plays QQ (Q if wimp, Q if strong) and Bob uses any mixture of DN (play D against B and play N against Q) and NN (play N no matter what) that places at least weight 1/2 on DN. The payoff to Alice is 21/10. The second is $\mathcal{B}$, in which Alice plays BB (B if wimp, B if strong) and Bob uses any mixture of ND (play N against B and play D against Q) and NN that places at least weight 1/2 on ND. The payoff to Alice is 29/10.

The Cho and Kreps (1987) intuitive criterion notes that by playing QQ, Alice earns 21/10, while by playing BB, she earns 29/10. Therefore, a rational Alice will choose BB. To find the LBR equilibrium, we note that both the $\mathcal{B}$ and the $\mathcal{Q}$ equilibria satisfy the first LBR conditions. Moreover, $\mathcal{B}$ is decisive with respect to $\mathcal{Q}$ and has a higher payoff for Alice. Thus, the only LBR equilibria are the $\mathcal{B}$.

9.11 An Unreasonable Perfect Equilibrium

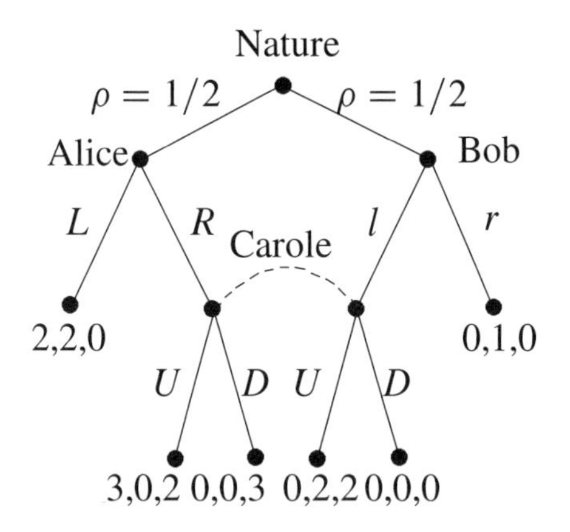

Figure 9.8. Perfection is not sensible, but the LBR criterion is

The game in figure 9.8, taken from McLennan (1985), has a strict Nash equilibrium at RlU and an interval of sequential equilibria $\mathcal{L}$ of the form Lr and D with a probability of at least 3/4. The $\mathcal{L}$ equilibria are "unintuitive" for many reasons. Perhaps the simplest is forward induction. If Carole gets to move, either Alice played R or Bob played l. In the former case, Alice must have expected Carole to play U, so that her payoff would be 3 instead of 2. If Bob moved, he must have moved l, in which case he also must have expected Carole to move U, so that his payoff would be 2 instead of 1. Thus, U is the only reasonable move for Carole.

For the LBR criterion, note that $\mathcal{L}$ is ruled out since the only Nash equilibrium reaching Carole's information set uses U with probability 1. The RlU equilibrium, however, satisfies the LBR criterion.

9.12 The Principle of Insufficient Reason

The figure to the right, due to Jeffery Ely (private communication) depicts a game for which the only reasonable Nash equilibria form a connected component $\mathcal{A}$ in which Alice plays A and Bob plays a with probability $\sigma_B(a) \in [1/3, 2/3]$, guaranteeing Alice a payoff of 0. There are two additional equilibria, Ca and Bb, and there are states at Alice's information set that are consistent with each of these equilibria. At Bob's choice information set, all conjectures σ_A for Alice with $\sigma_A(A) < 1$ are permitted. Thus, all Nash equilibria are LBR equilibria. However, by the principle of insufficient reason and the symmetry of the problem, if Bob gets to move, each of his moves is equally likely. Thus, it is only reasonable for Alice to assume $\sigma_B(a) = \sigma_B(b) = 1/2$, to which A is the only best response.

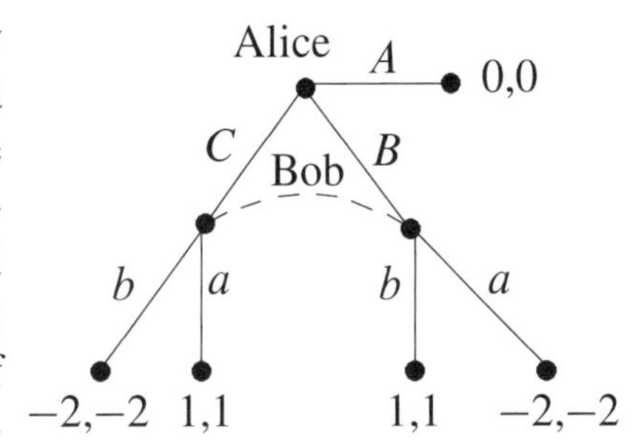

9.13 The Principle of Honest Communication

It is easy to check that the coordination game to the right has three Nash equilibria, each of which satisfies all traditional refinement criteria, and all are LBR equilibria as well. Clearly, however, only Ll is a reasonable equilibrium for rational players. One justification of this equilibrium is that if we add a preliminary round of communication, if each player communicates a promise to make a particular move, and if each player believes

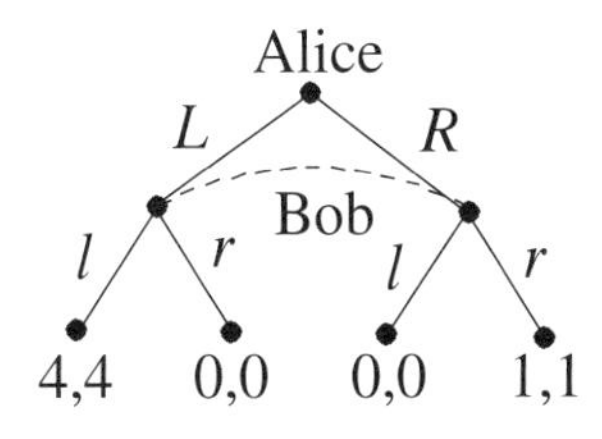

in the principle of honest communication (according to which players keep their promises unless they can benefit by violating these promises and being believed), then each will promise to play L or l, and they will keep their promises.

9.14 Induction: Forward is Robust, Backward is Fragile

The LBR criterion is an improvement of refinement criteria meant to render other standard refinement criteria superfluous. However, the LBR criterion is not an impediment to standard game-theoretic reasoning. If a game has a unique Nash equilibrium, this will be both a subgame perfect and an LBR equilibrium. If a game has a non-subgame perfect equilibrium with an incredible threat, this cannot be an LBR equilibrium because in an LBR equilibrium, players never conjecture incredible threats. In a repeated game, it is reasonable to argue, as does the LBR criterion, that players maximize at every information set they reach. We do not need the extra baggage of the sequential equilibrium concept to justify this assumption.

The LBR criterion supplies a rigorous and insightful equilibrium refinement criterion. Moreover, it clarifies the meaning of an "intuitively reasonable" equilibrium as being one in which players conjecture that other players use actions that are part of Nash equilibria and choose actions themselves that maximize their payoffs subject to such conjectures.

10

The Analytics of Human Sociality

> The whole earth had one language. Men said, "Come, let us build ourselves a city, and a tower with its top in the heavens." The Lord said, "Behold, they are one people, and they have all one language. Nothing will now be impossible for them. Let us go down and confuse their language." The Lord scattered them over the face of the earth, and they ceased building the city.
>
> Genesis 11:1

> An economic transaction is a solved political problem. Economics has gained the title of Queen of the Social Sciences by choosing solved political problems as its domain.
>
> Abba Lerner

10.1 Explaining Cooperation: An Overview

It is often said that sociology deals with cooperation and economics deals with competition. Game theory, however, shows that cooperation and competition are neither distinct nor antithetical. Cooperation involves aligning the beliefs and incentives of agents with distinct interests, competition among groups requires cooperation within these groups, and competition among individuals may be mutually beneficial.

A major goal of economic theory is to show the plausibility of wide-scale cooperation among self-regarding individuals. In an earlier period, this endeavor centered on the Walrasian model of general market equilibrium, culminating in the celebrated fundamental theorem of welfare economics (Arrow and Debreu 1954; Debreu 1959; Arrow and Hahn 1971). However, the theorem's key assumption that market exchange can be enforced at zero cost to the exchanging parties is often violated (Arrow 1971; Bowles and Gintis 1993; Gintis 2002; Bowles 2004).

The game theory revolution replaced reliance on exogenous enforcement with repeated game models in which punishment of defectors by cooperators secures cooperation among self-regarding individuals. Indeed, when a game $\mathcal{G}$ is repeated an indefinite number of times by the same players, many of the anomalies associated with finitely repeated games (4.11, §5.1,

5.7) disappear. Moreover, Nash equilibria of the repeated game arise that are not Nash equilibria of $\mathcal{G}$. The exact nature of these equilibria is the subject of the folk theorem (§10.3), which shows that when self-regarding individuals are Bayesian rational, have sufficiently long time horizons, and there is adequate public information concerning who obeyed the rules and who did not, efficient social cooperation can be achieved in a wide variety of cases.

The folk theorem requires that each action taken by each player carry a signal that is conveyed to the other players. We say a signal is *public* if all players receive the same signal. We say the signal is *perfect* if it accurately reports the player's action. The first general folk theorem that does not rely on incredible threats was proved by Fudenberg and Maskin (1986) for the case of perfect public signals (§10.3).

We say a signal is *imperfect* if it sometimes mis-reports a player's action. An imperfect public signal reports the same information to all players, but it is at times inaccurate. The folk theorem was extended to imperfect public signals by Fudenberg, Levine, and Maskin (1994), as will be analyzed in §10.4.

If different players receive different signals, or some receive no signal at all, we say the signal is *private*. The case of private signals has proved much more daunting than that of public signals, but folk theorems for private but near-public signals (i.e., where there is an arbitrarily small deviation ϵ from public signals) have been developed by several game theorists, including Sekiguchi (1997), Piccione (2002), Ely and Välimäki (2002), Bhaskar and Obara (2002), Hörner and Olszewski (2006), and Mailath and Morris (2006). It is difficult to assess how critical the informational requirements of these folk theorems are because generally the theorem is proved for "sufficiently small ϵ," with no discussion of the actual order of magnitude involved.

The question of the signal quality required for efficient cooperation to obtain is especially critical when the size of the game is considered. Generally, the folk theorem does not even mention the number of players, but in most situations in real life, the larger the number of players participating in a cooperative endeavor, the lower the average quality of the cooperation-vs.-defection signal because generally a player observes only a small number of other players with a high degree of accuracy, however large the group involved. We explore this issue in §10.4, which illustrates the problem by applying the Fudenberg, Levine, and Maskin (1994) framework to the Public Goods Game (§3.9) which in many respects is representative of contexts for cooperation in humans.

10.2 Bob and Alice Redux

Suppose Bob and Alice play the Prisoner's Dilemma shown on the right. In the one-shot game there is only one Nash equilibrium, in which both parties defect. However, suppose the same players play the game at times $t = 0, 1, 2, \ldots$. This is then a new game, called a *repeated game*, in which the payoff to each is the sum of the payoffs over all periods, weighted by a *discount factor* δ with $0 < \delta < 1$. We call the game played in each period the *stage game* of a *repeated game* in which at each period the players can condition their moves on the complete history of the previous stages. A strategy that dictates cooperating until a certain event occurs and then following a different strategy, involving defecting and perhaps otherwise harming one's partner, for the rest of the game is called a *trigger strategy*.

	C	D
C	5,5	$-3, 8$
D	$8, -3$	0,0

Note that we have exactly the same analysis if we assume that players do not discount the future, but in each period the probability that the game continues for at least one more period is δ. In general, we can think of δ as some combination of the discount factor and the probability of game continuation.

We show that the cooperative solution (5,5) can be achieved as a subgame perfect Nash equilibrium of the repeated game if δ is sufficiently close to unity and each player uses the trigger strategy of cooperating as long as the other player cooperates and defecting forever if the other player defects in one round. To see this, consider a repeated game that pays 1 now and in each future period to a certain player and the discount factor is δ. Let x be the value of the game to the player. The player receives 1 now and then gets to play exactly the same game in the next period. Because the value of the game in the next period is x, its present value is δx. Thus, $x = 1 + \delta x$, so $x = 1/(1 - \delta)$.

Now suppose both agents play the trigger strategy. Then, the payoff to each is $5/(1 - \delta)$. Suppose a player uses another strategy. This must involve cooperating for a number (possibly zero) of periods, then defecting forever; for once the player defects, his opponent will defect forever, the best response to which is to defect forever. Consider the game from the time t at which the first player defects. We can call this $t = 0$ without loss of generality. A player who defects receives 8 immediately and nothing thereafter. Thus, the cooperate strategy forms a Nash equilibrium if and only if $5/(1-\delta) \geq 8$, or $\delta \geq 3/8$. When δ satisfies this inequality, the pair of trigger strategies is also subgame perfect because the situation in which both parties defect forever is Nash subgame perfect.

This gives us an elegant solution to the problem, but in fact there are lots of other subgame perfect Nash equilibria for this game. For instance, Bob and Alice can trade off defecting on each other as follows. Consider the following trigger strategy for Alice: alternate $C, D, C, \ldots$ as long as Bob alternates $D, C, D, \ldots$. If Bob deviates from this pattern, defect forever. Suppose Bob plays the complementary strategy: alternate $D, C, D, \ldots$ as long as Alice alternates $C, D, C, \ldots$. If Alice deviates from this pattern, defect forever. These two strategies form a subgame perfect Nash equilibrium for δ sufficiently close to unity.

To see this, note that the payoffs are now $-3, 8, -3, 8, \ldots$ for Alice and $8, -3, 8, -3, \ldots$ for Bob. Let x be the payoffs to Alice. Alice gets -3 today and 8 in the next period and then gets to play the game all over again starting two periods from today. Thus, $x = -3 + 8\delta + \delta^2 x$. Solving this, we get $x = (8\delta - 3)/(1-\delta^2)$. The alternative is for Alice to defect at some point, the most advantageous time being when it is her turn to get -3. She then gets 0 in that and all future periods. Thus, cooperating forms a Nash equilibrium if and only if $x \geq 0$, which is equivalent to $8\delta - 3 \geq 0$, or $\delta \geq 3/8$.

For an example of a very unequal equilibrium, suppose Bob and Alice agree that Bob will play $C, D, D, C, D, D, \ldots$ and Alice will defect whenever Bob is supposed to cooperate, and vice versa. Let v_B be the value of the game to Bob when it is his turn to cooperate, provided he follows his strategy and Alice follows hers. Then, we have

$$v_B = -3 + 8\delta + 8\delta^2 + v_B\delta^3,$$

which we can solve, getting $v_B = (8\delta^2 + 8\delta - 3)/(1-\delta^3)$. The value to Bob of defecting is 8 now and 0 forever after. Hence, the minimum discount factor such that Bob will cooperate is the solution to the equation $v_B = 8$, which gives $\delta \approx 0.66$. Now let v_A be the value of the game to Alice when it is her first turn to cooperate, assuming both she and Bob follow their agreed strategies. Then we have

$$v_A = -3 - 3\delta + 8\delta^2 + v_A\delta^3,$$

which gives $v_A = (8\delta^2 - 3\delta - 3)/(1 - \delta^3)$. The value to Alice of defecting rather than cooperating when it is her first turn to do so is then given by $v_A = 8$, which we can solve for δ, getting $\delta \approx 0.94$. With this discount factor, the value of the game to Alice is 8, but $v_B \approx 72.47$, so Bob gains more than nine times as much as Alice.

10.3 The Folk Theorem

The *folk theorem* is so called because no one knows who first thought of it—it is just part of the folklore of game theory. We shall first present a stripped-down analysis of the folk theorem with an example and provide a more complete discussion in the next section.

Consider the stage game in §10.2. There is a subgame perfect Nash equilibrium in which each player gets zero. Moreover, neither player can be forced to receive a negative payoff in the repeated game based on this stage game because at least zero can be assured simply by playing D. Also, any point in the region $OEABCF$ in figure 10.1 can be attained in the stage game, assuming the players can agree on a mixed strategy for each. To see this, note that if Bob uses C with probability α and Alice uses C with probability β, then the expected payoff to the pair will be $(8\beta - 3\alpha, 8\alpha - 3\beta)$, which traces out every point in the quadrilateral $OEABCF$ for $\alpha, \beta \in [0, 1]$. Only the points in $OABC$ are superior to the universal defect equilibrium (0,0), however.

Consider the repeated game $\mathcal{R}$ based on the stage game $\mathcal{G}$ in §10.2. The folk theorem says that under the appropriate conditions concerning the cooperate/defect signal available to players, any point in the region $OABC$ can be sustained as the average per-period payoff of a subgame perfect Nash equilibrium of $\mathcal{R}$, provided the discount factors of the players are sufficiently near unity.

More formally, consider an n-player game with finite strategy sets S_i for $i = 1, \ldots, n$, so the set of strategy profiles for the game is $S = \prod_{i=1}^{n} S_i$. The payoff for player i is $\pi_i(s)$, where $s \in S$. For any $s \in S$, we write s_{-i} for the vector obtained by dropping the ith component of s, and for any $i = 1, \ldots, n$, we write $(s_i, s_{-i}) = s$. For a given player j, suppose the other players choose strategies m_{-j}^j such that j's best response m_j^j gives j the lowest possible payoff in the game. We call the resulting strategy profile m^j the *maximum punishment payoff* for j. Then, $\pi_j^* = \pi_j(m^j)$ is

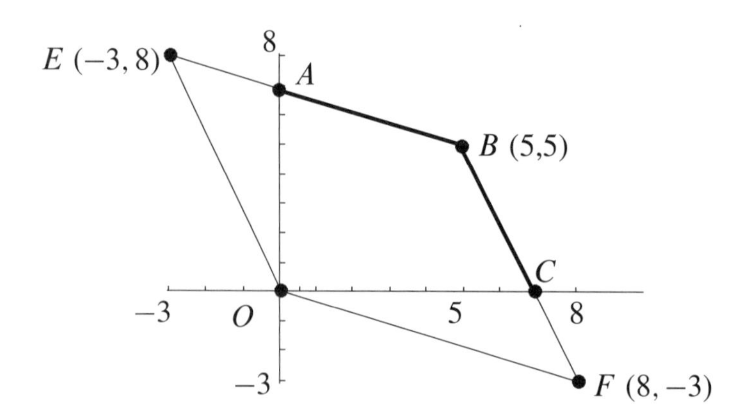

Figure 10.1. The folk theorem: any point in the region $OABC$ can be sustained as the average per-period payoff the subgame perfect Nash equilibrium of the repeated game based on the stage game in §10.2.

j's payoff when everyone else "gangs up on him." We call

$$\pi^* = (\pi_1^*, \ldots, \pi_n^*) \tag{10.1}$$

the *minimax point* of the game. Now define

$$\Pi = \{(\pi_1(s), \ldots, \pi_n(s)) | s \in S, \pi_i(s) \geq \pi_i^*, i = 1, \ldots, n\},$$

so Π is the set of strategy profiles in the stage game with payoffs at least as good as the maximum punishment payoff for each player.

This construction describes a stage game $\mathcal{G}$ for a repeated game $\mathcal{R}$ with discount factor δ, common to all the agents. If $\mathcal{G}$ is played in periods $t = 0, 1, 2, \ldots$, and if the sequence of strategy profiles used by the players is $s(1), s(2), \ldots$, then the payoff to player j is

$$\tilde{\pi}_j = \sum_{t=0}^{\infty} \delta^t \pi_j(s(t)).$$

Let us assume that information is *public* and *perfect*, so that when a player deviates from some agreed-upon action in some period, a signal to this effect is transmitted with probability 1 to the other players. If players can use mixed strategies, then any point in Π can be attained as payoffs to $\mathcal{R}$ by each player using the same mixed strategy in each period. However, it is not clear how a signal indicating deviation from a strictly mixed strategy should be interpreted. The simplest assumption guaranteeing the existence of such a signal is that there is a *public randomizing device* that can be seen by all players and that players use to decide which pure strategy to use, given that they have agreed to use a particular mixed strategy. Suppose, for instance, the randomizing device is a circular disk with a pointer that can be spun by a flick of the finger. Then, a player could mark off a number of regions around the perimeter of the disk, the area of each being proportional to the probability of using each pure strategy in a given mixed strategy to be used by that player. In each period, each player flicks his pointer and chooses the appropriate pure strategy, this behavior is recorded accurately by the signaling device, and the result is transmitted to all the players.

With these definitions, we have the following, where for $\pi \in \Pi$, $\sigma_i(\pi) \in \Delta S_i$ is a mixed strategy for player i such that $\pi_i(\sigma_1, \ldots, \sigma_n) = \pi_i$.

THEOREM 10.1 Folk Theorem. *Suppose players have a public randomizing device and the signal indicating cooperation or defection of each player is public and perfect. Then, for any $\pi = (\pi_1, \ldots, \pi_n) \in \Pi$, if δ is sufficiently close to unity, there is a Nash equilibrium of the repeated game such that π_j is j's payoff for $j = 1, \ldots, n$ in each period. The equilibrium is effected by each player i using $\sigma_i(\pi)$ as long as no player has been signaled as having defected, and by playing the minimax strategy m_i^j in all future periods after player j is first detected defecting.*

The idea behind this theorem is straightforward. For any such $\pi \in \Pi$, each player j uses the strategy $\sigma_j(\pi)$ that gives payoffs π in each period, provided the other players do likewise. If one player deviates, however, all the other players play the strategies that impose the maximum punishment payoff on j forever. Because $\pi_j \geq \pi_j^*$, player j cannot gain from deviating from $\sigma_j(\pi)$, so the profile of strategies is a Nash equilibrium.

Of course, unless the strategy profile $(m_1^j, \ldots, m_n^j)$ is a Nash equilibrium for each $j = 1, \ldots, n$, the threat to minimax even once, let alone forever, is not a credible threat. However, we do have the following theorem.

THEOREM 10.2 The folk theorem with subgame perfection. *Suppose $y = (y_1, \ldots, y_n)$ is the vector of payoffs in a Nash equilibrium of the underlying one-shot game and $\pi \in \Pi$ with $\pi_i \geq y_i$ for $i = 1, \ldots, n$. Then, if δ is sufficiently close to unity, there is a subgame perfect Nash equilibrium of the repeated game such that π_j is j's payoff for $j = 1, \ldots, n$ in each period.*

To see this, note that for any such $\pi \in \Pi$, each player j uses the strategy s_j that gives payoffs π in each period, provided the other players do likewise. If one player deviates, however, all the players play the strategies that give payoff vector y forever.

10.4 The Folk Theorem with Imperfect Public Information

An important model due to Fudenberg, Levine, and Maskin (1994) extends the Folk Theorem to many situations in which there is public imperfect signaling. Although their model does not discuss the n-player Public Goods Game, we shall here show that this game does satisfy the conditions for applying this theorem.

We shall see that the apparent power of the folk theorem in this case comes from letting the discount factor δ go to 1 *last*, in the sense that for any desired level of cooperation (by which we mean the level of *intended*, rather than *realized* cooperation), for any group size n and for any error rate ϵ, there is a δ sufficiently near unity that this level of cooperation can be realized. However, given δ, the level of cooperation may be quite low when n and ϵ are relatively small. Throughout this section, we shall assume that the signal imperfection takes the form of players defecting by accident with probability ϵ and hence failing to provide the benefit b to the group although they expend the cost c.

The Fudenberg, Levine, and Maskin stage game consists of players $i = 1, \ldots, n$, each with a finite set of pure actions $a_1, \ldots, a_{m_i} \in A_i$. A vector $a \in A \equiv \prod_{j=1}^{n} A_i$ is called a pure-action *profile*. For every profile $a \in A$, there is a probability distribution $y|a$ over the m possible public signals Y. Player i's payoff, $r_i(a_i, y)$, depends only on his own action and the resulting public signal. If $\pi(y|a)$ is the probability of $y \in Y$ given profile $a \in A$, i's expected payoff from a is given by

$$g_i(a) = \sum_{y \in Y} \pi(y|a) r_i(a_i, y).$$

Mixed actions and profiles, as well as their payoffs, are defined in the usual way and denoted by Greek letters, so α is a mixed-action profile and $\pi(y|\alpha)$ is the probability distribution generated by mixed action α.

Note that in the case of a simple Public Goods Game, in which each player can cooperate by producing b for the other players at a personal cost c, each action set consists of two elements $\{C, D\}$. We will assume that players choose only pure strategies. It is then convenient to represent the choice of C by 1 and D by 0. Let A be the set of strings of n zeros and ones, representing the possible pure strategy profiles of the n players, the kth entry representing the choice of the kth player. Let $\tau(a)$ be the number of ones in $a \in A$ and write a_i for the ith entry in $a \in A$. For any $a \in A$, the random variable $y \in Y$ represents the imperfect public information concerning $a \in A$. We assume defections are signaled correctly, but intended cooperation fails and appears as defection with probability $\epsilon > 0$. Let $\pi(y|a)$ be the probability that signal $y \in A$ is received by players when the actual strategy profile is $a \in A$. Clearly, if $y_i > a_i$ for some i, then $\pi(y|a) = 0$. Otherwise,

$$\pi(y|a) = \epsilon^{\tau(a)-\tau(y)}(1-\epsilon)^{\tau(y)} \qquad \text{for } \tau(y) \leq \tau(a). \tag{10.2}$$

The payoff to player i who chooses a_i and receives signal y is given by $r_i(a_i, y|a) = b\tau(y)(1-\epsilon) - a_i c$. The expected payoff to player i is just

$$g_i(a) = \sum_{y \in Y} \pi(y|a) r_i(a_i, y) = b\tau(a)(1-\epsilon) - a_i c. \tag{10.3}$$

Moving to the repeated game, we assume that in each period $t = 0, 1, \ldots,$ the stage game is played with public outcome $y^t \in Y$. The sequence $\{y^0, \ldots, y^t\}$ is thus the *public history* of the game through time t, and we assume that the strategy profile $\{\sigma^t\}$ played at time t depends only on this public history (Fudenberg, Levine, and Maskin show that allowing agents to condition their play on their previous private profiles does not add any additional equilibrium payoffs). We call a profile $\{\sigma^t\}$ of public strategies a *perfect public equilibrium* if, for any period t and any public history up to period t, the strategy profile specified for the rest of the game is a Nash equilibrium from that point on. Thus, a public perfect equilibrium is a subgame perfect Nash equilibrium implemented by public strategy profiles. The payoff to player i is then the discounted sum of the payoffs from each of the stage games.

The *minimax* payoff to player i is the largest payoff i can attain if all the other players collude to choose strategy profiles that minimize i's maximum payoff—see (10.1). In the Public Goods Game, the minimax payoff is zero for each player because the worst the other players can do is universally defect, in which case i's best action is to defect himself, giving payoff zero. Let V^* be the convex hull of stage game payoffs that dominate the minimax payoff for each player. A player who intends to cooperate and pays the cost c (which is not seen by the other players) can fail to produce the benefit b (which is seen by the other players) with probability $\epsilon > 0$. In the two-player case, V^* is the quadrilateral $ABCD$ in figure 10.2, where $b^* = b(1-\epsilon) - c$ is the expected payoff to a player if everyone cooperates.

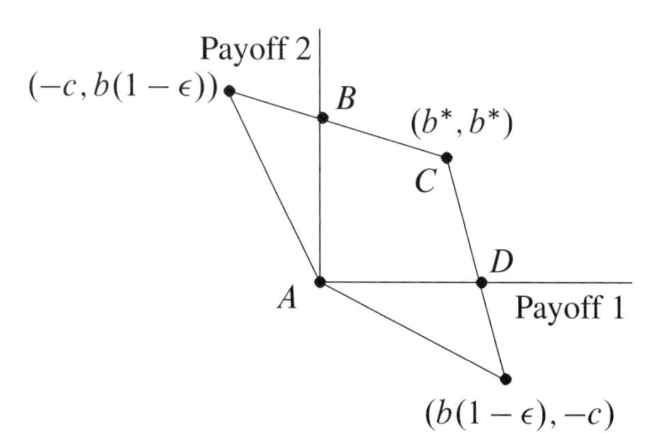

Figure 10.2. Two-player Public Goods Game

The folk theorem in Fudenberg, Levine, and Maskin(1994, p. 1025) is then as follows.[1] We say $W \subset V^*$ is *smooth* if W is closed and convex, has a nonempty interior, and is such that each boundary point $v \in W$ has a unique tangent hyperplane P_v that varies continuously with v (e.g., a closed ball with center interior to V^*). Then, if $W \subset V^*$ is smooth, there is a $\underline{\delta} < 1$ such that for all δ satisfying $\underline{\delta} \leq \delta < 1$, each point in W corresponds to a strict perfect public equilibrium with a discount factor δ, in which a pure-action profile is played in each period. In particular, we can choose W to have a boundary as close as we might desire to $\mathbf{v}^* \equiv (b^*, \ldots b^*)$, in which case the full cooperation payoff can be approximated as closely as desired.

The only condition of the theorem that must be verified in the case of the Public Goods Game is that the full cooperation payoff $\mathbf{v}^* = \{b^*, \ldots, b^*\}$ be on the boundary of an open set of payoffs in $\mathbf{R}^n$, assuming players can used mixed strategies. Suppose player i cooperates with probability x_i, so the payoff to player i is $v_i = \pi_i - cx_i$, where

$$\pi_i = b \sum_{j=1}^{n} x_j - x_i.$$

[1]I am suppressing two conditions on the signal y that are either satisfied trivially or are irrelevant in the case of a Public Goods Game.

If J is the Jacobian of the transformation $x \rightarrow v$, it is straightforward to show that

$$\det[J] = (-1)^{n+1}(b-c)\left(\frac{b}{n-1}+c\right)^{n-1},$$

which is nonzero, proving the transformation is not singular.

The method of recursive dynamic programming used to prove this theorem in fact offers an equilibrium construction algorithm, or rather a collection of such algorithms. Given a set $W \subset V^*$, a discount factor δ, and a strategy profile α, we say α is *enforceable* with respect to W and δ if there is a payoff vector $v \in \mathbf{R}^n$ and a *continuation function* $w: Y \rightarrow W$ such that, for all i,

$$v_i = (1-\delta)g_i(a_i,\alpha_{-i}) + \delta \sum_{y \in Y} \pi(y|a_i,\alpha_{-i})w_i(y) \quad \text{for all } a_i \text{ with } \alpha_i(a_i) > 0, \tag{10.4}$$

$$v_i \geq (1-\delta)g_i(a_i,\alpha_{-i}) + \delta \sum_{y \in Y} \pi(y|a_i,\alpha_{-i})w_i(y) \quad \text{for all } a_i \text{ with } \alpha_i(a_i) = 0. \tag{10.5}$$

We interpret the continuation function as follows. If signal $y \in Y$ is observed (the same signal will be observed by all, by assumption), each player switches to a strategy profile in the repeated game that gives player i the long-run average payoff $w_i(y)$. We thus say that $\{w(y)_{y \in Y}\}$ *enforces* α with respect to v and δ and that the payoff v is *decomposable* with respect to α, W, and δ. To render this interpretation valid, we must show that $W \subseteq E(\delta)$, where $E(\delta)$ is the set of average payoff vectors that correspond to equilibria when the discount factor is δ.

Equations (10.4) and (10.5) can be used to construct an equilibrium. First, we can assume that equations (10.4) and (10.5) are satisfied as equalities. There are then two equations for $|Y| = 2^n$ unknowns $\{w_i(y)\}$ for each player i. To reduce the underdetermination of the equations, we shall seek only pure strategies that are symmetric in the players, so no player can condition his behavior on having a particular index i. In this case, that $w_i(y)$ depends only on whether or not i signaled cooperate, and the number of other players who signaled cooperate. This reduces the number of strategies for a player at this point from 2^n to $2(n-1)$. In the interest of maximizing efficiency, we assume that in the first period all players cooperate, and as long as y indicates universal cooperation, players continue to play and all cooperate.

To minimize the amount of punishment meted out in the case of observed defections while satisfying (10.4) and (10.5), we first assume that if more than one agent signals defect, all continue to cooperate. If there is a single defection, this is punished by all the players defecting an amount that just satisfies the incentive compatibility equations (10.4) and (10.5). There is, of course, no assurance that this will be possible, but if so, there will be a unique punishment γ that is just sufficient to deter a self-regarding player from intentionally defecting. This level of γ is determined by using (10.2) and (10.3) to solve (10.4) and (10.5) using the parameters for the Public Goods Game. The calculations are quite tedious, but the solution for γ in terms of the model parameters is

$$\gamma = \frac{c(1-\delta)}{\delta(1-\epsilon)^{n-1}(1-n\epsilon)}. \tag{10.6}$$

Note that γ does not depend on b. This is because the amount of punishment must induce a player to expend only the cost c. The fraction of a period of production entailed by a given γ does, of course, depend on b. Note also that (10.6) holds only for $n\epsilon < 1$. We deal with the more general case below.

We can now calculate $v = v_i(\forall i)$, the expected one-period payoff. Again, the calculations are tedious, but we find

$$v = b(1-\epsilon) - c - \frac{n\epsilon c(1-\delta)}{1-n\epsilon}. \tag{10.7}$$

This shows that when $n\epsilon$ approaches unity, the efficiency of cooperation plummets.

The above solution is meaningful only when $n\epsilon < 1$. Suppose k is a positive integer such that $k-1 \leq n\epsilon < k$. An extension of the above argument shows that if no punishment is meted out unless exactly k defections are signaled, then (10.7) becomes

$$v = b(1-\epsilon) - c - \frac{n\epsilon c(1-\delta)}{k-n\epsilon}. \tag{10.8}$$

Again, for δ sufficiently close to unity, we can approximate Pareto-efficiency as closely as desired.

By inspecting (10.4) and (10.5), we can gain some insight into what the folk theorem is really saying in this case. When $n\epsilon$ is large, punishment is kept down by requiring several defections to trigger the punishment, in which case the punishment continues over several periods during which payoffs are zero. However, with positive probability when cooperation resumes there will be few defections, and if δ is near unity, the sum of the payoffs over these rare periods of cooperation will be high, so the second term in (10.4) will be large. Moreover, for δ near unity, the first term, representing the expected current period payoff, is near zero, so the present value of cooperation will be determined by the second term as $\delta \to 1$. There is clearly no sense in which this can be considered a solution to the problem of cooperation in large groups.

10.5 Cooperation with Private Signaling

Repeated game models with private signals, including Bhaskar and Obara (2002), Ely and Välimäki (2002), and Piccione (2002), are subject to the critique of the previous section, but private signaling models are complicated by the fact that no sequential equilibrium can support full cooperation, so strictly mixed strategies are necessary in equilibrium. To see this, consider the first period. If each player uses the full cooperation strategy, then if a player receives a defection signal from another player, with probability 1 this represents a bad signal rather than an intentional defection. Thus, with very high probability, no other member received a defection signal. Therefore, no player will react to a defect signal by defecting, and hence the always-defect strategy will have a higher payoff than the always-cooperate strategy. To deal with this problem, *all players defect with positive probability in the first period.*

Now, in any Nash equilibrium, the payoff for any two pure strategies used with positive probability by a player must have equal payoffs against the equilibrium strategies of the other players. Therefore, the probability of defecting must be chosen so that each player is indifferent between cooperating and defecting at least in the first round. Sekiguchi (1997) and Bhaskar and Obara (2002) accomplish this by assuming players randomize in the first round and play the grim trigger strategy in each succeeding round—cooperate as long as you receive a signal that your partner cooperated in the previous round, and after receiving a defect signal, defect yourself in each succeeding round. After the first round, it is possible that a defect signal really means a player defected because that player, who is also playing a trigger strategy, could have received a defect signal in the previous round.

This model is plausible when the number n of players is small, especially for $n=2$. However, when the error rate approaches $1/n$, the model becomes inefficient because the probability of at least one agent receiving a defect signal approaches unity, so the expected number of rounds in the game is close to 1, where the benefits of game repetition disappear. Moreover, it is true in most cases that the quality of the private signal deteriorates with increasing group size (e.g., if each individual receives a signal from a fixed maximum number of other players). As Bhaskar and Obara (2002) show, we can improve the performance of the model by "restarting" cooperation with a positive probability in each round after cooperation has ceased, but this only marginally improves efficiency because this process does not increase the incentive for players to cooperate in any given round.

Ely and Välimäki (2002) have developed a different approach to the problem, following the lead of Piccione (2002), who showed how to achieve coordination in a repeated game with private information without the need for the sort of belief updating and grim triggers used by Sekiguchi, Bhaskar, and Obara. They construct an equilibrium in which at every stage, each player is *indifferent* between cooperating and defecting no matter what his fellow members do. Such an individual is thus willing to follow an arbitrary mixed strategy in each period, and the authors show that there exists such a strategy for each player that ensures close to perfect cooperation provided individuals are sufficiently patient and the errors are small.

One problem with this approach is that it uses mixed strategies in every period, and unless the game can be purified (§6.2), there is no reason for players to play such strategies or to believe that their partners will do so. Bhaskar (2000) has shown that most repeated game models that use mixed strategies cannot be purified, and Bhaskar, Mailath, and Morris (2004) have shown that purification is generally impossible in the Ely-Välimäki approach to the Prisoner's Dilemma when the signal is public. The case of private signals is much more difficult, and there is no known example of purification in this case.

Without a choreographer, there is no mechanism that coordinates the activities of large numbers of people so as to implement a repeated game equilibrium with private information. It follows that the issue of whether or not such games, the Nash equilibria of which invariably require strictly mixed strategies, can be purified is not of fundamental importance. Nevertheless, it is useful to note that there are no examples of purification of such games, and at least two examples of the impossibility of purification. These examples, due to Bhaskar (1998, 2000), make it clear why purification is not likely: the Nash equilibria in repeated games with private information are engineered so that players are indifferent between acting on information concerning defection and ignoring such information. A slight change in payoffs, however, destroys this indifference, so players behave the same way whatever signal they receive. Such behavior is not compatible with a cooperative Nash equilibrium.

It might be thought that the lack of purification is not a fatal weakness, however, because we have already shown that the social instantiation of a repeated game model requires a choreographer, and there is no reason, at least in principle, that a social norm cannot implement a mixed strategy σ by suggesting each of the pure strategies in the support of σ with a probability equal to its weight in σ. This idea is, however, incorrect, as we exhibited in §6.3. Unless players have a sufficiently strong normative predisposition, small random changes in payoffs induce players to deviate from the choreographer's suggestion. The lack of purification for these models is virtually fatal—they cannot be socially instantiated.

10.6 One Cheer For the Folk Theorem

The folk theorem is the most promising analytically rigorous theory of human cooperation in the behavioral sciences. Its strength lies in its transformation of Adam Smith's "invisible hand" into an analytical model of elegance and clarity. The folk theorem's central weakness is that it is only an existence theorem with no consideration of how the Nash equilibria whose existence it demonstrates can actually be instantiated as a social process. Certainly, these equilibria cannot be implemented spontaneously or through a process of player learning. Rather, as we have stressed throughout this book, strategic interaction must be socially structured by a choreographer—a social norm with the status of common knowledge, as outlined in chapter 7.

This weakness is analytically trivial but scientifically monumental. Correcting it both strengthens repeated game models and suggests how they may be empirically tested—namely, by looking for the choreographer and, where it cannot be found, determining what premise of the repeated game model is violated and proposing an alternative model. Recognizing the normative dimension of social cooperation has the added benefit of explaining why repeated game models have virtually no relevance beyond our own species (Clements and Stephens 1995; Stephens, McLinn, and Stevens

2002; Hammerstein 2003), the reason being that normative behavior is extremely primitive, at best, for nonhuman species.

A second weakness of repeated game theory is its preoccupation with situations in which players are almost perfectly future-oriented (i.e., use a discount factor close to unity) and the noise in the system (e.g., signaling stochasticity or player error) is arbitrarily small. The reason for this preoccupation is simple: the folk theorem with self-regarding agents fails when agents are present-oriented, signals are imperfect, or players are likely to err.

The correct response to this weakness is to (a) observe how cooperation really occurs in society and (b) alter the characteristics of the repeated game model to incorporate what one has discovered. We learn from biology that there are huge gains in cooperation for an organism, but the challenges of coordinating behavior and keeping defection to manageable levels are extreme and are solved only by rare genetic innovation (Maynard Smith and Szathmáry 1997). The notion that human cooperation has a strong biological element, as stressed in chapter 3, is in line with this general biological point. We present in this chapter, for illustrative purposes, a model of cooperation based on observed characteristics of humans that are not captured by Bayesian rationality or by the social epistemology developed in earlier chapters (§10.7).

Both common observation and behavioral experiments suggest that humans are disposed to behave in prosocial ways when raised in appropriate cultural environments (Gintis 2003a). This disposition includes having other-regarding preferences, such as empathy for others, and the predisposition to embrace cooperative norms and to punish violators of these norms even at personal cost. It also includes upholding such character virtues as honesty, promise keeping, trustworthiness, bravery, group loyalty, and considerateness. Finally, it includes valuing self-esteem and recognizing that self-esteem depends on how one is evaluated by those with whom we strategically interact. Without these prosocial, biologically rooted traits, human language could not have developed, because there would then have been no means of maintaining veridical information transmission. Without high-quality information, efficient cooperation based on repeated game Nash equilibria would be impossible. Indeed, it is probably rare that information is of sufficient quality to sustain the cooperation of self-regarding actors.

10.7 Altruistic Punishing in the Public Goods Game

This section develops a model of cooperation in the Public Goods Game in which each agent is motivated by self-interest, unconditional altruism, and strong reciprocity, based on Carpenter et al. (2009). We investigate the conditions for a cooperative equilibrium, as well as how the efficiency of cooperation depends on the level of altruism and reciprocity. We show that if there is a stable interior equilibrium (i.e., including both cooperation and shirking), an increase in either altruism or reciprocity motives will generate higher efficiency.

Consider a group of size $n > 2$, where member i supplies an amount of effort $1 - \sigma_i \in [0, 1]$. We call σ_i the *level of shirking* of member i and write $\bar{\sigma} = \sum_{j=1}^{n} \sigma_j / n$ for the average level of shirking. We assume that working at level $1 - \sigma_i$ adds $q(1 - \sigma_i)$ dollars to the group output, where $q > 1$, while the cost of working is a quadratic function $s(1 - \sigma_i) = (1 - \sigma_i)^2/2$. We call q the *productivity of cooperation*. We assume the members of the group share their output equally, so member i's payoff is given by

$$\pi_i = q(1 - \bar{\sigma}) - (1 - \sigma_i)^2/2. \tag{10.9}$$

The payoff loss for each member of the group from one member shirking is $\beta = q/n$. We assume $1/n < \beta < 1$.

We assume member i can impose a cost on $j \neq i$ with monetary equivalent s_{ij} at cost $c_i(s_{ij})$ to himself. The cost s_{ij} results from public criticism, shunning, ostracism, physical violence, exclusion from desirable side deals, or another form of harm. We assume $c_i(0) = c_i'(0) = c_i''(0) = 0$ and $c_i(s_{ij})$ is increasing and strictly convex for all i, j when $s_{ij} > 0$.

Member j's cooperative behavior b_j depends on j's level of shirking and the harm that j inflicts on the group, which we assume is public knowledge. Specifically, we assume

$$b_j = \beta(1 - 2\sigma_j). \tag{10.10}$$

Thus, $\sigma_j = 1/2$ is the point at which i evaluates j's cooperative behavior as neither good nor bad.

To model cooperative behavior with social preferences, we say that individual i's utility depends on his own material payoff π_i and the payoff π_j to other individuals $j \neq i$ according to

$$u_i = \pi_i + \sum_{j \neq i} [(a_i + \lambda_i b_j)(\pi_j - s_{ij}) - c_i(s_{ij})] - s_i(\sigma_i) \tag{10.11}$$

where $s_i(\sigma_i) = \sum_{j \neq i} s_{ji}(\sigma_i)$ is the punishment inflicted upon i by other group members and $\lambda_i \geq 0$. The parameter a_i is i's level of unconditional altruism if $a_i > 0$ and unconditional spite if $a_i < 0$, and λ_i is i's strength of reciprocity motive, valuing j's payoffs more highly if j conforms to i's concept of good behavior, and conversely (Rabin 1993; Levine 1998). If λ_i and a_i are both positive, the individual is termed a strong reciprocator, motivated to reduce the payoffs of an individual who shirks even at a cost to himself.

Players maximize (10.11), and because b_j can be negative, this may lead i to increase his shirking σ_i and/or to punish j by increasing s_{ij} in response to a higher level of shirking by j. This motivation for punishing a shirker values the punishment per se rather than the benefits likely to accrue to the punisher if the shirker responds positively to the punishment. Moreover, members derive utility from punishing the shirker, not simply from observing that the shirker was punished. This means that punishing provides a warm glow rather than being instrumental towards affecting j's behavior (Andreoni 1995; Casari and Luini 2007).

This model requires only that a certain fraction of group members be reciprocators. This is in line with the evidence from behavioral game theory presented in chapter 3, which indicates that in virtually every experimental setting a certain fraction of the subjects do not act reciprocally, because they are self-regarding or they are purely altruistic. Note also that the punishment system could elicit a high level of cooperation, yet a low level of net material payoff. This is because punishment is not strategic in this model. In real societies, the amount of punishment of shirkers is generally socially regulated, and punishment beyond the level needed to secure compliance is sanctioned (Wiessner 2005).

In this model, i will choose $s^*_{ij}(\sigma_j)$ to maximize utility in (10.11), giving rise to the first-order condition (assuming an interior solution)

$$c'_i(s^*_{ij}) = \lambda_i \beta(2\sigma_j - 1) - a_i. \tag{10.12}$$

If $\lambda_i > 0$ and

$$\sigma_j \leq \sigma_i^0 = \frac{1}{2}\left[\frac{a_i}{\lambda_i \beta} + 1\right], \tag{10.13}$$

the maximization problem has a corner solution in which i does not punish. For $\lambda_i > 0$ and $\sigma_j > \sigma_i^0$, denoting the right-hand side of (10.12) by ϕ and differentiating (10.12) totally with respect to any parameter x, we get

$$\frac{ds_{ij}^*}{dx} = \frac{\partial\phi}{\partial x}\frac{1}{c_i''(s_{ij}^*)}. \tag{10.14}$$

In particular, setting $x = a_i$, $x = \lambda_i$, $x = \sigma_j$, $x = \beta$ and $x = n$ in turn in (10.14), we have the following theorem.

THEOREM 10.3 *For $\lambda_i > 0$ and $\sigma_j > \sigma_i^0$, the level of punishment by i imposed on j, s_{ij}^*, is (a) decreasing in i's unconditional altruism a_i; (b) increasing in i's reciprocity motive, λ_i; (c) increasing in the level σ_j of j's shirking; (d) increasing in the harm β that j inflicts upon i by shirking; and (e) decreasing in group size.*

The punishment $s_j(\sigma_j)$ inflicted upon j by the group is given by

$$s_j(\sigma_j) = \sum_{i \neq j} s_{ij}^*(\sigma_j), \tag{10.15}$$

which is then differentiable and strictly increasing in σ_j over some range, provided there is at least one reciprocator i ($\lambda_i > 0$).

The first-order condition on σ_i from (10.11) is given by

$$1 - \sigma_i - \beta = \beta \sum_{j \neq i}(a_i + \lambda_i b_j) + s_i'(\sigma_i), \tag{10.16}$$

so i shirks up to such a point that the net benefits of shirking (the left-hand side) equal i's valuation of the costs imposed on others by his shirking (the first term on the right-hand side) plus the marginal cost of shirking entailed by the increased level of punishment that i may expect. This defines i's optimal shirking level σ_i^* for all i and hence closes the model, assuming the second-order conditions $s_i''(\sigma_i) > -1$. Whether there is an interior solution depends on the array of parameters of the problem. For instance, if reciprocity is very weak, there could be complete shirking by every player, or if very strong, zero shirking by every player. We assume an interior solution to investigate the comparative statics of the problem.

The average shirking rate of i's partners is given by

$$\bar{\sigma}_{-i} = \frac{1}{n-1}\sum_{j \neq i}\sigma_j.$$

We say that i's partners *shirk on balance* if $\bar{\sigma}_i > 1/2$ and *work on balance* if the opposite inequality holds. We then have the following theorem, which is proved in Carpenter et al. (2009):

THEOREM 10.4 *Suppose there is an stable interior equilibrium under a best response dynamic. Then (a) an increase in i's unconditional altruism a_i leads i to shirk less; and (b) an increase in i's reciprocity motive λ_i leads i to shirk more when i's partners shirk on balance and to shirk less when i's partners work on balance.*

While this is a simple one-shot model, it could easily be developed into a repeated game model in which some of the parameters evolve endogenously and where reputation effects strengthen the other-regarding motives on which the above model depends.

10.8 The Failure of Models of Self-Regarding Cooperation

Providing a plausible game-theoretic model of cooperation among self-regarding agents would vindicate methodological individualism (§8.8), and render economic theory virtually independent of, and foundational to, the other behavioral disciplines. In fact, this project is not a success. A fully successful approach is likely to require a psychological model of social preferences and a social epistemology, as well as an analysis of social norms as correlating devices that choose among a plethora of Nash equilibria and choreograph the actions of heterogeneous agents into a harmonious operational system.

11

The Evolution of Property Rights

> Every Man has a property in his own Person. This no Body has any Right to but himself. The Labour of his Body, and the Work of his Hands, we may say, are properly his.
>
> John Locke

This chapter illustrates the synergy among the rational actor model, game theory, the socio-psychological theory of norms and gene-culture coevolution (§7.10), highlighting the gains that are possible when ossified disciplinary boundaries are shattered. The true power of game-theoretic analysis becomes manifest only when we cast our theoretical net beyond the strictures of methodological individualism (§8.8). The underlying model is taken from Gintis (2007b). A general case for the methodological approach followed in this chapter is presented in chapter 12.

Authors tracing back to the origins of political liberalism have treated property rights as a social norm the value of which lies in reducing conflict over rights of incumbency (Schlatter 1973). Our analysis of the bourgeois strategy as a social norm effecting an efficient correlated equilibrium embodies this classical notion (§7.3). However, we argued in chapter 7 that a social norm is likely to be fragile and unstable unless individuals generally have a normative predisposition to conform. We here interpret the well-known phenomena of loss-aversion and the endowment effect (§1.9) as highly rational forms of normative predisposition. In this case, the norm is shared with many species of animals as well, in the form of territoriality.

11.1 The Endowment Effect

The *endowment effect* is the notion that people value a good that they possess more highly than they value the same good when they do not possess it (§1.9). Experimental studies (§11.2) have shown that subjects exhibit a systematic endowment effect. The endowment effect is widely considered to be an instance of human irrationality. We suggest here that the endowment effect is not only rational, but also is the basis of key forms of human sociality, including respect for property rights.

Because the endowment effect is an aspect of prospect theory (§1.9), it can be modeled by amending the standard rational actor model to include an agent's current holdings as a parameter. The endowment effect gives rise to *loss-aversion*, according to which agents are more sensitive to losses than to gains (§1.7). We show here that the endowment effect can be modeled as respect for property rights in the absence of legal institutions ensuring third-party contract enforcement (Jones 2001; Stake 2004). In this sense, preinstitutional "natural" property rights have been observed in many species in the form of recognition of territorial possession. We develop a model loosely based on the Hawk-Dove Game (§2.9) and the War of Attrition (Maynard Smith and Price 1973) to explain the natural evolution of property rights.

We show that if agents in a group exhibit the endowment effect for an indivisible resource, then property rights for that resource can be established on the basis of incumbency, assuming incumbents and those who contest for incumbency are of equal perceived fighting ability.[1] The enforcement of these rights is then carried out by the agents themselves, so no third-party enforcement is needed. This is because the endowment effect leads the incumbent to be willing to expend more resources to protect his incumbency than an intruder will be willing to expend to expropriate the incumbent. For simplicity, we consider only the case where the marginal benefit of more than one unit of the resource is zero (e.g., a homestead, a spider's web, or a bird's nest).

The model assumes the agents know the present value π_g of incumbency, as well as the present value π_b of nonincumbency, measured in units of biological fitness. We assume utility and fitness coincide, except for one situation, described below: this situation explicitly involves *loss aversion*, where the disutility of loss exceeds the fitness cost of loss. When an incumbent faces an intruder, the intruder determines the expected value of attempting to seize the resource, and the incumbent determines the expected value of contesting vs. ceding incumbency when challenged. These conditions will not be the same, and in plausible cases there is a range of values of π_g/π_b

[1]The assumption of indivisibility is not very restrictive. In some cases it is naturally satisfied, as in a nest, a web, a dam, or a mate. In others, such as a hunter's kill, a fruit tree, a stretch of beach for an avian scavenger, it is simply the minimum size worth fighting over rather than dividing and sharing.

for which the intruder decides not to fight and the incumbent decides to fight if challenged. We call this a (natural) *property equilibrium*. In a property equilibrium, since the potential contestants are of equal power, it must be the case that individuals are *loss-averse*, the incumbent being willing to expend more resources to hold the resource than the intruder is to seize it.

Of course, π_g and π_b are generally endogenous in a fully specified model. Their values depend on the supply of the resource relative to the number of agents, the intrinsic value of the resource, the ease of finding an unowned unit of the resource, and the like.

In our model of decentralized property rights, agents contest for a unit of an indivisible resource, contests may be very costly, and in equilibrium, incumbency determines who holds the resource without costly contests. Our model, however, fills in critical gaps in the Hawk-Dove Game. The central ambiguity of the Hawk-Dove Game is that it treats the cost of contesting as exogenously given and taking on exactly two values, high for hawk and low for dove. Clearly, however, these costs are in large part under the control of the agents themselves and should not be considered exogenous. In our model, the level of resources devoted to a contest is endogenously determined, and the contest itself is modeled explicitly as a modified War of Attrition, the probability of winning being a function of the level of resources committed to combat. One critical feature of the War of Attrition is that the initial commitment of a level of resources to a contest must be *behaviorally ensured by the agent*, so that the agent will continue to contest even when the costs of doing so exceed the fitness benefits. Without this precommitment, the incumbent's threat of "fighting to the death" would not be credible (i.e., the agent would abandon the chosen best response when it came time to use it). From a behavioral point of view, this precommitment can be summarized as the incumbent having a degree of *loss aversion* leading his utility to differ from his fitness.

Our fuller specification of the behavioral underpinnings of the Hawk-Dove Game allows us to determine the conditions under which a property equilibrium will exist while its corresponding antiproperty equilibrium (in which a new arrival rather than the first entrant always assumes incumbency) does not exist. This aspect of our model is of some importance because the inability of the Hawk-Dove Game to favor property over antiproperty is a serious and rarely addressed weakness of the model (but see Mesterton-Gibbons 1992).

11.2 Territoriality

The endowment effect, according to which a good is more highly prized by an agent who is in possession of the good than by one who is not, was first documented by the psychologist Daniel Kahneman and his coworkers (Tversky and Kahneman 1991; Kahneman et al. 1991; Thaler 1992). Thaler describes a typical experimental verification of the phenomenon as follows. Seventy-seven students at Simon Fraser University were randomly assigned to one of three conditions, seller, buyer, or chooser. Sellers were given a mug with the university logo (selling for $6.00 at local stores) and asked whether they would be willing to sell at a series of prices ranging from $0.25 to $9.25. Buyers were asked whether they would be willing to purchase a mug at the same series of prices. For each price, choosers were asked to choose between receiving a mug or receiving that amount of money. The students were informed that a fraction of their choices, randomly chosen by the experimenter, would be carried out, thus giving the students a material incentive to reveal their true preferences. The average buyer price was $2.87, while the average seller price was $7.12. Choosers behaved like buyers, being on average indifferent between the mug and $3.12. The conclusion is that owners of the mug valued the object more than twice as highly as nonowners.

The aspect of the endowment effect that promotes natural property rights is known as *loss aversion*: the disutility of giving up something one owns is greater than the utility associated with acquiring it. Indeed, losses are commonly valued at about twice that of gains, so that to induce an individual to accept a lottery that costs $10 when one loses (which occurs with probability 1/2), it must offer a $20 payoff when one wins (Camerer 2003). Assuming that an agent's willingness to combat over possession of an object is increasing in the subjective value of the object, owners are prepared to fight harder to *retain* possession than non-owners are to *gain* possession. Hence there will be a predisposition in favor of recognizing property rights by virtue of incumbency, even where third-party enforcement institutions are absent.

We say an agent *owns* something, or *is incumbent*, if the agent has exclusive access to it and the benefits that flow from this privileged access. We say ownership (incumbency) is *respected* if it is rarely contested and, when contested, generally results in ownership remaining with the incumbent. The dominant view in Western thought, from Hobbes, Locke, Rousseau, and Marx to the present, is that property rights are a human social construction that emerged with the rise of modern civilization (Schlatter 1973). However, evidence from studies on animal behavior, gathered mostly in the past quarter-century, has shown this view to be incorrect. Various territorial claims are recognized in nonhuman species, including butterflies (Davies 1978), spiders (Riechert 1978), wild horses (Stevens 1988), finches (Senar, Camerino, and Metcalfe 1989), wasps (Eason, Cobbs, and Trinca 1999), nonhuman primates (Ellis 1985), lizards (Rand 1967), and many others (Mesterton-Gibbons and Adams 2003). There are, of course, some obvious forms of incumbent advantage that partially explain this phenomenon: the incumbent's investment in the territory may be idiosyncratically more valuable to the incumbent than to a contestant or the incumbent's familiarity with the territory may enhance its ability to fight. However, in the above-cited cases, these forms of incumbent advantage are unlikely to be important. Thus, a more general explanation of territoriality is needed.

In nonhuman species, that an animal owns a territory is generally established by the fact that the animal has occupied and altered the territory (e.g., by constructing a nest, burrow, hive, dam, or web, or by marking its limits with urine or feces). In humans there are other criteria of ownership, but physical possession and first to occupy remain of great importance, as expressed by John Locke in the epigraph for this chapter.

Since property rights in human society are generally protected by law and property rights are enforced by complex institutions (judiciary and police), it is natural to view property rights in animals as a categorically distinct phenomenon. In fact, however, decentralized, self-enforcing types of property rights, based on behavioral propensities akin to those found in nonhuman species (e.g., the endowment effect), are important for humans and arguably lay the basis for more institutional forms of property rights. For instance, many developmental studies indicate that toddlers and small children use behavioral rules similar to those of animals is recognizing and defending property rights (Furby 1980).

How respect for ownership has evolved and how it is maintained in an evolutionary context is a challenging puzzle. Why do loss aversion and the endowment effect exist? Why do humans fail to conform to the smoothly differentiable utility function assumed in most versions of the rational actor model? The question is equally challenging for nonhumans, although we are so used to the phenomenon that we rarely give it a second thought.

Consider, for instance, the sparrows that built a nest in a vine in my garden. The location is choice, and the couple spent days preparing the structure. The nest is quite as valuable to another sparrow couple. Why does another couple not try to evict the first? If they are equally strong, and both value the territory equally, each has a 50% chance of winning the territorial battle. Why bother investing if one can simply steal (Hirshleifer 1988)? Of course, if stealing were profitable, then there would be no nest building, and hence no sparrows, but that heightens rather than resolves the puzzle.

One common argument, borrowed from Trivers (1972), is that the original couple has more to lose since it has already put a good deal of effort into the improvement of the property. This, however, is a logical error that has come to be known as the *Concorde* or the *sunk cost* fallacy (Dawkins and Brockmann 1980; Arkes and Ayton 1999): to maximize future returns, an agent ought consider only the future payoffs of an entity, not how much the agent has expended on the entity in the past.

The Hawk-Dove Game was offered by Maynard Smith and Parker (1976) as a logically sound alternative to the sunk cost argument. In this game Hawks and Doves are phenotypically indistinguishable members of the same species, but they act differently in contesting ownership rights to a territory. When two Doves contest, they posture for a bit, and then each assumes the territory with equal probability. When a Dove and a Hawk contest, however, the Hawk takes the whole territory. Finally, when two Hawks contest, a terrible battle ensues, and the value of the territory is less than the cost of fighting for the contestants. Maynard Smith showed that, assuming that there is an unambiguous way to determine who first found the territory, there is an evolutionarily stable strategy in which all agents behave like Hawks when they are *first* to find the territory, and like Doves otherwise.

The Hawk-Dove Game is an elegant contribution to explaining the endowment effect, but the cost of contesting for Hawks and the cost of display for Doves cannot plausibly be taken as fixed and exogenously determined. Indeed, it is clear that Doves contest in the same manner as Hawks, except that they devote fewer resources to combat. Similarly, the value of the ownership is taken as exogenous, when in fact it depends on the frequency with which ownership is contested, as well as on other factors. As Grafen (1987) stresses, the costs and benefits of possession depend on the state of the population, the density of high-quality territories, the cost of search, and other variables that might well depend on the distribution of strategies in the population.

First, however, it is instructive to consider the evidence for a close association, as Locke suggested in his theory of property rights, between ownership and incumbency (physical contiguity and control) in children and nonhuman animals.

11.3 Property Rights in Young Children

Long before they become acquainted with money, markets, bargaining, and trade, children exhibit possessive behavior and recognize the property rights of others on the basis of incumbency.[2] In one study (Bakeman and Brownlee 1982), participant observers studied a group of 11 toddlers (12 to 24 months old) and a group of 13 preschoolers (40 to 48 months old) at a day care center. The observers found that each group was organized into a fairly consistent linear dominance hierarchy. They then cataloged *possession episodes*, defined as situations in which a *holder* touched or held an object and a *taker* touched the object and attempted to remove it from the holder's possession. Possession episodes averaged 11.7 per hour in the toddler group and 5.4 per hour in the preschool group.

For each possession episode, the observers noted (a) whether the taker had been playing with the object within the previous 60 seconds (prior possession), (b) whether the holder resisted the take attempt (resistance), and (c) whether the take was successful (success). They found that success was strongly and about equally associated with both dominance and prior possession. They also found that resistance was positively associated with dominance in the toddlers and negatively associated with prior possession in the preschoolers. They suggest that toddlers recognize possession as a basis for asserting control rights but do not respect the same rights in others. Preschoolers, more than twice the age of the toddlers, use physical proximity both to justify their own claims and to respect the claims of others. This study was replicated and extended by Weigel (1984).

11.4 Respect for Possession in Nonhuman Animals

In a famous paper, Maynard Smith and Parker noted that if two animals are competing for some resource (e.g., a territory), and if there is some discernible asymmetry (e.g., between an owner and a later-arriving animal), then it is evolutionarily stable for the asymmetry to settle the contest con-

[2] See Ellis (1985) for a review and an extensive bibliography of research in this area.

ventionally, without fighting. Among the findings of the many animal behaviorists who put this theory to the test, perhaps none is more elegant and unambiguous than that of Davies (1978), who studied the speckled wood (*Pararge aegeria*), a butterfly found in the Wytham Woods, near Oxford, England. Territories for this butterfly are shafts of sunlight breaking through the tree canopy. Males occupying these spots enjoyed heightened mating success, and on average only 60% of the males occupied the sunlit spots at any one time. A vacant spot was generally occupied within seconds, but an intruder at an already occupied spot was invariably driven away even if the incumbent had occupied the spot for only a few seconds. When Davies "tricked" two butterflies into thinking each had occupied the sunny patch first, the contest between the two lasted, on average, 10 times as long as the brief flurry that occurred when an incumbent chased off an intruder.

Stevens (1988) found a similar pattern of behavior among the feral horses occupying the sandy islands of the Rachel Carson Estuarine Sanctuary near Beaufort, North Carolina. In this case, it is freshwater that is scarce. After heavy rains, freshwater accumulates in many small pools in low-lying wooded areas, and bands of horses frequently stop to drink. Stevens found that there were frequent encounters between bands of horses competing for water at these temporary pools. If a band approached a water hole occupied by another band, a conflict ensued. During 76 hours of observation, Stevens observed 233 contests, of which the resident band won 178 (80%). In nearly all cases of usurpation, the intruding band was larger than the resident band. These examples, and many others like them, support the presence of an endowment effect and suggest that incumbents are willing to fight harder to maintain their positions than intruders are to usurp the owner.

Examples from nonhuman primates exhibit behavioral patterns in respecting property rights much closer to those of humans. In general, the taking of an object held by another individual is a rare event in primate societies (Torii 1974). A reasonable test of the respect for property in primates with a strong dominance hierarchy is the likelihood of a dominant individual refraining from taking an attractive object from a lower-ranking individual. In a study of hamadryas baboons (*Papio hamadryas*), for instance, Sigg and Falett (1985) handed a food can to a subordinate who was allowed to manipulate it and eat from it for 5 minutes before a dominant individual who had been watching from an adjacent cage was allowed to enter the subordinate's cage. A takeover was defined as the rival taking possession of the can before 30 minutes had elapsed. They found that (a) males never took the food can from other males; (b) dominant males took the can from subordinate females two-thirds of the time; and (c) dominant females took the can from subordinate females one-half of the time. With females, closer inspection showed that when the difference in rank was one or two, females showed respect for the property of other females, but when the rank difference was three or greater, takeovers tended to occur.

Kummer and Cords (1991) studied the role of proximity in respect for property in long-tailed macaques (*Macaca fascicularis*). As in the Sigg and Falett study, they assigned ownership to a subordinate and recorded the behavior of a dominant individual. The valuable object in all cases was a plastic tube stuffed with raisins. In one experiment, the tube was fixed to an object in half the trials and completely mobile in the other half. They found that with the fixed object, the dominant rival took possession in all cases and very quickly (median 1 minute), whereas in the mobile condition, the dominant rival took possession in only 10% of cases, and then only after a median delay of 18 minutes. The experiment took place in an enclosed area, so the relative success of the incumbent was not likely due to an ability to flee or hide. In a second experiment, the object was either mobile or attached to a fixed object by a stout 2- or 4-meter rope. The results were similar. A third case, in which the nonmobile object was attached to a long dragline that permitted free movement by the owner, produced the following results. Pairs of subjects were studied under two conditions, one where the rope attached to the dragline was 2 meters in length and a second where the rope was 4 meters in length. In 23 of 40 trials, the subordinate maintained ownership with both rope lengths, and in 6 trials the dominant rival took possession with both rope lengths. In the remaining 11 trials, the rival respected the subordinate's property in the short rope case but took possession in the long-rope case. The experimenters observed that when a dominant attempted to usurp a subordinate when other group members were around, the subordinate screamed, drawing the attention of third parties, who frequently forced the dominant individual to desist.

In *Wild Minds* (2000), Marc Hauser relates an experiment run by Kummer and his colleagues concerning mate property, using four hamadryas baboons, Joe, Betty, Sam, and Sue. Sam was let into Betty's cage while Joe looked on from an adjacent cage. Sam immediately began following Betty around and grooming her. When Joe was allowed entrance into the cage, he kept his distance, leaving Sam uncontested. The same experiment was repeated with Joe being allowed into Sue's cage. Joe behaved as Sam had in the previous experiment, and when Sam was let into the cage, he failed to challenge Joe's proprietary rights with respect to Sue.

No primate experiment, to my knowledge, has attempted to determine the probability that an incumbent will be contested for ownership by a rival who is, or could easily become, closely proximate to the desired object. This probability is likely very low in most natural settings, so the contests described in the papers cited in this section are probably rather rare in practice. At any rate, in the model of respect for property developed in the next section, we will make informational assumptions that render the probability of contestation equal to zero in equilibrium.

11.5 Conditions for a Property Equilibrium

Suppose that two agents, prior to fighting over possession, simultaneously precommit to expending a certain level of resources in the contest. As in the War of Attrition (Bishop and Cannings 1978), a higher level of resource commitment entails a higher fitness cost but increases the probability of winning. We assume throughout this chapter that the two contestants, an incumbent and an intruder, are ex ante equally capable contestants in that the costs and benefits of battle are symmetric in the resource commitments s_o (owner) and s_u (usurper) of the incumbent and the intruder, respectively, and $s_o, s_u \in [0, 1]$. To satisfy this requirement, we let $p_u = s_u^n/(s_u^n + s_o^n)$ be the probability that the intruder wins, where $n > 1$. Note that a larger n implies that resource commitments are more decisive in determining victory. We assume that combat leads to injury $\beta \in (0, 1]$ to the losing party with probability $p_d = (s_o + s_u)/2$, so $s = \beta p_d$ is the expected cost of combat for both parties.

We use a territorial analogy throughout, some agents being incumbents and others being migrants in search of either empty territories or occupied territories that they may be able to occupy by displacing current incumbents. Let π_g be the present value of being a currently uncontested incumbent and let π_b be the present value of being a migrant searching for a territory. We assume throughout that $\pi_g > \pi_b > 0$. Suppose a migrant comes upon an occupied territory. Should the migrant contest, the condition under which it pays an incumbent to fight back is then given by

$$\pi_c \equiv p_d(1 - p_u)\pi_g + p_d p_u(1 - \beta)(1 - c)\pi_b \\ +(1 - p_d)(1 - p_u)\pi_g + (1 - p_d)p_u\pi_b(1 - c) > \pi_b(1 - c).$$

The first term in π_c is the product of the probabilities that the intruder loses $(1 - p_u)$ and sustains an injury (p_d) times the value π_g of incumbency, which the incumbent then retains. The second term is the product of the probabilities that the incumbent loses (p_u), sustains an injury (p_d), survives the injury $(1 - \beta)$, and survives the passage to migrant status $(1 - c)$ times the present value π_b of being a migrant. The third and fourth terms are the parallel calculations when no injury is sustained. This inequality simplifies to

$$\frac{\pi_g}{\pi_b(1-c)} - 1 > \frac{s_u^n}{s_o^n} s. \tag{11.1}$$

The condition for a migrant refusing to contest for the territory, assuming the incumbent will contest if the migrant does, is

$$\pi_u \equiv p_d(p_u \pi_g + (1-p_u)(1-\beta)(1-c)\pi_b) \tag{11.2}$$

$$+(1-p_d)(p_u \pi_g + (1-p_u)\pi_b(1-c)) < \pi_b(1-c). \tag{11.3}$$

This inequality reduces to

$$\frac{s_o^n}{s_u^n} s > \frac{\pi_g}{\pi_b(1-c)} - 1. \tag{11.4}$$

A property equilibrium occurs when both inequalities obtain

$$\frac{s_o^n}{s_u^n} s > \frac{\pi_g}{\pi_b(1-c)} - 1 > \frac{s_u^n}{s_o^n} s. \tag{11.5}$$

An incumbent who is challenged chooses s_o to maximize π_c and then contests if and only if the resulting $\pi_c^* > \pi_b(1 - c)$, since the latter is the value of simply leaving the territory. It is easy to check that $\partial \pi_c / \partial s_o$ has the same sign as

$$\frac{\pi_g}{\pi_b(1-c)} - \left(\frac{s_o \beta}{2n(1-p_u)} + 1 - s \right).$$

The derivative of this expression with respect to s_o has the same sign as $(n - 1)\beta \pi_b / (1 - p_u)$, which is positive. Moreover, when $s_o = 0$, $\partial \pi_c / \partial s_o$ has the same sign as

$$\frac{\pi_g}{\pi_b(1-c)} - 1 + \frac{s_u \beta (1-c)}{2},$$

which is positive. Therefore, $\partial\pi_c/\partial s_o$ is always strictly positive, so $s_o = 1$ maximizes π_c.

In deciding whether or not to contest, the migrant chooses s_u to maximize π_u and then contests if this expression exceeds $\pi_b(1-c)$. But $\partial\pi_u/\partial s_u$ has the same sign as

$$\frac{\pi_g}{\pi_b(1-c)} - \left(s - 1 + \frac{s_u\beta}{2np_u}\right),$$

which is increasing in s_u and is positive when $s_u = 0$, so the optimal $s_u = 1$. The condition for not contesting the incumbent is then

$$\frac{\pi_g}{\pi_b(1-c)} - 1 < \beta. \tag{11.6}$$

In this case, the condition (11.4) for the incumbent contesting is the same as (11.6) with the inequality sign reversed.

By an *antiproperty* equilibrium we mean a situation where intruders always contest and incumbents always relinquish their possessions without a fight.

THEOREM 11.1 *If $\pi_g > (1+\beta)\pi_b(1-c)$, there is a unique equilibrium in which an migrant always fights for possession and an incumbent always contests. When the reverse inequality holds, there exists both a property equilibrium and an antiproperty equilibrium.*

Theorem 11.1 implies that property rights are more likely to be recognized when combatants are capable of inflicting great harm on one another, so β is close to its maximum of unity, or when migration costs are very high, so c is close to unity.

Theorem 11.1 may apply to a classic problem in the study of hunter-gatherer societies, which are important not only in their own right but also because our ancestors lived uniquely in such societies until about 10,000 years ago, and hence their social practices have doubtless been a major environmental condition to which the human genome has adapted (Cosmides and Tooby 1992). One strong uniformity across current-day hunter-gatherer societies is that low-value foodstuffs (e.g., fruits and small game) are consumed by the families that produce them, but high-value foodstuffs (e.g., large game and honey) are meticulously shared among all group members. The standard argument is that high-value foodstuffs exhibit a high variance, and sharing is a means of reducing individual variance. But an alternative with much empirical support is the *tolerated theft* theory that holds that high-value foodstuffs are worth fighting for (i.e., the inequality in theorem 11.1 is satisfied), and the sharing rule is a means of reducing the mayhem that would inevitably result from the absence of secure property rights to high-value foodstuffs (Hawkes 1993; Blurton Jones 1987; Betzig 1997; Bliege Bird and Bird 1997; Wilson 1998a).[3]

The only part of theorem 11.1 that remains to be proved is the existence of an antiproperty equilibrium. To see this, note that such an equilibrium exists when $\pi_c < \pi_b(1-c)$ and $\pi_u > \pi_b(1-c)$, which, by the same reasoning as above, occurs when

$$\frac{s_u^n}{s_o^n} > \frac{\pi_g}{\pi_b(1-c)} - 1 > \frac{s_o^n}{s_u^n}s. \tag{11.7}$$

It is easy to show that if the incumbent contests, then both parties will set $s_u = s_o = 1$, in which case the condition for the incumbent to do better by not contesting is exactly what it is in the property equilibrium.

The result that there exists an antiproperty equilibrium exactly when there is a property equilibrium is quite unrealistic since few, if any, antiproperty equilibria have been observed. Our model, of course, shares this anomaly with the Hawk-Dove Game, for which this weakness has never been analytically resolved. In our case, however, when we expand our model to determine π_g and π_g, the antiproperty equilibrium generally disappears. The problem with the above argument is that we cannot expect π_g and π_b to have the same values in a property and in an antiproperty equilibrium.

11.6 Property and Antiproperty Equilibria

To determine π_g and π_b, we must flesh out the above model of incumbents and migrants. Consider a field with many patches, each of which is indivisible and hence can have only one owner. In each time period, a fertile patch yields a benefit $b > 0$ to the owner and dies with probability $p > 0$, forcing its owner (should it have one) to migrate elsewhere in search of a fertile patch. Dead patches regain their fertility after a period of time, leaving the fraction of patches that are fertile constant from period to period. An agent who encounters an empty fertile patch invests an amount $v \in (0, 1/2)$ of

[3]For Theorem 11.1 to apply, the resource in question must be indivisible. In this case, the "territory" is the foodstuff that delivers benefits over many meals, and the individuals who partake of it are temporary occupiers of the territory.

fitness in preparing the patch for use and occupies the patch. An agent suffers a fitness cost $c > 0$ each period he is in the state of searching for a fertile patch. An agent who encounters an occupied patch may contest for ownership of the patch according to the War of Attrition structure analyzed in the previous section.

Suppose there are n_p patches and n_a agents. Let r be the probability of finding a fertile patch and let w be the probability of finding a fertile unoccupied patch. If the rate at which dead patches become fertile is q, which we assume for simplicity does not depend on how long a patch has been dead, then the equilibrium fraction f of patches that are fertile must satisfy $n_p f p = n_p(1 - f)q$, so $f = q/(p + q)$. Assuming that a migrant finds a new patch with probability ρ, we then have $r = f\rho$. If ϕ is the fraction of agents that are incumbents, then writing $\alpha = n_a/n_p$, we have

$$w = r(1 - \alpha\phi). \tag{11.8}$$

Assuming the system is in equilibrium, the number of incumbents whose patches die must be equal to the number of migrants who find empty patches, or $n_a\phi(1 - p) = n_a(1 - \phi)w$. Solving this equation gives ϕ, which is given by

$$\alpha r\phi^2 - (1 - p + r(1 + \alpha))\phi + r = 0. \tag{11.9}$$

It is easy to show that this equation has two positive roots, exactly one lying in the interval $(0, 1)$.

In a property equilibrium, we have

$$\pi_g = b + (1 - p)\pi_g + p\pi_b(1 - c), \tag{11.10}$$

and

$$\pi_b = w\pi_g(1 - v) + (1 - w)\pi_b(1 - c). \tag{11.11}$$

Note that the cost v of investing and the cost c of migrating are interpreted as fitness costs and hence as probabilities of death. Thus, the probability of a migrant becoming an incumbent in the next period is $w(1 - v)$, and the probability of remaining a migrant is $(1 - w)$. This explains (11.11). Solving these two equations simultaneously gives equilibrium values of incumbency and nonincumbency:

$$\pi_g^* = \frac{b(c(1 - w) + w)}{p(c(1 - vw) + vw)}, \tag{11.12}$$

$$\pi_b^* = \frac{b(1 - v)w}{p(c(1 - vw) + vw)}. \tag{11.13}$$

Note that $\pi_g, \pi_b > 0$ and

$$\frac{\pi_g^*}{\pi_b^*} - 1 = \frac{c(1-w) + wv}{w(1-v)}. \tag{11.14}$$

By theorem 11.1, the assumption that this is a property equilibrium is satisfied if and only if this expression is less than β, or

$$\frac{c(1-w) + wv}{w(1-v)} < \beta. \tag{11.15}$$

We have the following theorem.

THEOREM 11.2 *There is a strictly positive migration cost c^* and a cost of injury $\beta^*(c)$ for all $c < c^*$ such that a property equilibrium holds for all $c < c^*$ and $\beta > \beta^*(c)$.*

To see this, note that the left hand side of (11.15) is less than 1 precisely when $c < c^* =_{hboxdef} w(1-2v)/(1-w)$. We then set $\beta^*(c)$ equal to the left hand side of (11.15), ensuring that $\beta^*(c) < 1$.

This theorem shows that, in addition to our previous result, a low fighting cost and a high migration cost undermine the property equilibrium, a high probability w that a migrant encounters an incumbent undermines the property equilibrium, and a high investment v has the same effect.

Suppose, however, that the system is in an antiproperty equilibrium. In this case, letting q_u be the probability that an incumbent is challenged by an intruder, we have

$$\pi_g = b + (1-p)(1-q_u)\pi_g + (p(1-q_u) + q_u)\pi_b(1-c) \tag{11.16}$$

and

$$\pi_b = w\pi_g(1-v) + (r-w)\pi_g + (1-r)\pi_b(1-c). \tag{11.17}$$

Solving these equations simultaneously gives

$$\pi_g^* = \frac{b(c(1-r)+r)}{((p(1-q_u)+q_u))(vw + c(1-vw))}, \tag{11.18}$$

$$\pi_b^* = \frac{b(r-vw)}{(((p(1-q_u)+q_u))(vw + c(1-vw)))}. \tag{11.19}$$

Also, $\pi_g, \pi_b > 0$ and

$$\frac{\pi_g^*}{\pi_b^*} - 1 = \frac{c(1-r) + vw}{r - vw}. \tag{11.20}$$

Note that $r - vw = r(1 - v(1 - \alpha\phi)) > 0$. We must check whether a nonincumbent mutant who never invests, and hence passes up empty fertile patches, would be better off. In this case, the present value of the mutant, π_m, satisfies

$$\begin{aligned}\pi_m - \pi_b^* &= (r-w)\pi_g^* + (1-r+w)\pi_b^*(1-c) - \pi_b^* \\ &= \frac{bw(v(r-w) - c(1 - v(1-r+2)))}{(p(1-q_u) + q_u)(vw + c(1-vw))}.\end{aligned}$$

It follows that if

$$v \le \frac{c}{(r-w)(1-c) + c}, \tag{11.21}$$

then the mutant behavior (not investing) cannot invade, and we indeed have an anti-equilibrium. Note that (11.21) has a simple interpretation. The denominator in the fraction is the probability that a search ends either in death or in finding an empty patch. The right side is therefore the expected cost of searching for an occupied patch. If the cost v of investing in a empty patch is greater than the expected cost of waiting to usurp an already productive (fertile and invested in) patch, no agent will invest. We have the following theorem.

THEOREM 11.3 *There is an investment cost $v^* \in (0, 1)$ such that an antiproperty equilibrium exists if and only if $v \le v^*$. v^* is an increasing function of the migration cost c.*

To see this, note that the right hand side of (11.21) lies strictly between 0 and 1, and is strictly increasing in c.

If (11.21) is violated, then migrants will refuse to invest in an empty fertile patch. Then (11.9), which implicitly assumes that a migrant always occupies a vacant fertile patch, is violated. We argue as follows. Assume the system is in the antiproperty equilibrium as described above and, noting the failure of (11.21), migrants begin refusing to occupy vacant fertile patches. Then, as incumbents migrate from newly dead patches, ϕ falls, and hence w rises. This continues until (11.21) is satisfied as an equality. Thus, we must redefine an antiproperty equilibrium as one in which (11.9) is satisfied when (11.21) is satisfied; otherwise, (11.21) is satisfied as an equality and (11.9) is no longer satisfied. Note that in the latter case the equilibrium value of ϕ is strictly less than in the property equilibrium.

THEOREM 11.4 *Suppose (11.21) is violated when ϕ is determined by (11.9). Then the antiproperty equilibrium exhibits a lower average payoff than the property equilibrium.*

The reason is simply that the equilibrium value of ϕ is lower in the antiproperty equilibrium than in the property equilibrium, so there will be on average more migrants and fewer incumbents in the antiproperty equilibrium. But incumbents earn positive return b per period, while migrants suffer positive costs c per period.

Theorem 11.4 helps to explain why we rarely see antiproperty equilibria in the real world, If two groups differ only in that one plays the property equilibrium and the other plays the antiproperty equilibrium, the former will grow faster and hence displace the latter, provided that there is some scarcity of resources leading to a limitation on the combined size of the two groups.

This argument does not account for property equilibria in which there is virtually no investment by the incumbent. This includes the butterfly (Davies) and feral horse (Stevens) examples, among others. In such cases, the property and antiproperty equilibria differ in only one way: the identity of the patch owner changes in the latter more rapidly than in the former. It is quite reasonable to add to the model a small cost δ of ownership change, for instance, because the intruder must physically approach the patch and engage in some sort of display before the change in incumbency can be effected. With this assumption, the antiproperty equilibrium again has a lower average payoff than the property equilibrium, so it will be disadvantaged in a competitive struggle for existence.

The next section shows that if we respecify the ecology of the model appropriately, the unique equilibrium is precisely the antiproperty equilibrium.

11.7 An Antiproperty Equilibrium

Consider a situation in which agents die unless they have access to a fertile patch at least once every n days. While having access, they reproduce at rate b per period. A agent who comes upon a fertile patch that is already owned may value the patch considerably more than the current owner, since the intruder has, on average, less time to find another fertile patch than the current owner, who has a full n days. In this situation, the current owner may have no incentive to put up a sustained battle for the patch, whereas the intruder may. The newcomer may thus acquire the patch without a battle. Thus, there is a plausible antiproperty equilibrium.

To assess the plausibility of such a scenario, note that if π_g is the fitness of the owner of a fertile patch and $\pi_b(k)$ is the fitness of a nonowner who has k periods to find and exploit a fertile patch before dying, then we have the recursion equations

$$\pi_b(0) = 0, \tag{11.22}$$

$$\pi_b(k) = w\pi_g + (1-w)\pi_b(k-1) \qquad \text{for } k = 1, \ldots, n, \tag{11.23}$$

where r is the probability that a nonowner becomes the owner of a fertile patch, either because it is not owned or because the intruder costlessly evicts the owner. We can solve this, giving

$$\pi_b(k) = \pi_g(1-(1-r)^k) \qquad \text{for } k = 0, 1, \ldots n. \tag{11.24}$$

Note that the larger k and the larger r, the greater the fitness of a intruder. We also have the equation

$$\pi_g = b + (1-p)\pi_g + p\pi_g(n), \tag{11.25}$$

where p is the probability the patch dies or the owner is costlessly evicted by an intruder. We can solve this equation, finding

$$\pi_g = \frac{b}{p(1-r)^n}. \tag{11.26}$$

Note that the larger b, the smaller p, the larger r, and the larger n, the greater the fitness of the owner.

As in the previous model, assume the intruder devotes resources $s_u \in [0,1]$ and the incumbent devotes resources $s_o \in [0,1]$ to combat. With the same notation as above, we assume a fraction f_o of incumbents are contesters, and we derive the conditions for an incumbent and an intruder who has discovered the owner's fertile patch to conform to the antiproperty equilibrium. When these conditions hold, we have $f_o = 0$.

Let π_c be the fitness value of contesting rather than simply abandoning the patch. Then we have

$$\pi_c = s(1-p_u)\pi_g + (1-s)((1-p_u)\pi_g + p_u\pi_b(n)) - \pi_b(n),$$

which reduces to

$$\pi_c = \frac{\pi_g}{2}\left(\frac{s_u^2 + s_o(2+s_u)}{s_o+s_u}(1-r)^n - s_u\right). \tag{11.27}$$

Moreover, π_c is increasing in s_o, so if the owner contests, he will set $\sigma_o = 1$, in which case the condition for contesting being fitness-enhancing for the owner then becomes

$$\frac{s_u + 2/s_u + 1}{1+s_u}(1-r)^n > 1. \tag{11.28}$$

Now let $\pi_u(k)$ be the fitness of a nonowner who must own a patch before k periods have elapsed and who comes upon an owned fertile patch. The agent's fitness value of usurping is

$$\pi_u(k) = (1-f)\pi_g + f(sp_u\pi_g \\ +(1-s)(p_u\pi_g + (1-p_u)\pi_b(k-1))) - \pi_b(k-1).$$

The first term in this equation is the probability that the owner does not contest times the intruder's gain if this occurs. The second term is the probability that the owner does contest times the gain if the owner does contest. The final term is the fitness value of not usurping. We can simplify this equation to

$$\pi_u(k) = \pi_g\frac{s_o(1-f)+s_u}{s_o+s_u}. \tag{11.29}$$

This expression is always positive and is increasing in s_u and decreasing in s_o, provided $f_o > 0$. Thus, the intruder always sets $s_u = 1$. Also, as one might expect, if $f_o = 0$, the migrant usurps with probability 1, so $\pi_u(k) = \pi_g$. At any rate, the migrant always contests, whatever the value of f_o. The condition (11.28) for not contesting, and hence for there to be a globally stable antiproperty equilibrium, becomes

$$2(1-r)^n < 1, \tag{11.30}$$

which will be the case if either r or n is sufficiently large. When (11.30) does not hold, there is an antiproperty equilibrium.

The antiproperty equilibrium is not often entertained in the literature, although Maynard Smith (1982) describes the case of the spider *Oecibus civitas*, where intruders virtually always displace owners without a fight. More informally, I observe the model in action every summer's day at my bird feeders and bathers. A bird arrives, eats or bathes for a while, and if the feeder or bath is crowded, is then displaced, without protest, by another bird, and so on. It appears that, after having eaten or bathed for a while, it simply is not worth the energy to defend the territory.

11.8 Property Rights as Choreographer

Humans share with many other species a predisposition to recognize property rights. This takes the form of *loss aversion*: an incumbent is prepared to commit more vital resources to defending his property, ceteris paribus, than an intruder is willing to commit to taking the property. The major proviso is that if the property is sufficiently valuable, a property equilibrium will not exist (theorem 11.1).

History is written as though property rights are a product of modern civilization, a construction that exists only to the extent that it is defined and protected by judicial institutions operating according to legal notions of ownership. However, it is likely that property rights in the fruits of one's labor has existed for as long as humans have lived in small hunter-gatherer clans, unless the inequality in theorem 11.1 holds, as might plausibly be the case for big game. The true value of modern property rights, if the argument in this chapter is valid, lies in fostering the accumulation of property even when $\pi_g > (1+\beta)\pi_b(1-c)$. It is in this sense only that Thomas Hobbes may have been correct in asserting that life in an unregulated state of nature is "solitary, poor, nasty, brutish, and short." But even so, it must be recognized that modern notions of property are built on human behavioral propensities that we share with many species of nonhuman animals. Doubtless, an alien species with a genetic organization akin to that of our ants or termites would find our notions of individuality and privacy curious at best and probably incomprehensible.

12

The Unification of the Behavioral Sciences

> Each discipline of the social sciences rules comfortably within its own chosen domain... so long as it stays largely oblivious of the others.
>
> Edward O. Wilson

> The combined assumptions of maximizing behavior, market equilibrium, and stable preferences, used relentlessly and unflinchingly, form the heart of the economic approach
>
> Gary Becker

> While scientific work in anthropology, and sociology and political science will become increasingly indistinguishable from economics, economists will reciprocally have to become aware of how constraining has been their tunnel vision about the nature of man and social interaction.
>
> Jack Hirshleifer

The behavioral sciences include economics, anthropology, sociology, psychology, and political science, as well as biology insofar as it deals with animal and human behavior. These disciplines have distinct research foci, but they include four conflicting models of decision making and strategic interaction, as determined by what is taught in the graduate curriculum and what is accepted in journal articles without reviewer objection. The four are the psychological, the sociological, the biological, and the economic.

These four models are not only different, which is to be expected given their distinct explanatory aims, but are also *incompatible*. That is, each makes assertions concerning choice behavior that are denied by the others. This means, of course, that at least three of the four are certainly incorrect, and I will argue that in fact all four are flawed but can be modified to produce a unified framework for modeling choice and strategic interaction for all of the behavioral sciences. Such a framework would then be enriched in different ways to meet the particular needs of each discipline.

In the past, cross-disciplinary incoherence was tolerated because distinct disciplines dealt largely with distinct phenomena. Economics dealt with market exchange. Sociology dealt with stratification and social deviance. Psychology dealt with brain functioning. Biology, failing to follow up on Darwin's insightful monograph on human emotions (Darwin 1998), avoided dealing with human behavior altogether. In recent years, however, the value of transdisciplinary research in addressing questions of social theory has become clear, and sociobiology has become a major arena of scientific research. Moreover, contemporary social policy involves issues that fall squarely in the interstices of the behavioral disciplines, including substance abuse, crime, corruption, tax compliance, social inequality, poverty, discrimination, and the cultural foundations of market economies. Incoherence is now an impediment to progress.

My framework for unification includes five conceptual units: (a) gene-culture coevolution; (b) the sociopsychological theory of norms; (c) game theory, (d) the rational actor model; and (e) complexity theory. Gene-culture coevolution comes from the biological theory of social organization (sociobiology) and is foundational because *H. sapiens* is an evolved, highly social, biological species. The sociopsychological theory of norms includes fundamental insights from sociology and social psychology that apply to all forms of human social organization, from hunter-gatherer to advanced technological societies. These societies are the product of gene-culture coevolution but have *emergent properties* (§8.8), including social norms and their psychological correlates/prerequisites, that cannot be derived analytically from the component parts of the system—in this case the interacting agents (Morowitz 2002).

Game theory includes four related disciplines: classical, behavioral, epistemic, and evolutionary game theory, the first three of which have been developed in this book. The fourth, evolutionary game theory, is a macrolevel analytical apparatus allowing biological and cultural evolution to be mathematically modeled.

The rational actor model (§1.1, 1.5) is the most important analytical construct in the behavioral sciences operating at the level of the individual. While gene-culture coevolutionary theory is a form of ultimate explanation that does not predict, the rational actor model provides a proximate description of behavior that can be tested in the laboratory and in real life and is the basis of the explanatory success of economic theory. Classical, epistemic, and behavioral game theory make no sense without the rational actor model, and behavioral disciplines, such as anthropology and sociology, as well as social and cognitive psychology, that have abandoned this model have fallen into theoretical disarray.

Behavioral economists and psychologists have taken aim at the rational actor model in the belief that experimental results contradict rationality.

Showing that this view is wrong has been a constant theme of this book. The behaviorists' error is partly due to their having borrowed a flawed conception of rationality from classical game theory, partly due to their interpreting the rational actor model too narrowly, and partly due to an exuberant but unjustified irreverence for received wisdom.

Complexity theory is needed because human society is a complex adaptive system with *emergent properties* that cannot now be, and perhaps never will be, fully explained starting with more basic units of analysis. The hypothetico-deductive methods of game theory and the rational actor model, and even gene-culture coevolutionary theory, must therefore be complemented by the work of behavioral scientists who deal with society in more macrolevel, interpretive terms, and develop insightful schemas that shed light where analytical models cannot penetrate. Anthropological and historical studies fall into this category, as well as macroeconomic policy and comparative economic systems. Agent-based modeling of complex dynamical systems is also useful in dealing with emergent properties of complex adaptive systems.

The above principles are not meant to revolutionize research in any discipline. Indeed, they build on existing strengths, and they imply change only in the areas of overlap among disciplines. For instance, a psychologist working on visual processing, or an economist working on futures markets, or an anthropologist documenting food-sharing practices, or a sociologist gauging the effect of dual parenting on children's educational attainment might gain little from knowing that a unified model of decision making underlies all the behavioral disciplines. On the other hand, a unified model of human choice and strategic interaction might foster innovations that come to pervade the discipline, even in these relatively hermetically sealed areas.

12.1 Gene-Culture Coevolution: The Biological Model

The centrality of culture and complex social organization to the evolutionary success of *H. sapiens* implies that individual fitness in humans depends on the structure of social life. Since culture is limited and facilitated by human genetic propensities, it follows that human cognitive, affective, and moral capacities are the product of an evolutionary dynamic involving the interaction of genes and culture. This dynamic is known as *gene-culture coevolution* (Cavalli-Sforza and Feldman 1982; Boyd and Richerson 1985; Dunbar 1993; Richerson and Boyd 2004). This coevolutionary process has endowed us with preferences that go beyond the self-regarding concerns emphasized in traditional economic and biological theory and embrace a social epistemology facilitating the sharing of intentionality across minds, as well as such non-self-regarding values as a taste for cooperation, fairness, and retribution, the capacity to empathize, and the ability to value honesty, hard work, piety, toleration of diversity, and loyalty to one's reference group.

Gene-culture coevolution is the application of *sociobiology*, the general theory of the social organization of biological species, to species that transmit culture without informational loss across generations. An intermediate category is *niche construction*, which applies to species that transform their natural environment to facilitate social interaction and collective behavior (Odling-Smee, Laland, and Feldman 2003).

The genome encodes information that is used both to construct a new organism and to endow it with instructions for transforming sensory inputs into decision outputs. Because learning is costly and error-prone, efficient information transmission ensures that the genome encodes all aspects of the organism's environment that are constant or that change only slowly through time and space. By contrast, environmental conditions that vary rapidly can be dealt with by providing the organism with the capacity to *learn*.

There is an intermediate case, however, that is efficiently handled neither by genetic encoding nor by learning. When environmental conditions are positively but imperfectly correlated across generations, each generation acquires valuable information through learning that it cannot transmit genetically to the succeeding generation because such information is not encoded in the germ line. In the context of such environments, there is a fitness benefit to the transmission of *epigenetic* information concerning the current state of the environment.[1] Such epigenetic information is quite common (Jablonka and Lamb 1995) but achieves its highest and most flexible form in *cultural transmission* in humans and to a considerably lesser extent in other primates (Bonner 1984; Richerson and Boyd 1998). Cultural transmission takes the form of vertical (parents to children), horizontal (peer to peer), and oblique (elder to younger), as in Cavalli-Sforza and Feldman (1981), prestige (higher status influencing lower status), as in Henrich and Gil-White (2001), popularity-related as in Newman, Barabasi, and Watts

[1]An epigenetic mechanism is any nongenetic intergenerational information transmission mechanism, such a cultural transmission in humans.

(2006), and even random population-dynamic transmission, as in Shennan (1997) and Skibo and Bentley (2003).

The parallel between cultural and biological evolution goes back to Huxley (1955), Popper (1979), and James (1880)—see Mesoudi, Whiten, and Laland (2006) for details. The idea of treating culture as a form of epigenetic transmission was pioneered by Richard Dawkins, who coined the term "meme" in *The Selfish Gene* (1976) to represent an integral unit of information that could be transmitted phenotypically. There quickly followed several major contributions to a biological approach to culture, all based on the notion that culture, like genes, could evolve through replication (intergenerational transmission), mutation, and selection.

Cultural elements reproduce themselves from brain to brain and across time, mutate, and are subject to selection according to their effects on the fitness of their carriers (Parsons 1964; Cavalli-Sforza and Feldman 1982). Moreover, there are strong interactions between genetic and epigenetic elements in human evolution, ranging from basic physiology (e.g., transformation of the organs of speech with the evolution of language) to sophisticated social emotions, including empathy, shame, guilt, and revenge seeking (Zajonc 1980, 1984).

Because of their common informational and evolutionary character, there are strong parallels between genetic and cultural modeling (Mesoudi, Whiten, and Laland 2006). Like biological transmission, cultural transmission occurs from parents to offspring, and like cultural transmission, which occurs horizontally between unrelated individuals, in microbes and many plant species, genes are regularly transferred across lineage boundaries (Jablonka and Lamb 1995; Rivera and Lake 2004; Abbott et al. 2003). Moreover, anthropologists reconstruct the history of social groups by analyzing homologous and analogous cultural traits, much as biologists reconstruct the evolution of species by the analysis of shared characters and homologous DNA (Mace and Pagel 1994). Indeed, the same computer programs developed by biological systematists are used by cultural anthropologists (Holden 2002; Holden and Mace 2003). In addition, archaeologists who study cultural evolution have a similar modus operandi as paleobiologists who study genetic evolution (Mesoudi, Whiten, and Laland 2006). Both attempt to reconstruct lineages of artifacts and their carriers. Like paleobiology, archaeology assumes that when analogy can be ruled out, similarity implies causal connection by inheritance (O'Brian and Lyman 2000). Like biogeography's study of the spatial distribution of organisms (Brown and Lomolino 1998), behavioral ecology studies the interaction of ecological, historical, and geographical factors that determine the distribution of cultural forms across space and time (Smith and Winterhalder 1992).

Perhaps the most common critique of the analogy between genetic and cultural evolution is that the gene is a well-defined, discrete, independently reproducing and mutating entity, whereas the boundaries of the unit of culture are ill-defined and overlapping. In fact, however, this view of the gene is simply outdated. Overlapping, nested, and movable genes discovered in the past 35 years have some of the fluidity of cultural units, whereas quite often the boundaries of a cultural unit (a belief, icon, word, technique, stylistic convention) are quite delimited and specific. Similarly, alternative splicing, nuclear and messenger RNA editing, cellular protein modification, and genomic imprinting, which are quite common, quite undermine the standard view of the insular gene producing a single protein and support the notion of genes having variable boundaries and having strongly context-dependent effects.

Dawkins added a second fundamental mechanism of epigenetic information transmission in *The Extended Phenotype* (1982), noting that organisms can directly transmit environmental artifacts to the next generation in the form of such constructs as beaver dams, beehives, and even social structures (e.g., mating and hunting practices). The phenomenon of a species creating an important aspect of its environment and stably transmitting this environment across generations, known as *niche construction*, it a widespread form of epigenetic transmission (Odling-Smee, Laland, and Feldman 2003). Moreover, niche construction gives rise to what might be called a *gene-environment coevolutionary process* since a genetically induced environmental regularity becomes the basis for genetic selection, and genetic mutations that give rise to mutant niches will survive if they are fitness-enhancing for their constructors.

An excellent example of gene-environment coevolution is seen in the honey bee, which developed a complex division of labor in the hive, including a eusocial division of labor in which only a few individuals are permitted to reproduce on behalf of the whole social community, despite the fact that relatedness in the hive is generally quite low, because of multiple queen matings, multiple queens, queen deaths. The social structure of the hive is transmitted epigenetically across generations, and the honey bee genome is an adaptation to the social structure laid down in the distant past (Gadagkar 1991; Seeley 1997; Wilson and Holldobler 2005).[2]

Gene-culture coevolution in humans is a special case of gene-environment coevolution in which the environment is culturally constituted and transmitted (Feldman and Zhivotovsky 1992). The key to the success of our species in the framework of the hunter-gatherer social structure in which we evolved is the capacity of unrelated, or only loosely related, individuals to cooperate in relatively large egalitarian groups in hunting and territorial acquisition and defense (Boehm 2000; Richerson and Boyd 2004). While contemporary biological and economic theory have attempted to show that such cooperation can be effected by self-regarding rational agents (Trivers 1971; Alexander 1987; Fudenberg, Levine, and Maskin 1994), the conditions under which this is the case are highly implausible even for small groups (Boyd and Richerson 1988; Gintis 2005). Rather, the social environment of early humans was conducive to the development of prosocial traits, such as empathy, shame, pride, embarrassment, and reciprocity, without which social cooperation would be impossible.

Neuroscientific studies exhibit clearly the genetic basis for moral behavior. Brain regions involved in moral judgments and behavior include the prefrontal cortex, the orbitalfrontal cortex, and the superior temporal sulcus (Moll et al. 2005). These brain structures are virtually unique to, or most highly developed in humans and are doubtless evolutionary adaptations (Schulkin 2000). The evolution of the human prefrontal cortex is closely tied to the emergence of human morality (Allman, Hakeem, and Watson 2002). Patients with focal damage to one or more of these areas exhibit a variety of antisocial behaviors, including the absence of embarrassment, pride, and regret (Beer et al. 2003; Camille 2004), and sociopathic behavior (Miller et al. 1997). There is a likely genetic predisposition underlying sociopathy. Sociopaths comprise 3% to 4% of the male population, but they account for between 33% and 80% of the population of chronic criminal offenders in the United States (Mednick et al. 1977).

It is clear from this body of empirical information that culture is directly encoded in the human brain, which of course is the central claim of gene-culture coevolutionary theory.

[2]A *social* species is one that has a division of labor and cooperative behavior. A *eusocial* species is a social species that has a reproductive division of labor; i.e., some females, such as queen bees, produce offspring, while other females, such as worker bees, raise the queen's offspring.

12.2 Culture and Physiology of Human Communication

Consider, for instance, communication through language and complex facial expressions, which exists in more than rudimentary form only in humans. The gene-culture coevolutionary development of human communication is particularly clear because it left a strong fossil record. On an evolutionary time-scale, when a form of human communication became prevalent among hunter-gatherers, this new cultural form became the new environment within which new genetic mutations were evaluated for their fitness effects. Humans thus underwent massive physiological changes to facilitate speaking, understanding speech, and communicating with facial expressions.

To this end, regions in the human motor cortex expanded to carry out speech production in the evolution of *Homo sapiens*. Nerves and muscles to the mouth, larynx, and tongue became more numerous to handle the complexities of speech (Jurmain et al. 1997). Parts of the cerebral cortex, Broca's and Wernicke's areas, which do not exist or are small in other primates, evolved to permit grammatical speech and comprehension (Campbell, Loy and Cruz-Uribe 2005).

The most dramatic changes in human physiology involve speech production. Adult modern humans have a larynx low in the throat, "a position that allows the throat to serve as a resonating chamber capable of a greater number of sounds" (Relethford 2007). The first hominids that have skeletal structures supporting this laryngeal placement are the *Homo heidelbergensis*, who lived from 800,000 to 100,000 years ago. In addition, the production of consonants requires a short oral cavity, whereas our nearest primate relatives have much too long an oral cavity to produce most consonants. The position of the hyoid bone, which is a point of attachment for a tongue muscle, developed in *Homo sapiens* in a manner permitting highly precise and flexible tongue movements. Another indication that the tongue has evolved in hominids to facilitate speech is the size of the hypoglossal canal, an aperture that permits the hypoglossal nerve to reach the tongue muscles. This aperture is much larger in Neanderthals, and humans than in early hominids and non-human primates (Campbell, Loy, and Cruz-Uribe 2005).

Human facial nerves and musculature have also evolved to facilitate communication. This musculature is present in all vertebrates, but except in mammals, it serves feeding and respiratory functions (Burrows 2008). In mammals, this mimetic musculature attaches to skin in the face, thus permitting the subtle and accurate facial communication of such emotions as fear, surprise, disgust, and anger. In most mammals, however, a few wide sheet-like muscles are involved, rendering fine informational differentiation impossible. In primates, by contrast, this musculature divides into many independent muscles with distinct points of attachment to the epidermis and distinct ennervation, thus permitting higher bandwidth facial communication. Humans have the most highly developed facial musculature among vertebrates by far, with a degree of involvement of lips and eyes that is not present in any other species.

There is little doubt but that other human traits, such as empathy, shame, pride, embarrassment, reciprocity, and vengeance, traits without which social cooperation would be impossible, are the product of gene-culture coevolution. Unfortunately, such traits are less likely to leave a clear trace in the fossil record.

12.3 Biological and Cultural Dynamics

The analysis of living systems includes one concept that is not analytically represented in the natural sciences: that of a *strategic interaction* in which the behavior of agents is derived by assuming that each is choosing a *best response* to the actions of other agents. The study of systems in which agents choose best responses and in which such responses evolve dynamically is called *evolutionary game theory*.

A *replicator* is a physical system capable of drawing energy and chemical building blocks from its environment to make copies of itself. Chemical crystals, such as salt, have this property, but biological replicators have the additional ability to assume a myriad of physical forms based on the highly variable sequencing of their chemical building blocks. Biology studies the dynamics of such complex replicators using the evolutionary concepts of replication, variation, mutation, and selection (Lewontin 1974).

Biology plays a role in the behavioral sciences much like that of physics in the natural sciences. Just as physics studies the elementary processes that underlie all natural systems, so biology studies the general characteristics of survivors of the process of natural selection. In particular, genetic replicators, the epigenetic environments to which they give rise, and the effect of these environments on gene frequencies account for the characteristics of species, including the development of individual traits and the nature of intraspecific interaction. This does not mean, of course, that behavioral science in any sense can be *reduced* to biological laws. Just as one cannot deduce the character of natural systems (e.g., the principles of inorganic and organic chemistry, the structure and history of the universe, robotics, plate tectonics) from the basic laws of physics, similarly, one cannot deduce the structure and dynamics of complex life forms from basic biological principles. But, just as physical principles inform model creation in the natural sciences, so must biological principles inform all the behavioral sciences.

Within population biology, evolutionary game theory has become a fundamental tool. Indeed, evolutionary game theory is basically population biology with frequency-dependent fitnesses. Throughout much of the twentieth century, classical population biology did not employ a game-theoretic framework (Fisher 1930; Haldane 1932; Wright 1931). However, Moran (1964) showed that Fisher's fundamental theorem, which states that as long as there is positive genetic variance in a population, fitness increases over time, is false when more than one genetic locus is involved. Eshel and Feldman (1984) identified the problem with the population genetics model in its abstraction from mutation. But how do we attach a fitness value to a mutant? Eshel and Feldman (1984) suggested that payoffs be modeled game-theoretically on the phenotypic level and that a mutant gene be associated with a strategy in the resulting game. With this assumption, they showed that under some restrictive conditions, Fisher's fundamental theorem could be restored. Their results were generalized by Liberman (1988), Hammerstein and Selten (1994), Hammerstein (1996), Eshel, Feldman, and Bergman (1998), and others.

The most natural setting for genetic and cultural dynamics is game-theoretic. Replicators (genetic and/or cultural) endow copies of themselves with a repertoire of strategic responses to environmental conditions, including information concerning the conditions under which each is to be deployed in response to the character and density of competing replicators. Genetic replicators have been well understood since the rediscovery of Mendel's laws in the early twentieth century. Cultural transmission also apparently occurs at the neuronal level in the brain, in part through the action of *mirror neurons* (Williams et al. 2001; Rizzolatti et al. 2002; Meltzhoff and Decety 2003). Mutations include replacement of strategies by modified strategies, and the "survival of the fittest" dynamic (formally called a *replicator dynamic*) ensures that replicators with more successful strategies replace those with less successful strategies (Taylor and Jonker 1978).

Cultural dynamics, however, do not reduce to replicator dynamics. For one thing, the process of switching from lower- to higher-payoff cultural norms is subject to error, and with some positive frequency, lower-payoff forms can displace higher-payoff forms (Edgerton 1992). Moreover, cultural evolution can involve a conformist predisposition (Henrich and Boyd 1998; Henrich and Boyd 2001; Guzman, Sickert, and Rowthorn 2007), as well as oblique and horizontal transmission (Cavalli-Sforza and Feldman 1981; Gintis 2003b).

12.4 The Theory of Norms: The Sociological Model

Complex social systems generally have a division of labor, with distinct social positions occupied by individuals specially prepared for their roles. For instance, a beehive has workers, drones, and queens, and workers can be nurses, foragers, or scouts. Preparation for roles is by gender and larval nutrition. Modern human society has a division of labor characterized by dozens of specialized *roles*, appropriate behavior within which is given by *social norms*, and individuals are *actors* who are motivated to fulfill these roles through a combination of *material incentives* and *normative commitments*.

The centrality of culture in the social division of labor was clearly expressed by Emile Durkheim (1933 [1902]), who stressed that the great multiplicity of roles (which he called *organic solidarity*) required a commonality of beliefs (which he called *collective consciousness*) that would permit the smooth coordination of actions by distinct individuals. This theme was developed by Talcott Parsons (1937), who used his knowledge of economics to articulate a sophisticated model of the interaction between the situation (role) and its inhabitant (actor). The actor/role approach to social norms was filled out by Erving Goffman (1959), among others.

The social role has both normative and positive aspects. On the positive side, the payoffs—rewards and penalties—associated with a social role must provide the appropriate incentives for actors to carry out the duties associated with the role. This requirement is most easily satisfied when these payoffs are independent of the behavior of agents occupying other roles. However, this is rarely the case. In general, as developed in chapter 7, social roles are deeply interdependent and can be modeled as the strategy sets of players in an epistemic game, the payoffs to which are precisely these rewards and penalties, the choices of actors then forming a correlated equilibrium for which the required commonality of beliefs is provided by a society's common culture. This argument provides an analytical link uniting the actor/role framework in sociological theory with game-theoretic models of cooperation in economic theory.

Appropriate behavior in a social role is given by a *social norm* that specifies the duties, privileges, and normal behavior associated with the role. In the first instance, social norms have an instrumental character devoid of normative content, serving merely as informational devices that coordinate the behavior of rational agents (Lewis 1969; Gauthier 1986; Binmore 2005; Bicchieri 2006). However, in most cases, high level performance in a social role requires that the actor have a *personal commitment* to role performance that cannot be captured by the self-regarding "public" payoffs associated with the role (see chapter 7 and Conte and Castelfranchi, 1999). . This is because (a) actors may have private payoffs that conflict with the role's public payoffs, inducing them to behave counter to proper role-performance (e.g., corruption, favoritism, aversion to specific tasks); (b) the signal used to determine the public payoffs may be inaccurate and unreliable (e.g., the performance of a teacher, physician, scientist, or business executive cannot be fully objectively assessed at reasonable cost); and (c) the public payoffs required to gain compliance by self-regarding actors may be higher than those required when there is at least partial reliance upon the personal commitment of role incumbents; i.e., it may be less costly to use personally committed rather than purely materially motivated agents when performance cannot be easily measured (Bowles 2008). In such cases, self-regarding actors who treat social norms purely instrumentally behave in a socially inefficient manner (§6.3, 6.4).

The normative aspect of social roles flows from these considerations. First, to the extent that social roles are considered legitimate by incumbents, they place an intrinsic ethical value on role performance. We call this the *normative predisposition* associated with role occupancy (see chapter 7). Second, human ethical predispositions include *character virtues*, such as honesty, trustworthiness, promise keeping, and obedience, that may increase the value of conforming to the duties associated with role incumbency (§3.12). Third, humans are also predisposed to care about the esteem of others even when there can be no future reputational repercussions (Masclet et al. 2003) and take pleasure in punishing others who have violated social norms (Fehr and Fischbacher 2004). These ethical traits by no means contradict rationality (§12.6), because individuals trade off these values against material reward, and against each other, just as described in the economic theory of the rational actor (Andreoni and Miller 2002; Gneezy and Rustichini 2000).

The sociopsychological theory of norms can thus resolve the contradictions between the sociological and economic models of social cooperation, retaining the analytical clarity of game theory and the rational actor model while incorporating the collective, normative, and cultural characteristics stressed in psychosocial models of norm compliance.

12.5 Socialization and the Internalization of Norms

Society is held together by *moral values* that are transmitted from generation to generation by the process of *socialization*. These values are instantiated through the *internalization of norms* (Parsons 1967; Grusec and Kuczynski 1997; Nisbett and Cohen 1996; Rozin et al. 1999), a process in which the initiated instill values into the uninitiated (usually the younger generation) through an extended series of personal interactions, relying on a complex interplay of affect and authority. Through the internalization of norms, initiates are supplied with moral values that induce them to conform to the duties and obligations of the role positions they expect to occupy. The internalization of norms, of course, presupposes a genetic predisposition to moral cognition that can be explained only by gene-culture coevolution.

Internalized norms are accepted not as instruments for achieving other ends but rather as *arguments in the preference function that the individual maximizes*. For instance, an individual who has internalized the value of speaking truthfully does so even in cases where the net payoff to speaking truthfully would otherwise be negative. Such fundamental human emotions as shame, guilt, pride, and empathy are deployed by the well-socialized individual to reinforce these prosocial values when tempted by the immediate pleasures of such deadly sins as anger, avarice, gluttony, and lust. It is tempting to treat some norms as constraints rather than objectives, but virtually all norms are violated by individuals under some conditions, indicating that there are tradeoffs, such as those explored in §3.12 and §3.4, that could not exist were norms merely constraints on action.

The human openness to socialization is perhaps the most powerful form of epigenetic transmission found in nature. This epigenetic flexibility in considerable part accounts for the stunning success of the species *H. sapiens* because when individuals internalize a norm, the frequency of the desired behavior is higher than if people follow the norm only instrumentally—i.e., when they perceive it to be in their best interest to do so on other grounds. The increased incidence of prosocial behaviors is precisely what permits humans to cooperate effectively in groups (Gintis et al. 2005).

There are, of course, limits to socialization (Tooby and Cosmides 1992; Pinker 2002), and it is imperative to understand the dynamics of the emergence and abandonment of particular values, which in fact depend on their contribution to fitness and well-being, as economic and biological theory would suggest (Gintis 20031,b). Moreover, there are often swift, society-wide value changes that cannot be accounted for by socialization theory (Wrong 1961; Gintis 1975). However, socialization theory has an important place in the general theory of culture, strategic learning, and moral development.

One of the more stunning indications of the disarray of the behavioral sciences is the fact that the internalization of norms does not appear in the economic and biological models of human behavior.

12.6 Rational Choice: The Economic Model

General evolutionary principles suggest that individual decision making for members of a species can be modeled as optimizing a preference function. Natural selection leads the content of preferences to reflect biological fitness. The principle of expected utility extends this optimization to stochastic outcomes. The resulting model is called the *rational actor model* in economics, although there is some value to referring to it as the *beliefs, preferences, and constraints* (BPC) model, thus avoiding the often misleading connotations attached to the term "rational."

For every constellation of sensory inputs, each decision taken by an organism generates a probability distribution over outcomes, the expected value of which is the *fitness* associated with that decision. Since fitness is a scalar variable, for each constellation of sensory inputs, each possible action the organism might take has a specific fitness value, and organisms whose decision mechanisms are optimized for this environment choose the available action that maximizes this value. This argument was presented verbally by Darwin (1872) and is implicit in the standard notion of "survival of the fittest," but formal proof is recent (Grafen 1999, 2000, 2002). The case with frequency-dependent (nonadditive genetic) fitness has yet to be formally demonstrated, but the informal arguments are compelling.

Given the state of its sensory inputs, if an organism with an optimized brain chooses action A over action B when both are available, and chooses action B over action C when both are available, then it will also choose action A over action C when both are available. Thus, choice consistency follows from basic evolutionary dynamics. The rational actor model is often presented as though it applies only when actors possess extremely powerful information-processing capacities. As we saw in chapter 1, in fact, the basic model depends only on choice consistency, the expected utility theorem being considerably more demanding.

Four *caveats* are in order. First, individuals do not consciously maximize something called "utility," or anything else. Second, individual choices, even if they are self-regarding (e.g., personal consumption) are not necessarily welfare-enhancing. Third, preferences must have some stability across time to be theoretically useful, but preferences are ineluctably a function of an individual's *current state*, and beliefs can change dramatically in response to immediate sensory and social experiences. Finally, beliefs need not be correct nor need they be updated correctly in the face of new evidence, although Bayesian assumptions concerning updating can be made part of consistency in elegant and compelling ways (Jaynes 2003).

The rational actor model is the cornerstone of contemporary economic theory and in the past few decades has become the heart of the biological modeling of animal behavior (Real 1991; Alcock 1993; Real and Caraco 1986). Economic and biological theory thus have a natural affinity: the choice consistency on which the rational actor model of economic theory depends is rendered plausible by evolutionary theory, and the optimization techniques pioneered in economics are routinely applied and extended by biologists in modeling the behavior of nonhuman organisms. I suggest below that this is due to the *routine* choice paradigm that applies in economics and biology, as opposed to the *deliberative* choice paradigm that applies in cognitive psychology.

Perhaps the most pervasive critique of the BPC model is that put forward by Herbert Simon (1982), holding that because information processing is costly and humans have a finite information-processing capacity, individuals *satisfice* rather than *maximize* and hence are only *boundedly rational*. There is much substance to this view, including the importance of including information-processing costs and limited information in modeling choice behavior and recognizing that the decision on how much information to collect depends on unanalyzed subjective priors at some level (Winter 1971; Heiner 1983). Indeed, from basic information theory and quantum mechanics, it follows that *all rationality is bounded*. However, the popular message taken from Simon's work is that we should reject the BPC model. For instance, the mathematical psychologist D. H. Krantz (1991) asserts, "The normative assumption that individuals *should* maximize *some* quantity may be wrong. ...People do and should act as *problem solvers*, not *maximizers*." This is incorrect. In fact, as long as individuals are involved in routine choice (see §12.14) and hence have consistent preferences, they can be modeled as maximizing an objective function subject to constraints.

This point is lost on even such capable researchers as Gigerenzer and Selten (2001), who reject the "optimization subject to constraints" method on the grounds that individuals do not in fact solve optimization problems. However, just as billiards players do not solve differential equations in choosing their shots, so decision makers do not solve Lagrangian equations, even though in both cases we may use such optimization models to describe their behavior. Of course, as stressed by Gigerenzer and Selten (2001), from an analytical standpoint, generalizing the rational actor model may not be the best way to capture the heuristics of decision making in particular areas.

12.7 Deliberative Choice: The Psychological Model

The psychological literature on decision making is rich and multifaceted, traditional approaches being augmented in recent years by neural net theory and evidence from neuroscientific data on brain functioning (Kahneman, Slovic, and Tversky 1982; Baron 2007; Oaksford and Chater 2007; Hinton and Sejnowski 1999; Newell, Lagnado, and Shanks 2007; Juslin and Montgomery 1999; Bush and Mosteller 1955; Gigerenzer and Todd 1999; Betch and Haberstroh 2005; Koehler and Harvey 2004). There does not yet exist a unitary model underlying the psychological understanding of judgment and decision making, doubtless because the mental processes involved are so varied and complex.

The sorts of decision making studied by psychologists include the formation of long-term goals, which are evaluated according to the value if attained, the range of probable costs, and the probability of goal attainment. All three dimensions of goal formation have inherent uncertainties, so among the strategies of goal choice is the formation of subgoals with the aim of reducing these uncertainties. The most complex of human decisions tend to involve goals that arise infrequently in the course of a life, such as choosing a career, whether to marry and to whom, how many children to have, and how to deal with a health threat, where the scope for learning from mistakes is narrow. Psychologists also study how people make decisions based on noisy single- or multidimensional data under conditions of trial-and-error learning.

The difficulty in modeling a deliberative choice is exacerbated by the fact that, because of the complexity of such decisions, much human decision making has a distinctly group dynamic, in which some individuals experiment and other imitate the more successful of the experimenters (Bandura 1977). This dynamic cannot be successfully modeled on the individual level.

By contrast, the rational actor model applies to choice situations where ambiguities are absent, the choice set is clearly delineated, and the payoffs are unmediated, so that no deliberation is involved beyond the comparison of feasible alternatives. Accordingly, most psychologists working in this area accept the rational actor model as the appropriate model of choice behavior in this realm of routine choice, yet recognize that there is no obvious way to extend the model to the more complex situations they study. For instance, Newell, Lagnado, and Shanks (2007) assert, "We view judgment and decision making as often exquisitely subtle and well-tuned to the world, especially in situations where we have the opportunity to respond repeatedly under similar conditions where we can learn from feedback." (p. 2)

There is thus no deep conceptual divide between the psychological approach to decision making and the economic approach. While in some important areas, human decision makers appear to violate the consistency condition for rational choice, in virtually all such cases, as we suggested in §1.9, consistency can be restored by assuming that the current state of the agent is an argument of the preference structure. Another possible challenge to preference consistency is preference reversal in the choice of lotteries. Lichtenstein and Slovic (1971) were the first to find that in many cases, individuals who prefer lottery A to lottery B are nevertheless willing to take less money for A than for B. Reporting this to economists several years later, Grether and Plott (1979) asserted, "A body of data and theory has been developed... [that] are simply inconsistent with preference theory... (p. 623). These preference reversals were explained several years later by Tversky, Slovic, and Kahneman (1990) as a bias toward the higher probability of winning in a lottery choice and toward the higher maximum amount of winnings in monetary valuation. However, the phenomenon has been documented only when the lottery pairs A and B are so close in expected value that one needs a calculator (or a quick mind) to determine which would be preferred by an expected value maximizer. For instance, in Grether and Plott (1979) the average difference between expected values of comparison pairs was 2.51% (calculated from table 2, p. 629). The corresponding figure for Tversky, Slovic, and Kahneman (1990) was 13.01%. When the choices are so close to indifference, it is not surprising that inappropriate cues are relied upon to determine choice, as would be suggested by the heuristics and biases model (Kahneman, Slovic, and Tversky 1982) favored by behavioral economists and psychologists.

The expected utility model (§1.5) is closer to the concerns of psychologists because it deals with uncertainty in a fundamental way, and applying Bayes' rule certainly may involve complex deliberations. The Ellsberg paradox is an especially clear example of the failure of the probability reasoning behind the expected utility model. Nevertheless the model has a considerable body of empirical support, so the basic modeling issue is to be able to say clearly when the expected utility theorem is likely to be violated, and to supply an alternative model outside this range (Newell, Lagnado, and Shanks 2007; Oaksford and Chater 2007).

I conclude that there should be a basic synergy between the rational actor model when dealing with routine choice and the sorts of models developed by psychologists to explain complex human deliberation, goal formation, and learning.

12.8 Application: Addictive Behavior

Substance abuse appears to be irrational. Abusers are time inconsistent and their behavior is welfare-reducing. Moreover, even draconian increases in the penalties for illicit substance use lead to the swelling of prison populations rather than abandonment of the sanctioned activity. Because rational actors generally trade off among desired goals, this curious phenomenon has led some researchers to reject the BPC model out of hand.

However, the BPC model remains the most potent tool for analyzing substance abuse on a societywide level. The most salient target of the critics has been the "rational addiction" model of Becker and Murphy (1988). While this model does have some shortcomings, its use of the rational actor model is not among them. Indeed, empirical research supports the contention that illicit drugs respond normally to market forces. For instance, Saffer and Chaloupka (1999) estimated the price elasticities of heroin and cocaine using a sample of 49,802 individuals from the National Household Survey of Drug Abuse to be 1.70 and 0.96, respectively. These elasticities are in fact quite high. Using these estimates, the authors judge that lower prices flowing from the legalization of these drugs would lead to an increase of about 100% and 50% in the quantities of heroin and cocaine consumed, respectively.

How does this square with the observation that draconian punishments do not squelch the demand altogether? Gruber and Köszegi (2001), who use the rational actor model but do not assume time consistency, explain this by showing that drug users exhibit the commitment and self-control problems typical of time-inconsistent agents, for whom the possible future penalties have a highly attenuated deterrent value in the present. This behavior may be welfare-reducing, but the rational actor model does not presume that preferred outcomes are necessarily welfare-improving.

12.9 Game Theory: The Universal Lexicon of Life

Game theory is a logical extension of evolutionary theory. To see this, suppose there is only one replicator, deriving its nutrients and energy from nonliving sources. The replicator population will then grow at a geometric rate until it presses upon its environmental inputs. At that point, mutants that exploit the environment more efficiently will outcompete their less efficient conspecifics and with input scarcity, mutants will emerge that "steal" from conspecifics who have amassed valuable resources. With the rapid growth of such predators, mutant prey will devise means of avoiding predation, and predators will counter with their own novel predatory capacities. In this manner, strategic interaction is born from elemental evolutionary forces. It is only a conceptual short step from this point to cooperation and competition among cells in a multicellular body, among conspecifics who cooperate in social production, between males and females in a sexual species, between parents and offspring, and among groups competing for territorial control.

Historically, game theory did not emerge from biological considerations but rather from strategic concerns in World War II (Von Neumann and Morgenstern 1944; Poundstone 1992). This led to the widespread caricature of game theory as applicable only to static confrontations of rational self-regarding agents possessed of formidable reasoning and information-processing capacity. Developments within game theory in recent years, however, render this caricature inaccurate.

Game theory has become the basic framework for modeling animal behavior (Maynard Smith 1982; Alcock 1993; Krebs and Davies 1997) and thus has shed its static and hyperrationalistic character, in the form of evolutionary game theory (Gintis 2009). Evolutionary and behavioral game theory do not require the formidable information-processing capacities of classical game theory, so disciplines that recognize that cognition is scarce and costly can make use of game-theoretic models (Young 1998; Gintis 2009; Gigerenzer and Selten 2001). Thus, agents may consider only a restricted subset of strategies (Winter 1971; Simon 1972), and they may use rule-of-thumb heuristics rather than maximization techniques (Gigerenzer and Selten 2001). Game theory is thus a generalized schema that permits the precise framing of meaningful empirical assertions but imposes no particular structure on the predicted behavior.

12.10 Epistemic Game Theory and Social Norms

Economics and sociology have highly contrasting models of human interaction. Economics traditionally considers individuals to be rational, self-regarding payoff maximizers, while sociology considers individuals to be highly socialized, other-regarding, moral agents who strive to fill social roles and whose self-esteem depends on the approbation of others. The project of unifying the behavioral sciences must include a resolution of these inconsistencies in a manner that preserves the key insights of each.

Behavioral game theory helps us adjudicate these disciplinary differences, providing experimental data supporting the sociological stress on moral values and other-regarding preferences, and also supports the economic stress on rational payoff maximization. For instance, most individuals care about reciprocity and fairness as well as personal gain (Gintis et al. 2005), value such character virtues as honesty for their own sake (Gneezy 2005), care about the esteem of others even when there can be no future reputational repercussions (Masclet et al. 2003), and take pleasure in punishing others who have hurt them (deQuervain et al. 2004). Moreover, as suggested by socialization theory, individuals have consistent values, based on their particular sociocultural situations, that they apply in the laboratory even in one-shot games under conditions of anonymity (Henrich et al.

2004; Henrich et al. 2006). This body of evidence suggests that sociologists would benefit from reincorporating the rational actor model into sociological theory, and economists broaden their concept of human preferences.

A second discrepancy between economics and sociology concerns the contrasting claims of game theory and the sociopsychological theory of norms in explaining social cooperation. Our exposition of this area in chapter 7 can be interpreted in the larger context of the unity of the behavioral sciences as follows.

The basic model of the division of labor in economic theory is the Walrasian general equilibrium model, according to which a system of flexible prices induces firms and individuals to supply and demand goods and services in such amounts that all markets clear in equilibrium (Arrow and Debreu 1954). However, this model assumes that all contracts among individuals can be costlessly enforced by a third party, such as the judicial system. In fact, however, many critical forms of social cooperation are not mediated by a third-party enforceable contract but rather take the form of repeated interactions in which an informal, but very real, threat of rupturing the relationship is used to induce mutual cooperation (Fudenberg, Levine, and Maskin 1994; Ely and Välimäki 2002). For instance, an employer hires a worker who works hard under the threat of dismissal not the threat of an employer lawsuit.

Repeated game theory thus steps in for economists to explain forms of face-to-face cooperation that do not reduce to simple price-mediated market exchanges. Repeated game theory shows that in many cases the activity of many individuals can be coordinated, in the sense that there is a Nash equilibrium ensuring that no self-regarding player can gain by deviating from the strategy assigned to him by the equilibrium, assuming other players also use the strategies assigned to them (§10.4). If this theory were adequate, which most economists believe is the case, then there would be no role for the sociopsychological theory of norms, and sociological theory would be no more that a thick description of a social mechanism analytically accounted for by repeated game theory.

However, repeated game theory with self-regarding agents does not solve the problem of social cooperation (§10.6). When the group consists of more than two individuals and the signal indicating how well a player is performing his part is imperfect and private (i.e., players receive imperfectly correlated signals about another player's behavior), the efficiency of cooperation may be quite low, and the roles assigned to each player will be extremely complex mixed strategies that players have no incentive to use (§10.5). As we suggested in chapter 7, the sociopsychology of norms can step in at this point to provide mechanisms that induce individuals to play their assigned parts. A social norm may provide the rules for each individual in the division of labor, players may have a general predilection for honesty that allows them to consolidate their private signals concerning another player's behavior into a public signal that can be the bases for coordinated collective punishment and reward, and players may have a personal normative predisposition towards following the social roles assigned to them. The sociological and economic forces thus complement rather than contradict one another.

A central analytical contribution to this harmonization of economics and sociology was provided by Robert Aumann (1987), who showed that the natural concept of equilibrium in game theory for rational actors who share common beliefs is not the Nash equilibrium but the correlated equilibrium. A correlated equilibrium is the Nash equilibrium in the game formed by adding to the original game a new player, whom I call the *choreographer* (Aumann calls this simply a "correlating device"), who samples the probability distribution given by the players (common) beliefs and then instructs each player what action to take. The actions recommended by the choreographer are all best responses to one another, conditional on their having been simultaneously ordered by the choreographer, so self-regarding players can do no better than to follow the choreographer's advice.

Sociology, and more generally sociobiology (see chapter 11), then come in not only by supplying the choreographer, in the form of a complex of social norms, but also by supplying cultural theory to explain why players might have a common set of beliefs, without which the correlated equilibrium would not exist. Cognitive psychology explains the normative predisposition that induces players to take the advice of the choreographer (i.e., to follow the social norm) when in fact there might be many other actions with equal, or even higher, payoff that the player might have an inclination to choose.

12.11 Society as a Complex Adaptive System

The behavioral sciences advance not only by developing analytical and quantitative models but also by accumulating historical, descriptive, and ethnographic evidence that pays heed to the detailed complexities of life in the sweeping array of wondrous forms that nature reveals to us. Historical contingency is a primary focus for many students of sociology, anthropology, ecology, biology, politics, and even economics. By contrast, the natural sciences have found little use for narrative alongside analytical modeling.

The reason for this contrast between the natural and the behavioral sciences is that *living systems are generally complex, dynamic adaptive systems* with emergent properties that cannot be fully captured in analytical models that attend only to local interactions. The hypothetico-deductive methods of game theory, the BPC model, and even gene-culture coevolutionary theory must therefore be complemented by the work of behavioral scientists who adhere to more historical and interpretive traditions, as well as that of researchers who use agent-based programming techniques to explore the dynamic behavior of approximations to real-world complex adaptive systems.

A *complex system* consists of a large population of similar entities (in our case, human individuals) who interact through regularized channels (e.g., networks, markets, social institutions) with significant stochastic elements, without a system of centralized organization and control (i.e., if there is a state, it controls only a fraction of all social interactions and is itself a complex system). A complex system is *adaptive* if it undergoes an evolutionary (genetic, cultural, agent-based, or other) process of reproduction, mutation, and selection (Holland 1975). To characterize a system as complex adaptive does not explain its operation and does not solve any problems. However, it suggests that certain modeling tools are likely to be effective that have little use in a noncomplex system. In particular, the traditional mathematical methods of physics and chemistry must be supplemented by other modeling tools such as agent-based simulation and network theory.

The stunning success of modern physics and chemistry lies in their ability to avoid or control emergence. The experimental method in natural science is to create highly simplified laboratory conditions under which modeling becomes analytically tractable. Physics is no more effective than economics or biology in analyzing complex real-world phenomena in situ. The various branches of engineering (electrical, chemical, mechanical) are effective because they re-create in everyday life artificially controlled, noncomplex, nonadaptive environments in which the discoveries of physics and chemistry can be directly applied. This option is generally not open to most behavioral scientists, who rarely have the opportunity of "engineering" social institutions and cultures.

12.12 Counterpoint: Biology

Biologists are generally comfortable with three of the five principles laid out in the introduction to this chapter. Only gene-culture coevolution and the sociopsychology of norms have generated significant opposition.

Gene-culture coevolutionary theory has been around only since the 1980s and applies to only one species—*H. sapiens*. Not surprisingly, many sociobiologists have been slow to adopt it and have deployed a formidable array of population biology concepts toward explaining human sociality in more familiar terms—especially kin selection (Hamilton 1964) and reciprocal altruism (Trivers 1971). The explanatory power of these models convinced a generation of researchers that what appears to be altruism— personal sacrifice on the behalf of others—is really just long-run self-interest, and that elaborate theories drawn from anthropology, sociology, and economics are unnecessary to explain human cooperation and conflict.

Richard Dawkins, for instance, in *The Selfish Gene* (1989 [1976]), asserts, "We are survival machines—robot vehicles blindly programmed to preserve the selfish molecules known as genes.... This gene selfishness will usually give rise to selfishness in individual behavior." Similarly, in *The Biology of Moral Systems* (1987), R. D. Alexander asserts, "Ethics, morality, human conduct, and the human psyche are to be understood only if societies are seen as collections of individuals seeking their own self-interest...." (p. 3) In a similar vein, Michael Ghiselin (1974) writes, "No hint of genuine charity ameliorates our vision of society, once sentimentalism has been laid aside. What passes for cooperation turns out to be a mixture of opportunism and exploitation.... Scratch an altruist, and watch a hypocrite bleed" (p. 247)

Evolutionary psychology, which has been a major contributor to human sociobiology, has incorporated the kin selection/reciprocal altruism perspective into a broadside critique of the role of culture in society (Barkow, Cosmides, and Tooby 1992) and of the forms of group dynamics upon which gene-culture coevolution depends (Price, Cosmides, and Tooby 2002). I believe these claims have been effectively refuted (Richerson and Boyd 2004; Gintis et al. 2009), although the highly interesting debate in population biology concerning group selection has been clarified but not completely resolved (Lehmann and Keller 2006; Lehmann et al. 2007; Wilson and Wilson 2007).

12.13 Counterpoint: Economics

Economists generally believe in methodological individualism, a doctrine claiming that all social behavior can be explained by strategic interactions among agents. Were this correct, gene-culture coevolution would be unnecessary, complexity theory would be irrelevant, and the sociopsychological theory of norms could be derived from game theory. We concluded in chapter 8, however, that methodological individualism is contradicted by the evidence.

Economists also generally reject the idea of society as a complex adaptive system, on grounds that we may yet be able to tweak the Walrasian general equilibrium framework, suitably fortified by sophisticated mathematical methods, so as to explain macroeconomic activity. In fact, there has been virtually no progress in general equilibrium theory since the mid-twentieth-century existence proofs (Arrow and Debreu 1954). Particularly noteworthy has been the absence of any credible stability model (Fisher 1983). Indeed, the standard models predict price instability and chaos (Saari 1985; Bala and Majumdar 1992). Moreover, analysis of excess demand functions suggests that restrictions on preferences are unlikely to entail the stability of Walrasian price dynamics (Sonnenschein 1972, 1973; Debreu 1974; Kirman and Koch 1986).

My response to this sad state of affairs has been to show that agent-based models of generalized exchange, based on the notion that the economy is a complex nonlinear dynamical system, exhibit a high degree of stability and efficiency (Gintis 2006, 2007a). There does not appear to be any serious doctrinal impediment to the use of agent-based modeling in economics.

12.14 Counterpoint: Psychology

Decision theory, based on the rational actor model, represents one of the great scientific achievements of all time, beginning with Bernoulli and Pascal in the seventeenth and eighteenth centuries and culminating in the work of Ramsey, de Finetti, Savage, and von Neumann and Morgenstern in the early and middle years of the twentieth century. Its preeminence in the behavioral disciplines that deal with human choice, especially its position as the keystone of modern economic theory, however, has led to an extreme level of empirical scrutiny of decision theory. Because I include the rational actor model as one of my five organizing principles for the unification of the behavioral sciences, this critique deserves careful consideration.

The most salient critique has taken inspiration from the brilliant series of experiments by Daniel Kahneman and Amos Tversky. These researchers have documented several key and systematic divergences between the normative principle of decision theory and the actual choices of intelligent, educated individuals (see chapter 1). Such phenomena as loss aversion, the base rate fallacy, framing effects, and the conjunction fallacy must be added to the traditional paradoxes of Allais and Ellsberg as representing fundamental aspects of human decision making that fall outside the purview of traditional decision theory (§1.7).

Psychologists have used these contributions improperly to mount a sustained attack on the rational actor model, leading many researchers to reject traditional decision theory and seek alternatives lying quite outside the rational actor tradition, in such areas as computer modeling of neural nets and neuroscientific studies of brain functioning. This dismissal of traditional decision theory may be emotionally satisfying, but it is immature, short-sighted, and scientifically destructive. There is no alternative to the traditional decision-theoretic model on the horizon, and there is not likely to be one, for one simple reason: the theory is mostly correct, and where it fails, the principles accounting for failure are complementary to, rather than destructive of, the standard theory. For instance, the documented inconsistencies in the traditional rational actor model can be handled effectively by assuming that the preference function has the current state of the individual as an argument, so all assessments are of deviations from the status quo ante. Prospect theory, for which Kahneman was awarded the Nobel prize, is precisely of this form, as is the treatment of time inconsistency and regret phenomena. In other cases, by assuming that individuals have other-regarding preferences (which laboratory evidence strongly supports), we rupture the traditional prejudice that rationality implies selfishness.

My suggestion for resolving the conflict between psychological and economic models of decision making has four points. First, the two disciplines should recognize the distinction between deliberative and routine decision making. Second, psychology should introduce the evolution of routine decision making into its core framework, based on the principle that the brain is a fitness-enhancing adaptation. Third, deliberative decision making is an adaptation to the increased social complexity of primates and hominid groups. Finally, routine decision making shades into deliberative decision making under conditions that are only imperfectly known but are of great potential importance for understanding human choice.

12.15 The Behavioral Disciplines Can Be Unified

In this chapter, I have proposed five analytical tools that together serve to provide a common basis for the behavioral sciences. These are gene-culture coevolution, the sociopsychological theory of norms, game theory, the rational actor model, and complexity theory. While there are doubtless formidable scientific issues involved in providing the precise articulations between these tools and the major conceptual tools of the various disciplines, as exhibited, for instance, in harmonizing the socio-psychological theory of norms and repeated game theory, these intellectual issues are likely to be dwarfed by the sociological issues surrounding the semifeudal nature of modern behavioral disciplines, which renders even the most pressing reform a monumental enterprise. If these institutional obstacles can be overcome, the behavioral disciplines can be rendered mutually consistent and reinforcing.

13

Summary

> This above all: to thine own self be true,
> And it must follow, as the night the day,
> Thou canst not then be false to any man.
>
> Shakespeare

In a long book with many equations, it is easy to become mired in details and hence miss the big picture. This chapter is a summary of the book's main points.

- Game theory is an indispensable tool in modeling human behavior. Behavioral disciplines that reject or peripheralize game theory are theoretically handicapped.
- The traditional equilibrium concept in game theory, the Nash equilibrium, is implemented by rational actors only if they share beliefs as to how the game will be played.
- The rational actor model includes no principles entailing the communality of beliefs across individuals. For this reason, the complex Nash equilibria that arise in modeling the coordination of behavior in groups do not emerge spontaneously from the interaction of rational agents. Rather, they require a higher-level correlating device, or choreographer.
- Hence, the Nash equilibrium is not the appropriate equilibrium concept for social theory.
- The correlated equilibrium is the appropriate equilibrium concept for a set of rational individuals having common priors. The appropriate correlating devices may be broadly identified with social norms.
- Social systems are complex adaptive dynamical systems. Social norms are among the emergent properties of such systems. Social norms range from simple conventions (e.g., vocabulary and traffic signals) to complex products of gene-culture coevolution (e.g., territoriality and property rights). Complex norms may be taught, learned, and internalized, but individuals must be genetically predisposed to recognize and obey social norms.
- There is thus a social epistemology based on the specific character of the evolved human brain, as well as the operation of culturally

specific social institutions that effect the commonality of beliefs in humans.

- Even with a commonality of beliefs and a social norm choreographing a correlated equilibrium, self-regarding individuals do not have incentives to play correlated equilibria. Rather, humans are other-regarding: they are predisposed to obey social norms even when it is costly to do so. We term this a *normative predisposition*.
- The behavioral disciplines today have four incompatible models of human behavior. The behavioral sciences must develop a unified model of choice that eliminates these incompatibilities and that can be specialized in different ways to meet the heterogeneous needs of the various disciplines.
- *The Bounds of Reason* contributes to the task of unifying the behavioral sciences by showing that game theory needs a broader social theory to have explanatory power, and that social theory without game theory is seriously compromised.
- The bounds of reason are not forms of irrationality but rather forms of sociality.

14

Table of Symbols

$\{a,b,x\}$	Set with members a, b and x
$\{x\|p(x)\}$	The set of x for which $p(x)$ is true
$p \wedge q, p \vee q, \neg p$	p and q, p or q, not p
iff	If and only if
$p \Rightarrow q$	p implies q
$p \Leftrightarrow q$	p if and only if q
(a,b)	Ordered pair: $(a,b) = (c,d)$ iff $a = c$ and $b = d$
$a \in A$	a is a member of the set A
$A \times B$	$\{(a,b)\|a \in A \text{ and } b \in B\}$
$\mathbf{R}$	The real numbers
$\mathbf{R}^n$	The n-dimensional real vector space
$(x_1,\ldots,x_n) \in \mathbf{R}^n$	An n-dimensional vector
$f{:}A \rightarrow B$	A function $b = f(a)$, where $a \in A$ and $b \in B$
$f(\cdot)$	A function f where we suppress its argument
$f^{-1}(y)$	The inverse of function $y = f(x)$
$\sum_{x=a}^{b} f(x)$	$f(a) + \cdots + f(b)$
$S_1 \times \cdots \times S_n$	$\{(s_1,\ldots s_n)\|s_i \in S_i, i = 1,\ldots n\}$
$\prod_{i=1}^{n} S_i$	$S_1 \times \cdots \times S_n$
ΔS	Set of probability distributions (lotteries) over S
$\Delta^* \prod_i S_i$	$\Pi_i \Delta S_i$ (set of mixed strategies)
$[a,b]$,(a,b)	$\{x \in \mathbf{R}\|a \leq x \leq b\}$,$\{x \in \mathbf{R}\|a < x < b\}$
$[a,b)$,$(a,b]$	$\{x \in \mathbf{R}\|a \leq x < b\}$,$\{x \in \mathbf{R}\|a < x \leq b\}$
$A \cup B$	$\{x\|x \in A \text{ or } x \in B\}$
$A \cap B$	$\{x\|x \in A \text{ and } x \in B\}$
$\cup_\alpha A_\alpha$	$\{x\|x \in A_\alpha \text{ for some } \alpha\}$
$\cap_\alpha A_\alpha$	$\{x\|x \in A_\alpha \text{ for all } \alpha\}$
$A \subset B$	$A \neq B \wedge (x \in A \Rightarrow x \in B)$
$A \subseteq B$	$x \in A \Rightarrow x \in B$
$=_{\text{def}}$	Equal by definition
$[\psi]$	$\{\omega \in \Omega\|\psi(\omega) \text{ is true}\}$
$f \circ g(x)$	$f(g(x))$

Symbols for Chapter 11

$\beta \in (0, 1]$	Amount of injury from combat
$\phi \in (0, 1]$	Fraction of agents who are incumbents
π_g	Present value of being a currently uncontested incumbent
π_b	Present value of being a migrant searching for a territory
$\rho \in (0, 1]$	Probability that a migrant locates a patch
b	Benefit from incumbency
$c \in (0, 1]$	Fitness cost associated with territorial search
$f \in (0, 1]$	Fraction of patches that are fertile
$f_o \in (0, 1]$	Fraction of incumbents who contest
n	Number of days agent can live without incumbency
n_p	Number of patches
n_a	Number of agents
$p \in (0, 1]$	Probability of patch death
$q \in (0, 1]$	Probability of a dead patch becoming fertile
$q_u \in (0, 1]$	Probability incumbency is challenged by an intruder
$r \in (0, 1]$	Probability of finding a fertile patch
$v \in (0, 1]$	Cost of investing in a newly fertile patch
$w \in (0, 1]$	Probability of finding a fertile unoccupied patch
$p_d \in (0, 1]$	Probability of combat leading to injury
$p_u \in (0, 1]$	Probability that intruder wins the contest
$s =\in (0, 1]$	Expected injury from combat
$s_o \in (0, 1]$	Resources committed to combat by incumbent
$s_u \in (0, 1]$	Resources committed to combat by intruder

References

Abbink, Klaus, Jordi Brandts, Benedikt Herrmann, and Henrik Orzen, "Inter-Group Conflict and Intra-Group Punishment in an Experimental Contest Game," 2007. CREED, University of Amsterdam.

Abbott, R. J., J. K. James, R. I. Milne, and A. C. M. Gillies, "Plant Introductions, Hybridization and Gene Flow," *Philosophical Transactions of the Royal Society of London B* 358 (2003):1123–1132.

Ahlbrecht, Martin, and Martin Weber, "Hyperbolic Discounting Models in Prescriptive Theory of Intertemporal Choice," *Zeitschrift für Wirtschafts- und Sozialwissenschaften* 115 (1995):535–568.

Ainslie, George, "Specious Reward: A Behavioral Theory of Impulsiveness and Impulse Control," *Psychological Bulletin* 82 (July 1975):463–496.

Ainslie, George, and Nick Haslam, "Hyperbolic Discounting," in George Loewenstein and Jon Elster (eds.) *Choice over Time* (New York: Russell Sage, 1992) pp. 57–92.

Akerlof, George A., "Labor Contracts as Partial Gift Exchange," *Quarterly Journal of Economics* 97,4 (November 1982):543–569.

Alcock, John, *Animal Behavior: An Evolutionary Approach* (Sunderland, MA: Sinauer, 1993).

Alexander, R. D., *The Biology of Moral Systems* (New York: Aldine, 1987).

Allais, Maurice, "Le comportement de l'homme rationnel devant le risque, critique des postulats et axiomes de l'école Américaine," *Econometrica* 21 (1953):503–546.

Allman, J., A. Hakeem, and K. Watson, "Two Phylogenetic Specializations in the Human Brain," *Neuroscientist* 8 (2002):335–346.

Anderson, Christopher, and Louis Putterman, "Do Non-strategic Sanctions Obey the Law of Demand? The Demand for Punishment in the Voluntary Contribution Mechanism," *Games and Economic Behavior* 54,1 (2006):1–24.

Andreoni, James, "Cooperation in Public Goods Experiments: Kindness or Confusion," *American Economic Review* 85,4 (1995):891–904.

Andreoni, James, and John H. Miller, "Rational Cooperation in the Finitely Repeated Prisoner's Dilemma: Experimental Evidence," *Economic Journal* 103 (May 1993):570–585.

—, "Giving According to GARP: An Experimental Test of the Consistency of Preferences for Altruism," *Econometrica* 70,2 (2002):737–753.

Andreoni, James, Brian Erard, and Jonathan Feinstein, "Tax Compliance," *Journal of Economic Literature* 36,2 (June 1998):818–860.

Anscombe, F., and Robert J. Aumann, "A Definition of Subjective Probability," *Annals of Mathematical Statistics* 34 (1963):199–205.

Arkes, Hal R., and Peter Ayton, "The Sunk Cost and *Concorde* Effects: Are Humans Less Rational Than Lower Animals?" *Psychological Bulletin* 125,5 (1999):591–600.

Arrow, Kenneth J., "An Extension of the Basic Theorems of Classical Welfare Economics," in J. Neyman (ed.) *Proceedings of the Second Berkeley Symposium on Mathematical Statistics and Probability* (Berkeley: University of California Press, 1951) pp. 507–532.

—, "Political and Economic Evaluation of Social Effects and Externalities," in M. D. Intriligator (ed.) *Frontiers of Quantitative Economics* (Amsterdam: North Holland, 1971) pp. 3–23.

Arrow, Kenneth J., and Frank Hahn, *General Competitive Analysis* (San Francisco: Holden-Day, 1971).

Arrow, Kenneth J., and Gerard Debreu, "Existence of an Equilibrium for a Competitive Economy," *Econometrica* 22,3 (1954):265–290.

Ashraf, Nava, Dean S. Karlan, and Wesley Yin, "Tying Odysseus to the Mast: Evidence from a Commitment Savings Product in the Philippines," *Quarterly Journal of Economics* 121,2 (2006):635–672.

Aumann, Robert J., "Subjectivity and Correlation in Randomizing Strategies," *Journal of Mathematical Economics* 1 (1974):67–96.

Aumann, Robert J., "Agreeing to Disagree," *Annals of Statistics* 4,6 (1976):1236–1239.

—, "Correlated Equilibrium and an Expression of Bayesian Rationality," *Econometrica* 55 (1987):1–18.

—, "Backward Induction and Common Knowledge of Rationality," *Games and Economic Behavior* 8 (1995):6–19.

Aumann, Robert J., and Adam Brandenburger, "Epistemic Conditions for Nash Equilibrium," *Econometrica* 65,5 (September 1995):1161–1180.

Bakeman, Roger, and John R. Brownlee, "Social Rules Governing Object Conflicts in Toddlers and Preschoolers," in Kenneth H. Rubin and Hildy S. Ross (eds.) *Peer Relationships and Social Skills in Childhood* (New York: Springer-Verlag, 1982) pp. 99–112.

Bala, V., and M. Majumdar, "Chaotic Tatonnement," *Economic Theory* 2 (1992):437–445.

Bandura, Albert, *Social Learning Theory* (Englewood Cliffs, NJ: Prentice Hall, 1977).

Barkow, Jerome H., Leda Cosmides, and John Tooby, *The Adapted Mind: Evolutionary Psychology and the Generation of Culture* (New York: Oxford University Press, 1992).

Baron, James, *Thinking and Deciding* (Cambridge: Cambridge University Press, 2007).

Basu, Kaushik, "On the Non-Existence of a Rationality Definition for Extensive Games," *International journal of Game Theory* 19 (1990):33–44.

—, "The Traveler's Dilemma: Paradoxes of Rationality in Game Theory," *American Economic Review* 84,2 (May 1994):391–395.

Battigalli, Pierpallo, "On Rationalizability in Extensive Form Games," *Journal of Economic Theory* 74 (1997):40–61.

Becker, Gary S., *Accounting for Tastes* (Cambridge, MA: Harvard University Press, 1996).

Becker, Gary S., and Casey B. Mulligan, "The Endogenous Determination of Time Preference," *Quarterly Journal of Economics* 112,3 (August 1997):729–759.

Becker, Gary S., and Kevin M. Murphy, "A Theory of Rational Addiction," *Journal of Political Economy* 96,4 (August 1988):675–700.

Beer, J. S., E. A. Heerey, D. Keltner, D. Skabini, and R. T. Knight, "The Regulatory Function of Self-conscious Emotion: Insights from Patients with Orbitofrontal Damage," *Journal of Personality and Social Psychology* 65 (2003):594–604.

Ben-Porath, Elchanan, "Rationality, Nash Equilibrium and Backward Induction in Perfect-Information Games," *Review of Economic Studies* 64 (1997):23–46.

Ben-Porath, Elchanan, and Eddie Dekel, "Signaling Future Actions and the Potential for Self-sacrifice," *Journal of Economic Theory* 57 (1992):36–51.

Berg, Joyce, John Dickhaut, and Kevin McCabe, "Trust, Reciprocity, and Social History," *Games and Economic Behavior* 10 (1995):122–142.

Bernheim, B. Douglas, "Rationalizable Strategic Behavior," *Econometrica* 52,4 (July 1984):1007–1028.

Betch, T., and H. Haberstroh, *The Routines of Decision Making* (Mahwah, NJ: Lawrence Erlbaum Associates, 2005).

Betzig, Laura, "Delated Reciprocity and Tolerated Theft," *Current Anthropology* 37 (1997):49–78.

Bewley, Truman F., *Why Wages Don't Fall During a Recession* (Cambridge: Cambridge University Press, 2000).

Bhaskar, V., "Informational Constraints and the Overlapping Generations Model: Folk and Anti-Folk Theorems," *Review of Economic Studies* 65,1 (January 1998):135–149.

—, "Noisy Communication and the Evolution of Cooperation," *Journal of Economic Theory* 82,1 (September 1998):110–131.

—, "The Robustness of Repeated Game Equilibria to Incomplete Payoff Information," 2000. University of Essex.

Bhaskar, V., and Ichiro Obara, "Belief-Based Equilibria: The Repeated Prisoner's Dilemma with Private Monitoring," *Journal of Economic Theory* 102 (2002):40–69.

Bhaskar, V., George J. Mailath, and Stephen Morris, "Purification in the Infinitely Repeated Prisoner's Dilemma," 2004. University of Essex.

Bicchieri, Cristina, "Self-Refuting Theories of Strategic Interaction: A Paradox of Common Knowledge," *Erkenntniss* 30 (1989):69–85.

—, *The Grammar of Society: The Nature and Dynamics of Social Norms* (Cambridge: Cambridge University Press, 2006).

Binmore, Kenneth G., "Modeling Rational Players: I," *Economics and Philosophy* 3 (1987):179–214.

—, *Game Theory and the Social Contract: Playing Fair* (Cambridge, MA: MIT Press, 1993).

—, "A Note on Backward Induction," *Games and Economic Behavior* 18 (1996):135–137.

—, *Game Theory and the Social Contract: Just Playing* (Cambridge, MA: MIT Press, 1998).

—, *Natural Justice* (Oxford: Oxford University Press, 2005).

Binmore, Kenneth G., and Larry Samuelson, "The Evolution of Focal Points," *Games and Economic Behavior* 55,1 (April 2006):21–42.

Binswanger, Hans, "Risk Attitudes of Rural Households in Semi-Arid Tropical India," *American Journal of Agricultural Economics* 62,3 (1980):395–407.

Binswanger, Hans, and Donald Sillers, "Risk Aversion and Credit Constraints in Farmers' Decision-Making: A Reinterpretation," *Journal of Development Studies* 20,1 (1983):5–21.

Bishop, D. T., and C. Cannings, "The Generalised War of Attrition," *Advances in Applied Probability* 10,1 (March 1978):6–7.

Black, Fisher, and Myron Scholes, "The Pricing of Options and Corporate Liabilities," *Journal of Political Economy* 81 (1973):637–654.

Bliege Bird, Rebecca L., and Douglas W. Bird, "Delayed Reciprocity and Tolerated Theft," *Current Anthropology* 38 (1997):49–78.

Blount, Sally, "When Social Outcomes Aren't Fair: The Effect of Causal Attributions on Preferences," *Organizational Behavior & Human Decision Processes* 63,2 (August 1995):131–144.

Blume, Lawrence E., and William R. Zame, "The Algebraic Geometry of Perfect and Sequential Equilibrium," *Econometrica* 62 (1994):783–794.

Blurton Jones, Nicholas G., "Tolerated Theft: Suggestions about the Ecology and Evolution of Sharing, Hoarding, and Scrounging," *Social Science Information* 26,1 (1987):31–54.

Bochet, Olivier, Talbot Page, and Louis Putterman, "Communication and Punishment in Voluntary Contribution Experiments," *Journal of Economic Behavior and Organization* 60,1 (2006):11–26.

Boehm, Christopher, "The Evolutionary Development of Morality as an Effect of Dominance Behavior and Conflict Interference," *Journal of Social and Biological Structures* 5 (1982):413–421.

—, *Hierarchy in the Forest: The Evolution of Egalitarian Behavior* (Cambridge, MA: Harvard University Press, 2000).

Boles, Terry L., Rachel T. A. Croson, and J. Keith Murnighan, "Deception and Retribution in Repeated Ultimatum Bargaining," *Organizational Behavior and Human Decision Processes* 83,2 (2000):235–259.

Bolton, Gary E., and Rami Zwick, "Anonymity versus Punishment in Ultimatum Games," *Games and Economic Behavior* 10 (1995):95–121.

Boldon, Gary E., Elena Katok, and Rami Zwick, "Dictator Game Giving: Rules of Fairness versus Acts of Kindness," *International Journal of Game Theory* 27,2 (July 1998):269–299.

Bonner, John Tyler, *The Evolution of Culture in Animals* (Princeton, NJ: Princeton University Press, 1984).

Börgers, Tillman, "Weak Dominance and Approximate Common Knowledge," *Journal of Economic Theory* 64 (1994):265–276.

Bowles, Samuel, *Microeconomics: Behavior, Institutions, and Evolution* (Princeton: Princeton University Press, 2004).

—, "Policies Designed for Self-interested Citizens May Undermine "the Moral Sentiments": Evidence from Economic Experiments," *Science* 320,5883 (2008).

Bowles, Samuel, and Herbert Gintis, "The Revenge of Homo economicus: Contested Exchange and the Revival of Political Economy," *Journal of Economic Perspectives* 7,1 (Winter 1993):83–102.

—, "The Origins of Human Cooperation," in Peter Hammerstein (ed.) *Genetic and Cultural Origins of Cooperation* (Cambridge, MA: MIT Press, 2004).

Boyd, Robert, and Peter J. Richerson, *Culture and the Evolutionary Process* (Chicago: University of Chicago Press, 1985).

—, "The Evolution of Reciprocity in Sizable Groups," *Journal of Theoretical Biology* 132 (1988):337–356.

Brosig, J., A. Ockenfels, and J. Weimann, "The Effect of Communication Media on Cooperation," *German Economic Review* 4 (2003):217–242.

Brown, J. H., and M. V. Lomolino, *Biogeography* (Sunderland, MA: Sinauer, 1998).

Burks, Stephen V., Jeffrey P. Carpenter, and Eric Verhoogen, "Playing Both Roles in the Trust Game," *Journal of Economic Behavior and Organization* 51 (2003):195–216.

Burrows, Anne M., "The Facial Expression Musculature in Primates and its Evolutionary Significance," *BioEssays* 30,3 (2008):212–225.

Bush, R. R., and F. Mosteller, *Stochastic Models for Learning* (New York: John Wiley & Sons, 1955).

Cabrales, Antonio, Rosemarie Nagel, and Roc Armenter, "Equilibrium Selection Through Incomplete Information in Coordination Games: An Experimental Study," *Experimental Economics* 10,3 (September 2007):221–234.

Camerer, Colin, "Prospect Theory in the Wild: Evidence from the Field," in Daniel Kahneman and Amos Tversky (eds.) *Choices, Values, and Frames* (Cambridge: Cambridge University Press, 2000) pp. 288–300.

—, *Behavioral Game Theory: Experiments in Strategic Interaction* (Princeton, NJ: Princeton University Press, 2003).

Camerer, Colin, and Richard Thaler, "Ultimatums, Dictators, and Manners," *Journal of Economic Perspectives* 9,2 (1995):209–219.

Camille, N., "The Involvement of the Orbitofrontal Cortex in the Experience of Regret," *Science* 304 (2004):1167–1170.

Campbell, Bernard G., James D. Loy, and Katherine Cruz-Uribe, *Humankind Emerging* (New York: Allyn and Bacon, 2005).

Carlsson, Hans, and Eric van Damme, "Global Games and Equilibrium Selection," *Econometrica* 61,5 (September 1993):989–1018.

Carpenter, Jeffrey, and Peter Matthews, "Norm Enforcement: Anger, Indignation, or Reciprocity," 2005. Department of Economics, Middlebury College, Working Paper 0503.

Carpenter, Jeffrey, Samuel Bowles, Herbert Gintis, and Sung Ha Hwang, "Strong Reciprocity and Team Production," 2009. Journal of Economic Behavior and Organization.

Casari, Marco, and Luigi Luini, "Group Cooperation Under Alternative Peer Punishment Technologies: An Experiment," 2007. Department of Economics, University of Siena.

Cavalli-Sforza, L., and M. W. Feldman, "Models for Cultural Inheritance: Within Group Variation," *Theoretical Population Biology* 42,4 (1973):42–55.

Cavalli-Sforza, Luca L., and Marcus W. Feldman, "Theory and Observation in Cultural Transmission," *Science* 218 (1982):19–27.

Cavalli-Sforza, Luigi L., and Marcus W. Feldman, *Cultural Transmission and Evolution* (Princeton, NJ: Princeton University Press, 1981).

Chaitin, Gregory, *Algorithmic Information Theory* (Cambridge: Cambridge University Press, 2004).

Charness, Gary, and Ernan Haruvy, "Altruism, Equity, and Reciprocity in a Gift-Exchange Experiment: An Encompassing Approach," *Games and Economic Behavior* 40 (2002):203–231.

Charness, Gary, and Martin Dufwenberg, "Promises and Partnership," October 2004. University of California at Santa Barbara.

Cho, In-Koo, and David M. Kreps, "Signalling Games and Stable Equilibria," *Quarterly Journal of Economics* 102,2 (May 1987):180–221.

Chow, Timothy Y., "The Surprise Examination or Unexpected Hanging Paradox," *American Mathematical Monthly* 105 (1998):41–51.

Chung, Shin-Ho, and Richard J. Herrnstein, "Choice and Delay of Reinforcement," *Journal of Experimental Analysis of Behavior* 10,1 (1967):67–74.

Cinyabuguma, Matthias, Talbott Page, and Louis Putterman, "On Perverse and Second-Order Punishment in Public Goods Experiments with Decentralized Sanctions," 2004. Department of Economics, Brown University.

Clements, Kevin C., and David W. Stephens, "Testing Models of Non-kin Cooperation: Mutualism and the Prisoner's Dilemma," *Animal Behaviour* 50 (1995):527–535.

Collins, John, "How We Can Agree to Disagree," 1997. Department of Philosophy, Columbia University.

Conte, Rosaria, and Cristiano Castelfranchi, "From Conventions to Prescriptions. Towards an Integrated View of Norms," *Artificial Intelligence and Law* 7 (1999):323–340.

Cooper, W. S., "Decision Theory as a Branch of Evolutionary Theory," *Psychological Review* 4 (1987):395–411.

Cosmides, Leda, and John Tooby, "Cognitive Adaptations for Social Exchange," in Jerome H. Barkow, Leda Cosmides, and John Tooby (eds.) *The Adapted Mind: Evolutionary Psychology and the Generation of Culture* (New York: Oxford University Press, 1992 pp. 163–228).

Cox, James C., "How to Identify Trust and Reciprocity," *Games and Economic Behavior* 46 (2004):260–281.

Cubitt, Robin P., and Robert Sugden, "Common Knowledge, Salience and Convention: A Reconstruction of David Lewis' Game Theory," *Economics and Philosophy* 19 (2003):175–210.

Damasio, Antonio R., *Descartes' Error: Emotion, Reason, and the Human Brain* (New York: Avon Books, 1994).

Dana, Justin, Daylian M. Cain, and Robyn M. Dawes, "What You Don't Know Won't Hurt Me: Costly (But Quiet) Exit in Dictator Games," *Organizational Behavior and Human Decision Processes* 100 (2006):193–201.

Darwin, Charles, *The Origin of Species by Means of Natural Selection* 6th Edition (London: John Murray, 1872).

—, *The Expression of Emotions in Man and Animals* Paul Eckman (ed.) (Oxford: Oxford University Press, 1998).

Davies, N. B., "Territorial Defence in the Speckled Wood Butterfly (*Pararge Aegeria*): The Resident Always Wins," *Animal Behaviour* 26 (1978):138–147.

Davis, Douglas D., and Charles A. Holt, *Experimental Economics* (Princeton, NJ: Princeton University Press, 1993).

Dawes, R. M., A. J. C Van de Kragt, and J. M. Orbell, "Not me or Thee but We: The Importance of Group Identity in Eliciting Cooperation in Dilemma Situations: Experimental Manipulations," *Acta Psychologica* 68 (1988):83–97.

Dawkins, Richard, *The Selfish Gene* (Oxford: Oxford University Press, 1976).

—, *The Extended Phenotype: The Gene as the Unit of Selection* (Oxford: Freeman, 1982).

—, *The Selfish Gene*, 2nd Edition (Oxford: Oxford University Press, 1989).

Dawkins, Richard., and H. J. Brockmann, "Do Digger Wasps Commit the *Concorde* Fallacy?" *Animal Behaviour* 28 (1980):892–896.

de Laplace, Marquis, *A Philosophical Essay on Probabilities* (New York: Dover, 1996).

Debreu, Gérard, *Theory of Value* (New York: John Wiley & Sons, 1959).

Debreu, Gerard, "Excess Demand Function," *Journal of Mathematical Economics* 1 (1974):15–23.

Dekel, Eddie, and Faruk Gul, "Rationality and Knowledge in Game Theory," in David M. Kreps and K. F. Wallis (eds.) *Advances in Economics and Econometrics*, Vol. I (Cambridge: Cambridge University Press, 1997) pp. 87–172.

Denant-Boemont, Laurent, David Masclet, and Charles Noussair, "Punishment, Counterpunishment and Sanction Enforcement in a Social Dilemma Experiment," *Economic Theory* 33,1 (October 2007):145–167.

deQuervain, Dominique J.-F., Urs Fischbacher, Valerie Treyer, Melanie Schellhammer, Ulrich Schnyder, Alfred Buck, and Ernst Fehr, "The Neural Basis of Altruistic Punishment," *Science* 305 (27 August 2004):1254–1258.

di Finetti, Benedetto, *Theory of Probability* (Chichester: John Wiley & Sons, 1974).

Dunbar, R. I. M., "Coevolution of Neocortical Size, Group Size and Language in Humans," *Behavioral and Brain Sciences* 16,4 (1993):681–735.

Durkheim, Emile, *The Division of Labor in Society* (New York: The Free Press, 1933 [1902]).

Eason, P. K., G. A. Cobbs, and K. G. Trinca, "The Use of Landmarks to Define Territorial Boundaries," *Animal Behaviour* 58 (1999):85–91.

Easterlin, Richard A., "Does Economic Growth Improve the Human Lot? Some Empirical Evidence," in *Nations and Households in Economic Growth: Essays in Honor of Moses Abramovitz* (New York: Academic Press, 1974).

—, "Will Raising the Incomes of All Increase the Happiness of All?" *Journal of Economic Behavior and Organization* 27,1 (June 1995):35–47.

Edgerton, Robert B., *Sick Societies: Challenging the Myth of Primitive Harmony* (New York: The Free Press, 1992).

Edgeworth, Francis Ysidro, *Papers Relating to Political Economy I* (London: Macmillan, 1925).

Ellis, Lee, "On the Rudiments of Possessions and Property," *Social Science Information* 24,1 (1985):113–143.

Ellsberg, Daniel, "Risk, Ambiguity, and the Savage Axioms," *Quarterly Journal of Economics* 75 (1961):643–649.

Elster, Jon, *The Cement of Society* (Cambridge: Cambridge University Press, 1989).

—, "Social Norms and Economic Theory," *Journal of Economic Perspectives* 3,4 (1989):99–117.

Ely, Jeffrey C., and Juuso Välimäki, "A Robust Folk Theorem for the Prisoner's Dilemma," *Journal of Economic Theory* 102 (2002):84–105.

Ertan, Arhan, Talbot Page, and Louis Putterman, "Can Endogenously Chosen Institutions Mitigate the Free-Rider Problem and Reduce Perverse Punishments?" 2005. Working Paper 2005-13, Department of Economics, Brown University.

Eshel, Ilan, and Marcus W. Feldman, "Initial Increase of New Mutants and Some Continuity Properties of ESS in two Locus Systems," *American Naturalist* 124 (1984):631–640.

Eshel, Ilan, Marcus W. Feldman, and Aviv Bergman, "Long-term Evolution, Short-term Evolution, and Population Genetic Theory," *Journal of Theoretical Biology* 191 (1998):391–396.

Fagin, Ronald, Joseph Y. Halpern, Yoram Moses, and Moshe Y. Vardi, *Reasoning about Knowledge* (Cambridge, MA: MIT Press, 1995).

Fehr, Ernst, and Klaus M. Schmidt, "A Theory of Fairness, Competition, and Cooperation," *Quarterly Journal of Economics* 114 (August 1999):817–868.

Fehr, Ernst, and Lorenz Goette, "Do Workers Work More If Wages Are High? Evidence from a Randomized Field Experiment," *American Economic Review* 97,1 (March 2007):298–317.

Fehr, Ernst, and Peter Zych, "The Power of Temptation: Irrationally Myopic Excess Consumption in an Addiction Experiment," September 1994. University of Zurich.

Fehr, Ernst, and Simon Gächter, "How Effective Are Trust- and Reciprocity-Based Incentives?" in Louis Putterman and Avner Ben-Ner (eds.) *Economics, Values and Organizations* (New York: Cambridge University Press, 1998) pp. 337–363.

—, "Cooperation and Punishment," *American Economic Review* 90,4 (September 2000):980–994.

—, "Altruistic Punishment in Humans," *Nature* 415 (10 January 2002):137–140.

Fehr, Ernst, and Urs Fischbacher, "Third Party Punishment and Social Norms," *Evolution & Human Behavior* 25 (2004):63–87.

Fehr, Ernst, Georg Kirchsteiger, and Arno Riedl, "Does Fairness Prevent Market Clearing?" *Quarterly Journal of Economics* 108,2 (1993):437–459.

—, "Gift Exchange and Reciprocity in Competitive Experimental Markets," *European Economic Review* 42,1 (1998):1–34.

Fehr, Ernst, Simon Gächter, and Georg Kirchsteiger, "Reciprocity as a Contract Enforcement Device: Experimental Evidence," *Econometrica* 65,4 (July 1997):833–860.

Feldman, Marcus W., and Lev A. Zhivotovsky, "Gene-Culture Coevolution: Toward a General Theory of Vertical Transmission," *Proceedings of the National Academy of Sciences* 89 (December 1992):11935–11938.

Feller, William, *An Introduction to Probability Theory and Its Applications* Vol. 1 (New York: John Wiley & Sons, 1950).

Fisher, Franklin M., *Disequilibrium Foundations of Equilibrium Economics* (Cambridge, UK: Cambridge University Press, 1983).

Fisher, Ronald A., *The Genetical Theory of Natural Selection* (Oxford: Clarendon Press, 1930).

Fong, Christina M., Samuel Bowles, and Herbert Gintis, "Reciprocity and the Welfare State," in Herbert Gintis, Samuel Bowles, Robert Boyd, and Ernst Fehr (eds.) *Moral Sentiments and Material Interests: On the Foundations of Cooperation in Economic Life* (Cambridge, MA: MIT Press, 2005).

Forsythe, Robert, Joel Horowitz, N. E. Savin, and Martin Sefton, "Replicability, Fairness and Pay in Experiments with Simple Bargaining Games," *Games and Economic Behavior* 6,3 (May 1994):347–369.

Frederick, S., George F. Loewenstein, and T. O'Donoghue, "Time Discounting: A Critical Review," *Journal of Economic Literature* 40 (2002):351–401.

Friedman, Milton, and Leonard J. Savage, "The Utility Analysis of Choices Involving Risk," *Journal of Political Economy* 56 (1948):279–304.

Fudenberg, Drew, and Eric Maskin, "The Folk Theorem in Repeated Games with Discounting or with Incomplete Information," *Econometrica* 54,3 (May 1986):533–554.

Fudenberg, Drew, and Jean Tirole, "Perfect Bayesian Equilibrium and Sequential Equilibrium," *journal of Economic Theory* 53 (1991):236–260.

Fudenberg, Drew, David K. Levine, and Eric Maskin, "The Folk Theorem with Imperfect Public Information," *Econometrica* 62 (1994):997–1039.

Fudenberg, Drew, David M. Kreps, and David Levine, "On the Robustness of Equilibrium Refinements," *Journal of Economic Theory* 44 (1988):354–380.

Furby, Lita, "The Origins and Early Development of Possessive Behavior," *Political Psychology* 2,1 (1980):30–42.

Gächter, Simon, and Ernst Fehr, "Collective Action as a Social Exchange," *Journal of Economic Behavior and Organization* 39,4 (July 1999):341–369.

Gadagkar, Raghavendra, "On Testing the Role of Genetic Asymmetries Created by Haplodiploidy in the Evolution of Eusociality in the Hymenoptera," *Journal of Genetics* 70,1 (April 1991):1–31.

Gauthier, David, *Morals by Agreement* (Oxford: Clarendon Press, 1986).

Genesove, David, and Christopher Mayer, "Loss Aversion and Seller Behavior: Evidence from the Housing Market," *Quarterly Journal of Economics* 116,4 (November 2001):1233–1260.

Ghiselin, Michael T., *The Economy of Nature and the Evolution of Sex* (Berkeley: University of California Press, 1974).

Gigerenzer, Gerd, and P. M. Todd, *Simple Heuristics That Make Us Smart* (New York: Oxford University Press, 1999).

Gigerenzer, Gerd, and Reinhard Selten, *Bounded Rationality* (Cambridge, MA: MIT Press, 2001).

Gillies, Donald, *Philosophical Theories of Probability* (London: Routledge, 2000).

Gintis, Herbert, "Consumer Behavior and the Concept of Sovereignty," *American Economic Review* 62,2 (May 1972):267–278.

—, "A Radical Analysis of Welfare Economics and Individual Development," *Quarterly Journal of Economics* 86,4 (November 1972):572–599.

—, "Welfare Criteria with Endogenous Preferences: The Economics of Education," *International Economic Review* 15,2 (June 1974):415–429.

—, "Welfare Economics and Individual Development: A Reply to Talcott Parsons," *Quarterly Journal of Economics* 89,2 (February 1975):291–302.

—, "The Nature of the Labor Exchange and the Theory of Capitalist Production," *Review of Radical Political Economics* 8,2 (Summer 1976):36–54.

—, "Some Implications of Endogenous Contract Enforcement for General Equilibrium Theory," in Fabio Petri and Frank Hahn (eds.) *General Equilibrium: Problems and Prospects* (London: Routledge, 2002) pp. 176–205.

—, "The Hitchhiker's Guide to Altruism: Genes, Culture, and the Internalization of Norms," *Journal of Theoretical Biology* 220,4 (2003):407–418.

—, "Solving the Puzzle of Human Prosociality," *Rationality and Society* 15,2 (May 2003):155–187.

—, "Behavioral Game Theory and Contemporary Economic Theory," *Analyze & Kritik* 27,1 (2005):48–72.

—, "The Emergence of a Price System from Decentralized Bilateral Exchange," *Contributions to Theoretical Economics* 6,1,13 (2006). Available at www.bepress.com/bejte/contributions/vol6/iss1/art13.

—, "The Dynamics of General Equilibrium," *Economic Journal* 117 (October 2007):1289–1309.

—, "The Evolution of Private Property," *Journal of Economic Behavior and Organization* 64,1 (September 2007):1–16.

—, "A Framework for the Unification of the Behavioral Sciences," *Behavioral and Brain Sciences* 30,1 (2007):1–61.

—, *Game Theory Evolving* 2nd Edition, (Princeton, NJ: Princeton University Press, 2009).

Gintis, Herbert, Joseph Henrich, Samuel Bowles, Robert Boyd, and Ernst Fehr, "Strong Reciprocity and the Roots of Human Morality," *Social Justice Research* (2009).

Gintis, Herbert, Samuel Bowles, Robert Boyd, and Ernst Fehr, *Moral Sentiments and Material Interests: On the Foundations of Cooperation in Economic Life* (Cambridge: MIT Press, 2005).

Glaeser, Edward, David Laibson, Jose A. Scheinkman, and Christine L. Soutter, "Measuring Trust," *Quarterly Journal of Economics* 65 (2000):622–846.

Glimcher, Paul W., *Decisions, Uncertainty, and the Brain: The Science of Neuroeconomics* (Cambridge, MA: MIT Press, 2003).

Glimcher, Edward, and Aldo Rustichini, "Neuroeconomics: The Consilience of Brain and Decision," *Science* 306 (15 October 2004):447–452.

Glimcher, Edward, Michael C. Dorris, and Hannah M. Bayer, "Physiological Utility Theory and the Neuroeconomics of Choice," 2005. Center for Neural Science, New York University.

Gneezy, Uri, "Deception: The Role of Consequences," *American Economic Review* 95,1 (March 2005):384–394.

Gneezy, Uri, and Aldo Rustichini, "A Fine Is a Price," *Journal of Legal Studies* 29 (2000):1–17.

Goffman, Erving, *The Presentation of Self in Everyday Life* (New York: Anchor, 1959).

Govindan, Srihari, Phillip J. Reny, and Arthur J. Robson, "A Short Proof of Harsanyi's PUrification Theorem," 2003. University of Western Ontario and University of Chicago.

Grafen, Alan, "The Logic of Divisively Asymmetric Contests: Respect for Ownership and the Desperado Effect," *Animal Behavior* 35 (1987):462–467.

—, "Formal Darwinism, the Individual-as-Maximizing-Agent: Analogy, and Bet-hedging," *Proceedings of the Royal Society of London B* 266 (1999):799–803.

—, "Developments of Price's Equation and Natural Selection Under Uncertainty," *Proceedings of the Royal Society of London B* 267 (2000):1223–1227.

—, "A First Formal Link between the Price Equation and an Optimization Program," *Journal of Theoretical Biology* 217 (2002):75–91.

Green, Leonard, Joel Myerson, Daniel D. Holt, John R. Slevin, and Sara J. Estle, "Discounting of Delayed Food Rewards in Pigeons and Rats: Is There a Magnitude Effect?" *Journal of the Experimental Analysis of Behavior* 81 (2004):31–50.

Greenberg, M. S., and D. M. Frisch, "Effect of Intentionality on Willingness to Reciprocate a Favor," *Journal of Experimental Social Psychology* 8 (1972):99–111.

Grether, David, and Charles Plott, "Economic Theory of Choice and the Preference Reversal Phenomenon," *American Economic Review* 69,4 (September 1979):623–638.

Gruber, J., and B. Köszegi, "Is Addiction Rational? Theory and Evidence," *Quarterly Journal of Economics* 116,4 (2001):1261–1305.

Grusec, Joan E., and Leon Kuczynski, *Parenting and Children's Internalization of Values: A Handbook of Contemporary Theory* (New York: John Wiley & Sons, 1997).

Gul, Faruk, "A Comment on Aumann's Bayesian View," *Econometrica* 66,4 (1998):923–928.

Gunnthorsdottir, Anna, Kevin McCabe, and Vernon Smith, "Using the Machiavellianism Instrument to Predict Trustworthiness in a Bargaining Game," *Journal of Economic Psychology* 23 (2002):49–66.

Güth, Werner, and Reinhard Tietz, "Ultimatum Bargaining Behavior: A Survey and Comparison of Experimental Results," *Journal of Economic Psychology* 11 (1990):417–449.

Güth, Werner, R. Schmittberger, and B. Schwarze, "An Experimental Analysis of Ultimatum Bargaining," *Journal of Economic Behavior and Organization* 3 (May 1982):367–388.

Guzman, R. A., Carlos Rodriguez Sickert, and Robert Rowthorn, "When in Rome Do as the Romans Do: The Coevolution of Altruistic Punishment, Conformist Learning, and Cooperation," *Evolution and Human Behavior* 28 (2007):112–117.

Haldane, J. B. S., *The Causes of Evolution* (London: Longmans, Green & Co., 1932).

Hamilton, William D., "The Genetical Evolution of Social Behavior, I & II," *Journal of Theoretical Biology* 7 (1964):1–16,17–52.

Hammerstein, Peter, "Darwinian Adaptation, Population Genetics and the Streetcar Theory of Evolution," *Journal of Mathematical Biology* 34 (1996):511–532.

—, "Why Is Reciprocity So Rare in Social Animals?" in Peter Hammerstein (ed.) *Genetic and Cultural Evolution of Cooperation* (Cambridge, MA: MIT Press, 2003) pp. 83–93.

Hammerstein, Peter, and Reinhard Selten, "Game Theory and Evolutionary Biology," in Robert J. Aumann and Sergiu Hart (eds.) *Handbook of Game Theory with Economic Applications* (Amsterdam: Elsevier, 1994) pp. 929–993.

Harsanyi, John C., "Games with Incomplete Information Played by Bayesian Players, Parts I, II, and III," *Behavioral Science* 14 (1967):159–182, 320–334, 486–502.

—, "Games with Randomly Disturbed Payoffs: A New Rationale for Mixed-Strategy Equilibrium Points," *International Journal of Game Theory* 2 (1973):1–23.

Harsanyi, John C., and Reinhard Selten, *A General Theory of Equilibrium Selection in Games* (Cambridge, MA: MIT Press, 1988).

Hauser, Marc, *Wild Minds* (New York: Henry Holt, 2000).

Hawkes, Kristen, "Why Hunter-Gatherers Work: An Ancient Version of the Problem of Public Goods," *Current Anthropology* 34,4 (1993):341–361.

Hayashi, N., E. Ostrom, J. Walker, and T. Yamagishi, "Reciprocity, Trust, and the Sense of Control: A Cross-societal Study," *Rationality and Society* 11 (1999):27–46.

Heinemann, Frank, Rosemarie Nagel, and Peter Ockenfels, "The Theory of Global Games on Test: Experimental Analysis of Coordination Games with Public and Private Information," *Econometrica* 72,5 (September 2004):1583–1599.

Heiner, Ronald A., "The Origin of Predictable Behavior," *American Economic Review* 73,4 (1983):560–595.

Helson, Harry, *Adaptation Level Theory: An Experimental and Systematic Approach to Behavior* (New York: Harper and Row, 1964).

Henrich, Joseph, and Francisco Gil-White, "The Evolution of Prestige: Freely Conferred Status as a Mechanism for Enhancing the Benefits of Cultural Transmission," *Evolution and Human Behavior* 22 (2001):165–196.

Henrich, Joseph, and Robert Boyd, "The Evolution of Conformist Transmission and the Emergence of Between-Group Differences," *Evolution and Human Behavior* 19 (1998):215–242.

—, "Why People Punish Defectors: Weak Conformist Transmission Can Stabilize Costly Enforcement of Norms in Cooperative Dilemmas," *Journal of Theoretical Biology* 208 (2001):79–89.

Henrich, Joseph, Richard McElreath, Abigail Barr, Jean Ensminger, Clark Barrett, Alexander Bolyanatz, Juan Camilo Cardenas, Michael Gurven, Edwins Gwako, Natalie Henrich, Carolyn Lesorogol, Frank Marlowe, David Tracer, , and John Ziker, "Costly Punishment Across Human Societies," *Science* 312 (2006):1767–1770.

Henrich, Joseph, Robert Boyd, Samuel Bowles, Colin Camerer, Ernst Fehr, and Herbert Gintis, *Foundations of Human Sociality: Economic Experiments and Ethnographic Evidence from Fifteen Small-scale Societies* (Oxford: Oxford University Press, 2004).

Herrmann, Benedikt, Christian Thöni, and Simon Gächter, "Anti-social Punishment Across Societies," *Science* 319 (7 March 2008):1362–1367.

Herrnstein, Richard J., and Drazen Prelec, "A Theory of Addiction," in George Loewenstein and Jon Elster (eds.) *Choice over Time* (New York: Russell Sage, 1992) pp. 331–360.

Herrnstein, Richard J., David Laibson, and Howard Rachlin, *The Matching Law: Papers on Psychology and Economics* (Cambridge, MA: Harvard University Press, 1997).

Hinton, Geoffrey, and Terrence J. Sejnowski, *Unsupervised Learning: Fundation of Neural Computation* (Cambridge, MA: MIT Press, 1999).

Hirshleifer, Jack, "The Analytics of Continuing Conflict," *Synthése* 76 (1988):201–233.

Hobbes, Thomas, *Leviathan* (New York: Penguin, 1968[1651]). Edited by C. B. MacPherson.

Holden, C. J., "Bantu Language Trees Reflect the Spread of Farming Across Sub-Saharan Africa: A Maximum-parsimony Analysis," *Proceedings of the Royal Society of London B* 269 (2002):793–799.

Holden, C. J., and Ruth Mace, "Spread of Cattle Led to the Loss of Matrilineal Descent in Africa: A Coevolutionary Analysis," *Proceedings of the Royal Society of London B* 270 (2003):2425–2433.

Holland, John H., *Adaptation in Natural and Artificial Systems* (Ann Arbor: University of Michigan Press, 1975).

Holt, Charles A., *Industrial Organization: A Survey of Laboratory Research* (Princeton, NJ: Princeton University Press, 1995).

Holt, Charles A., Loren Langan, and Anne Villamil, "Market Power in an Oral Double Auction," *Economic Inquiry* 24 (1986):107–123.

Hörner, Johannes, and Wojciech Olszewski, "The Folk Theorem for Games with Private Almost-Perfect Monitoring," *Econometrica* 74,6 (2006):1499–1545.

Huxley, Julian S., "Evolution, Cultural and Biological," *Yearbook of Anthropology* (1955):2–25.

Jablonka, Eva, and Marion J. Lamb, *Epigenetic Inheritance and Evolution: The Lamarckian Case* (Oxford: Oxford University Press, 1995).

James, William, "Great Men, Great Thoughts, and the Environment," *Atlantic Monthly* 46 (1880):441–459.

Jaynes, E. T., *Probability Theory: The Logic of Science* (Cambridge: Cambridge University Press, 2003).

Jones, Owen D., "Time-Shifted Rationality and the Law of Law's Leverage: Behavioral Economics Meets Behavioral Biology," *Northwestern University Law Review* 95 (2001):1141–1206.

Jurmain, Robert, Harry Nelson, Lynn Kilgore, and Wenda Travathan, *Introduction to Physical Anthropology* (Cincinatti: Wadsworth Publishing Company, 1997).

Juslin, P., and H. Montgomery, *Judgment and Decision Making: New-Burswikian and Process-Tracing Approaches* (Hillsdale, NJ: Lawrence Erlbaum Associates, 1999).

Kachelmaier, S. J., and M. Shehata, "Culture and Competition: A Laboratory Market Comparison between China and the West," *Journal of Economic Behavior and Organization* 19 (1992):145–168.

Kagel, John H., and Alvin E. Roth, *Handbook of Experimental Economics* (Princeton, NJ: Princeton University Press, 1995).

Kagel, John H., Raymond C. Battalio, and Leonard Green, *Economic Choice Theory: An Experimental Analysis of Animal Behavior* (Cambridge: Cambridge University Press, 1995).

Kahneman, Daniel, and Amos Tversky, *Choices, Values, and Frames* (Cambridge: Cambridge University Press, 2000).

Kahneman, Daniel, Jack L. Knetsch, and Richard H. Thaler, "Experimental Tests of the Endowment Effect and the Coase Theorem," *Journal of Political Economy* 98,6 (December 1990):1325–1348.

—, "The Endowment Effect, Loss Aversion, and Status Quo Bias," *Journal of Economic Perspectives* 5,1 (Winter 1991):193–206.

Kahneman, Daniel, Paul Slovic, and Amos Tversky, *Judgment under Uncertainty: Heuristics and Biases* (Cambridge, UK: Cambridge University Press, 1982).

Karlan, Dean, "Using Experimental Economics to Measure Social Capital and Predict Real Financial Decisions," *American Economic Review* 95,5 (December 2005):1688–1699.

Keynes, John Maynard, *A Treatise on Probability* (New York: Dover, 2004).

Kirby, Kris N., and Richard J. Herrnstein, "Preference Reversals Due to Myopic Discounting of Delayed Reward," *Psychological Science* 6,2 (March 1995):83–89.

Kirman, Alan P., and K. J. Koch, "Market Excess Demand in Exchange Economies with Identical Preferences and Collinear Endowments," *Review of Economic Studies* LIII (1986):457–463.

Kiyonari, Toko, Shigehito Tanida, and Toshio Yamagishi, "Social Exchange and Reciprocity: Confusion or a Heuristic?," *Evolution and Human Behavior* 21 (2000):411–427.

Koehler, D., and N. Harvey, *Blackwell Handbook of Judgment and Decision Making* (New York: Blackwell, 2004).

Kohlberg, Elon, and Jean-Françis Mertens, "On the Strategic Stability of Equilibria," *Econometrica* 54,5 (September 1986):1003–1037.

Kolmogorov, A. N., *Foundations of the Theory of Probability* (New York: Chelsea, 1950).

Konow, James, and Joseph Earley, "The Hedonistic Paradox: Is Homo Economicus Happier?" *Journal of Public Economics* 92 (2008):1–33.

Koopmans, Tjalling, "Allocation of Resources and the Price System," in *Three Essays on the State of Economic Science* (New York: McGraw-Hill, 1957) pp. 4–95.

Krantz, D. H., "From Indices to Mappings: The Representational Approach to Measurement," in D. Brown and J. Smith (eds.) *Frontiers of Mathematical Psychology* (Cambridge: Cambridge University Press, 1991) pp. 1–52.

Krebs, J. R., and N. B. Davies, *Behavioral Ecology: An Evolutionary Approach*, 4th Edition, (Oxford: Blackwell Science, 1997).

Kreps, David M., *Notes on the Theory of Choice* (London: Westview, 1988).

Kreps, David M., and Robert Wilson, "Sequential Equilibria," *Econometrica* 50,4 (July 1982):863–894.

Kummer, Hans, and Marina Cords, "Cues of Ownership in Long-tailed Macaques, *Macaca fascicularis*," *Animal Behavior* 42 (1991):529–549.

Kurz, Mordecai, "Endogenous Economic Fluctuations and Rational Beliefs: A General Perspective," in Mordecai Kurz (ed.) *Endogenous Economic Fluctuations: Studies in the Theory of Rational Beliefs* (Berlin: Springer-Verlag, 1997) pp. 1–37.

Laibson, David, "Golden Eggs and Hyperbolic Discounting," *Quarterly Journal of Economics* 112,2 (May 1997):443–477.

Lane, Robert E., *The Market Experience* (Cambridge: Cambridge University Press, 1991).

Lane, Robert E., "Does Money Buy Happiness?" *The Public Interest* 113 (Fall 1993):56–65.

Ledyard, J. O., "Public Goods: A Survey of Experimental Research," in John H. Kagel and Alvin E. Roth (eds.) *The Handbook of Experimental Economics* (Princeton, NJ: Princeton University Press, 1995) pp. 111–194.

Lehmann, Laurent, and Laurent Keller, "The Evolution of Cooperation and Altruism—A General Framework and a Classification of Models," *Journal of Evolutionary Biology* 19 (2006):1365–1376.

Lehmann, Laurent, F. Rousset, D. Roze, and Laurent Keller, "Strong Reciprocity or Strong Ferocity? A Population Genetic View of the Evolution of Altruistic Punishment," *American Naturalist* 170,1 (July 2007):21–36.

Lerner, Abba, "The Economics and Politics of Consumer Sovereignty," *American Economic Review* 62,2 (May 1972):258–266.

Levine, David K., "Modeling Altruism and Spitefulness in Experiments," *Review of Economic Dynamics* 1,3 (1998):593–622.

Levy, Haim, "First Degree Stochastic Dominance Violations: Decision-Weights and Bounded Rationality," *Economic Journal* 118 (April 2008):759–774.

Lewis, David, *Conventions: A Philosophical Study* (Cambridge, MA: Harvard University Press, 1969).

Lewontin, Richard C., *The Genetic Basis of Evolutionary Change* (New York: Columbia University Press, 1974).

Liberman, Uri, "External Stability and ESS Criteria for Initial Increase of a New Mutant Allele," *Journal of Mathematical Biology* 26 (1988):477–485.

Lichtenstein, Sarah, and Paul Slovic, "Reversals of Preferences Between Bids and Choices in Gambling Decisions," *Journal of Experimental Psychology* 89 (1971):46–55.

Loewenstein, George, "Anticipation and the Valuation of Delayed Consumption," *Economic Journal* 97 (1987):666–684.

Loewenstein, George, and Daniel Adler, "A Bias in the Prediction of Tastes," *Economic Journal* 105 (431) (July 1995):929–937.

Loewenstein, George, and Drazen Prelec, "Anomalies in Intertemporal Choice: Evidence and an Interpretation," *Quarterly Journal of Economics* 57 (May 1992):573–598.

Loewenstein, George, and Nachum Sicherman, "Do Workers Prefer Increasing Wage Profiles?" *Journal of Labor Economics* 91,1 (1991):67–84.

Loewenstein, George F., Leigh Thompson, and Max H. Bazerman, "Social Utility and Decision Making in Interpersonal Contexts," *Journal of Personality and Social Psychology* 57,3 (1989):426–441.

Loomes, Graham, "When Actions Speak Louder than Prospects," *American Economic Review* 78,3 (June 1988):463–470.

Lorini, Emiliano, Luca Tummolini, and Andreas Herzig, "Establishing Mutual Beliefs by Joint Attention: Towards and Formal Model of Public Events," 2005. Institute of Cognitive Sciences, Rome.

Lucas, Robert, *Studies in Business Cycle Theory* (Cambridge, MA: MIT Press, 1981).

Mace, Ruth, and Mark Pagel, "The Comparative Method in Anthropology," *Current Anthropology* 35 (1994):549–564.

Machina, Mark J., "Choice Under Uncertainty: Problems Solved and Unsolved," *Journal of Economic Perspectives* 1,1 (Summer 1987):121–154.

Mailath, George J., and Stephen Morris, "Coordination Failure in Repeated Games with Almost-public Monitoring," *Theoretical Economics* 1 (2006):311–340.

Mandeville, Bernard, *The Fable of the Bees: Private Vices, Publick Benefits* (Oxford: Clarendon, 1924 [1705]).

Mas-Colell, Andreu, Michael D. Whinston, and Jerry R. Green, *Microeconomic Theory* (New York: Oxford University Press, 1995).

Masclet, David, Charles Noussair, Steven Tucker, and Marie-Claire Villeval, "Monetary and Nonmonetary Punishment in the Voluntary Contributions Mechanism," *American Economic Review* 93,1 (March 2003):366–380.

Maynard Smith, John, *Evolution and the Theory of Games* (Cambridge, UK: Cambridge University Press, 1982).

Maynard Smith, John, and Eors Szathmáry, *The Major Transitions in Evolution* (Oxford: Oxford University Press, 1997).

Maynard Smith, John, and G. A. Parker, "The Logic of Asymmetric Contests," *Animal Behaviour* 24 (1976):159–175.

Maynard Smith, John, and G. R. Price, "The Logic of Animal Conflict," *Nature* 246 (2 November 1973):15–18.

McClure, Samuel M., David I. Laibson, George Loewenstein, and Jonathan D. Cohen, "Separate Neural Systems Value Immediate and Delayed Monetary Rewards," *Science* 306 (15 October 2004):503–507.

McKelvey, R. D., and T. R. Palfrey, "An Experimental Study of the Centipede Game," *Econometrica* 60 (1992):803–836.

McLennan, Andrew, "Justifiable Beliefs in Sequential Equilibrium," *Econometrica* 53,4 (July 1985):889–904.

Mednick, S. A., L. Kirkegaard-Sorenson, B. Hutchings, J Knop, R. Rosenberg, and F. Schulsinger, "An Example of Bio-social Interaction Research: The Interplay of Socio-environmental and Individual Factors in the Etiology of Criminal Behavior," in S. A. Mednick and K. O. Christiansen (eds.) *Biosocial Bases of Criminal Behavior* (New York: Gardner Press, 1977) pp. 9–24.

Meltzhoff, Andrew N., and J. Decety, "What Imitation Tells Us About Social Cognition: A Rapprochement Between Developmental Psychology and Cognitive Neuroscience," *Philosophical Transactions of the Royal Society of London B* 358 (2003):491–500.

Mesoudi, Alex, Andrew Whiten, and Kevin N. Laland, "Towards a Unified Science of Cultural Evolution," *Behavioral and Brain Sciences (*2006).

Mesterton-Gibbons, Mike, "Ecotypic Variation in the Asymmetric Hawk-Dove Game: When Is Bourgeois an ESS?" *Evolutionary Ecology* 6 (1992):1151–1186.

Mesterton-Gibbons, Mike, and Eldridge S. Adams, "Landmarks in Territory Partitioning," *American Naturalist* 161,5 (May 2003):685–697.

Miller, B. L., A. Darby, D. F. Benson, J. L. Cummings, and M. H. Miller, "Aggressive, Socially Disruptive and Antisocial Behaviour Associated with Fronto-temporal Dementia," *British Journal of Psychiatry* 170 (1997):150–154.

Moll, Jorge, Roland Zahn, Ricardo di Oliveira-Souza, Frank Krueger, and Jordan Grafman, "The Neural Basis of Human Moral Cognition," *Nature Neuroscience* 6 (October 2005):799–809.

Montague, P. Read, and Gregory S. Berns, "Neural Economics and the Biological Substrates of Valuation," *Neuron* 36 (2002):265–284.

Moran, P. A. P., "On the Nonexistence of Adaptive Topographies," *Annals of Human Genetics* 27 (1964):338–343.

Morowitz, Harold, *The Emergence of Everything: How the World Became Complex* (Oxford: Oxford University Press, 2002).

Morris, Stephen, "The Common Prior Assumption in Economic Theory," *Economics and Philosophy* 11 (1995):227–253.

Moulin, Hervé, *Game Theory for the Social Sciences* (New York: New York University Press, 1986).

Myerson, Roger B., "Refinements of the Nash Equilibrium Concept," *International Journal of Game Theory* 7,2 (1978):73–80.

Nagel, Rosemarie, "Unravelling in Guessing Games: An Experimental Study," *American Economic Review* 85 (1995):1313–1326.

Nash, John F., "Equilibrium Points in n-Person Games," *Proceedings of the National Academy of Sciences* 36 (1950):48–49.

Nerlove, Marc, and Tjeppy D. Soedjiana, "Slamerans and Sheep: Savings and Small Ruminants in Semi-Subsistence Agriculture in Indonesia," 1996. Department of Agriculture and Resource Economics, University of Maryland.

Newell, Benjamin R., David A. Lagnado, and David R. Shanks, *Straight Choices: The Psychology of Decision Making* (New York: Psychology Press, 2007).

Newman, Mark, Albert-Laszlo Barabasi, and Duncan J. Watts, *The Structure and Dynamics of Networks* (Princeton, NJ: Princeton University Press, 2006).

Nikiforakis, Nikos S., "Punishment and Counter-punishment in Public Goods Games: Can we Still Govern Ourselves?" *Journal of Public Economics* 92,1–2 (2008):91–112.

Nisbett, Richard E., and Dov Cohen, *Culture of Honor: The Psychology of Violence in the South* (Boulder, CO: Westview Press, 1996).

Oaksford, Mike, and Nick Chater, *Bayesian Rationality: The Probabilistic Approach to Human Reasoning* (Oxford: Oxford University Press, 2007).

O'Brian, M. J., and R. L. Lyman, *Applying Evolutionary Archaeology* (New York: Kluwer Academic, 2000).

Odling-Smee, F. John, Kevin N. Laland, and Marcus W. Feldman, *Niche Construction: The Neglected Process in Evolution* (Princeton, NJ: Princeton University Press, 2003).

O'Donoghue, Ted, and Matthew Rabin, "Doing It Now or Later," *American Economic Review* 89,1 (March 1999):103–124.

—, "Incentives for Procrastinators," *Quarterly Journal of Economics* 114,3 (August 1999):769–816.

—, "The Economics of Immediate Gratification," *Journal of Behavioral Decision-Making* 13,2 (April/June 2000):233–250.

—, "Choice and Procrastination," *Quarterly Journal of Economics* 116,1 (February 2001):121–160.

Ok, Efe A., and Yusufcan Masatlioglu, "A General Theory of Time Preference," 2003. Economics Department, New York University.

Orbell, John M., Robyn M. Dawes, and J. C. Van de Kragt, "Organizing Groups for Collective Action," *American Political Science Review* 80 (December 1986):1171–1185.

Osborne, Martin J., and Ariel Rubinstein, *A Course in Game Theory* (Cambridge, MA: MIT Press, 1994).

Ostrom, Elinor, James Walker, and Roy Gardner, "Covenants with and without a Sword: Self-Governance Is Possible," *American Political Science Review* 86,2 (June 1992):404–417.

Oswald, Andrew J., "Happiness and Economic Performance," *Economic Journal* 107,445 (November 1997):1815–1831.

Page, Talbot, Louis Putterman, and Bulent Unel, "Voluntary Association in Public Goods Experiments: Reciprocity, Mimicry, and Efficiency," *Economic Journal* 115 (October 2005):1032–1053.

Parker, A. J., and W. T. Newsome, "Sense and the Single Neuron: Probing the Physiology of Perception," *Annual Review of Neuroscience* 21 (1998):227–277.

Parsons, Talcott, *The Structure of Social Action* (New York: McGraw-Hill, 1937).

—, "Evolutionary Universals in Society," *American Sociological Review* 29,3 (June 1964):339–357.

—, *Sociological Theory and Modern Society* (New York: Free Press, 1967).

Pearce, David, "Rationalizable Strategic Behavior and the Problem of Perfection," *Econometrica* 52 (1984):1029–1050.

Pettit, Philip, and Robert Sugden, "The Backward Induction Paradox," *The Journal of Philosophy* 86,4 (1989):169–182.

Piccione, Michele, "The Repeated Prisoner's Dilemma with Imperfect Private Monitoring," *Journal of Economic Theory* 102 (2002):70–83.

Pinker, Steven, *The Blank Slate: The Modern Denial of Human Nature* (New York: Viking, 2002).

Plott, Charles R., "The Application of Laboratory Experimental Methods to Public Choice," in Clifford S. Russell (ed.) *Collective Decision Making: Applications from Public Choice Theory* (Baltimore, MD: Johns Hopkins University Press, 1979) pp. 137–160.

Popper, Karl, "The Propensity Interpretation of Probability," *British Journal of the Philosophy of Science* 10 (1959):25–42.

Popper, Karl, *Objective knowledge: An Evolutionary Approach* (Oxford: Clarendon Press, 1979).

Poundstone, William, *Prisoner's Dilemma* (New York: Doubleday, 1992).

Premack, D. G., and G. Woodruff, "Does the Chimpanzee Have a Theory of Mind?" *Behavioral and Brain Sciences* 1 (1978):515–526.

Price, Michael, Leda Cosmides, and John Tooby, "Punitive Sentiment as an Anti-Free Rider Psychological Device," *Evolution & Human Behavior* 23,3 (May 2002):203–231.

Rabbie, J. M., J. C. Schot, and L. Visser, "Social Identity Theory: A Conceptual and Empirical Critique from the Perspective of a Behavioral Interaction Model," *European Journal of Social Psychology* 19 (1989):171–202.

Rabin, Matthew, "Incorporating Fairness into Game Theory and Economics," *American Economic Review* 83,5 (1993):1281–1302.

Rabin, Matthew, "Risk Aversion and Expected-Utility Theory: A Calibration Theorem," *Econometrica* 68,5 (2000):1281–1292.

Rand, A. S., "Ecology and Social Organization in the Iguanid Lizard *Anolis lineatopus*," *Proceedings of the US National Museum* 122 (1967):1–79.

Real, Leslie A., "Animal Choice Behavior and the Evolution of Cognitive Architecture," *Science* 253 (30 August 1991):980–986.

Real, Leslie, and Thomas Caraco, "Risk and Foraging in Stochastic Environments," *Annual Review of Ecology and Systematics* 17 (1986):371–390.

Relethford, John H., *The Human Species: An Introduction to Biological Anthropology* (New York: McGraw-Hill, 2007).

Reny, Philip J., "Common Belief and the Theory of Games with Perfect Information," *Journal of Economic Theory* 59 (1993):257–274.

Reny, Philip J., and Arthur J. Robson, "Reinterpreting Mixed Strategy Equilibria: A Unification of the Classical and Bayesian Views," *Games and Economic Behavior* 48 (2004):355–384.

Richerson, Peter J., and Robert Boyd, "The Evolution of Ultrasociality," in I. Eibl-Eibesfeldt and F.K. Salter (eds.) *Indoctrinability, Idology and Warfare* (New York: Berghahn Books, 1998) pp. 71–96.

—, *Not By Genes Alone* (Chicago: University of Chicago Press, 2004).

Riechert, S. E., "Games Spiders Play: Behavioural Variability in Territorial Disputes," *Journal of Theoretical Biology* 84 (1978):93–101.

Rivera, M. C., and J. A. Lake, "The Ring of Life Provides Evidence for a Genome Fusion Origin of Eukaryotes," *Nature* 431 (2004):152–155.

Rizzolatti, G., L. Fadiga, L Fogassi, and V. Gallese, "From Mirror Neurons to Imitation: Facts and Speculations," in Andrew N. Meltzhoff and Wolfgang Prinz (eds.) *The Imitative Mind: Development, Evolution and Brain Bases* (Cambridge: Cambridge University Press, 2002) pp. 247–266.

Robson, Arthur J., "A Biological Basis for Expected and Non-Expected Utility," March 1995. Department of Economics, University of Western Ontario.

Rosenthal, Robert W., "Games of Perfect Information, Predatory Pricing and the Chain-Store Paradox," *Journal of Economic Theory* 25 (1981):92–100.

Rosenzweig, Mark R., and Kenneth I. Wolpin, "Credit Market Constraints, Consumption Smoothing, and the Accumulation of Durable Production Assets in

Low-Income Countries: Investment in Bullocks in India," *Journal of Political Economy* 101,2 (1993):223–244.

Roth, Alvin E., Vesna Prasnikar, Masahiro Okuno-Fujiwara, and Shmuel Zamir, "Bargaining and Market Behavior in Jerusalem, Ljubljana, Pittsburgh, and Tokyo: An Experimental Study," *American Economic Review* 81,5 (December 1991):1068–1095.

Rozin, Paul, L. Lowery, S. Imada, and Jonathan Haidt, "The CAD Triad Hypothesis: A Mapping Between Three Moral Emotions (Contempt, Anger, Disgust) and Three Moral Codes (Community, Autonomy, Divinity)," *Journal of Personality & Social Psychology* 76 (1999):574–586.

Rubinstein, Ariel, "Comments on the Interpretation of Game Theory," *Econometrica* 59,4 (July 1991):909–924.

—, "Dilemmas of an Economic Theorist," *Econometrica* 74,4 (July 2006):865–883.

Saari, Donald G., "Iterative Price Mechanisms," *Econometrica* 53 (1985):1117–1131.

Saffer, Henry, and Frank Chaloupka, "The Demand for Illicit Drugs," *Economic Inquiry* 37,3 (1999):401–411.

Saha, Atanu, Richard C. Shumway, and Hovav Talpaz, "Joint Estimation of Risk Preference Structure and Technology Using Expo-Power Utility," *American Journal of Agricultural Economics* 76,2 (May 1994):173–184.

Sally, David, "Conversation and Cooperation in Social Dilemmas," *Rationality and Society* 7,1 (January 1995):58–92.

Samuelson, Paul, *The Foundations of Economic Analysis* (Cambridge: Harvard University Press, 1947).

Sato, Kaori, "Distribution and the Cost of Maintaining Common Property Resources," *Journal of Experimental Social Psychology* 23 (January 1987):19–31.

Savage, Leonard J., *The Foundations of Statistics* (New York: John Wiley & Sons, 1954).

Schall, J. D., and K. G. Thompson, "Neural Selection and Control of Visually Guided Eye Movements," *Annual Review of Neuroscience* 22 (1999):241–259.

Schelling, Thomas C., *The Strategy of Conflict* (Cambridge, MA: Harvard University Press, 1960).

Schlatter, Richard Bulger, *Private Property: History of an Idea* (New York: Russell & Russell, 1973).

Schulkin, J., *Roots of Social Sensitivity and Neural Function* (Cambridge, MA: MIT Press, 2000).

Schultz, W., P. Dayan, and P. R. Montague, "A Neural Substrate of Prediction and Reward," *Science* 275 (1997):1593–1599.

Seeley, Thomas D., "Honey Bee Colonies are Group-Level Adaptive Units," *American Naturalist* 150 (1997):S22–S41.

Sekiguchi, Tadashi, "Efficiency in Repeated Prisoner's Dilemma with Private Monitoring," *Journal of Economic Theory* 76 (1997):345–361.

Selten, Reinhard, "Re-examination of the Perfectness Concept for Equilibrium Points in Extensive Games," *International Journal of Game Theory* 4 (1975):25–55.

—, "A Note on Evolutionarily Stable Strategies in Asymmetric Animal Conflicts," *Journal of Theoretical Biology* 84 (1980):93–101.

Senar, J. C., M. Camerino, and N. B. Metcalfe, "Agonistic Interactions in Siskin Flocks: Why are Dominants Sometimes Subordinate?" *Behavioral Ecology and Sociobiology* 25 (1989):141–145.

Shafir, Eldar, and Amos Tversky, "Thinking Through Uncertainty: Nonconsequential Reasoning and Choice," *Cognitive Psychology* 24,4 (October 1992):449–474.

—, "Decision Making," in Edward E. Smith and Daniel N. Osherson (eds.) *Thinking: An Invitation to Cognitive Science*, Vol. 3, 2nd Edition (Cambridge, MA: MIT Press, 1995) pp. 77–100.

Shennan, Stephen, *Quantifying Archaeology* (Edinburgh: Edinburgh University Press, 1997).

Shizgal, Peter, "On the Neural Computation of Utility: Implications from Studies of Brain Stimulation Reward," in Daniel Kahneman, Edward Diener, and Norbert Schwarz (eds.) *Well-Being: The Foundations of Hedonic Psychology* (New York: Russell Sage, 1999) pp. 502–526.

Sigg, Hans, and Jost Falett, "Experiments on Respect of Possession and Property in Hamadryas Baboons *Papio hamadryas)*," *Animal Behaviour* 33 (1985):978–984.

Simon, Herbert, "Theories of Bounded Rationality," in C. B. McGuire and Roy Radner (eds.) *Decision and Organization* (New York: American Elsevier, 1972) pp. 161–176.

—, *Models of Bounded Rationality* (Cambridge, MA: MIT Press, 1982).

—, "A Mechanism for Social Selection and Successful Altruism," *Science* 250 (1990):1665–1668.

Skibo, James M., and R. Alexander Bentley, *Complex Systems and Archaeology* (Salt Lake City: University of Utah Press, 2003).

Sloman, S. A., "Two Systems of Reasoning," in Thomas Gilovich, Dale Griffin, and Daniel Kahneman (eds.) *Heuristics and Biases: The Psychology of Intuitive Judgment* (Cambridge: Cambridge University Press, 2002) pp. 379–396.

Smith, Adam, *The Theory of Moral Sentiments* (New York: Prometheus, 2000 [1759]).

Smith, Eric Alden, and B. Winterhalder, *Evolutionary Ecology and Human Behavior* (New York: Aldine de Gruyter, 1992).

Smith, Vernon, "Microeconomic Systems as an Experimental Science," *American Economic Review* 72 (December 1982):923–955.

Smith, Vernon, and Arlington W. Williams, "Experimental Market Economics," *Scientific American* 267,6 (December 1992):116–121.

Sonnenschein, Hugo, "Market Excess Demand Functions," *Econometrica* 40 (1972):549–563.

—, "Do Walras' Identity and Continuity Characterize the Class of Community Excess Demand Functions?" *Journal of Ecomonic Theory* 6 (1973):345–354.

Spence, A. Michael, "Job Market Signaling," *Quarterly Journal of Economics* 90 (1973):225–243.

Stake, Jeffrey Evans, "The Property Instinct," *Philosophical Transactions of the Royal Society of London B* 359 (2004):1763–1774.

Stephens, W., C. M. McLinn, and J. R. Stevens, "Discounting and Reciprocity in an Iterated Prisoner's Dilemma," *Science* 298 (13 December 2002):2216–2218.

Stevens, Elisabeth Franke, "Contests Between Bands of Feral Horses for Access to Fresh Water: The Resident Wins," *Animal Behaviour* 36,6 (1988):1851–1853.

Strotz, Robert H., "Myopia and Inconsistency in Dynamic Utility Maximization," *Review of Economic Studies* 23,3 (1955):165–180.

Sugden, Robert, *The Economics of Rights, Co-operation and Welfare* (Oxford: Basil Blackwell, 1986).

—, "An Axiomatic Foundation for Regret Theory," *Journal of Economic Theory* 60,1 (June 1993):159–180.

—, "Reference-dependent Subjective Expected Utility," *Journal of Economic Theory* 111 (2003):172–191.

Sugrue, Leo P., Gregory S. Corrado, and William T. Newsome, "Choosing the Greater of Two Goods: Neural Currencies for Valuation and Decision Making," *Nature Reviews Neuroscience* 6 (2005):363–375.

Sutton, R., and A. G. Barto, *Reinforcement Learning* (Cambridge, MA: The MIT Press, 2000).

Tajfel, Henri, "Experiments in Intercategory Discrimination," *Annual Review of Psychology* 223,5 (1970):96–102.

Tajfel, Henri, M. Billig, R.P. Bundy, and Claude Flament, "Social Categorization and Intergroup Behavior," *European Journal of Social Psychology* 1 (1971):149–177.

Tan, Tommy Chin-Chiu, and Sergio Ribeiro da Costa Werlang, "The Bayesian Foundations of Solution Concepts of Games," *Journal of Economic Theory* 45 (1988):370–391.

Taylor, Peter, and Leo Jonker, "Evolutionarily Stable Strategies and Game Dynamics," *Mathematical Biosciences* 40 (1978):145–156.

Thaler, Richard H., *The Winner's Curse* (Princeton: Princeton University Press, 1992).

Tomasello, Michael, *The Cultural Origins of Human Cognition* (Cambridge, MA: Harvard University Press, 1999).

Tooby, John, and Leda Cosmides, "The Psychological Foundations of Culture," in Jerome H. Barkow, Leda Cosmides, and John Tooby (eds.) *The Adapted Mind: Evolutionary Psychology and the Generation of Culture* (New York: Oxford University Press, 1992) pp. 19–136.

Torii, M., "Possession by Non-human Primates," *Contemporary Primatology* (1974):310–314.

Trivers, Robert L., "The Evolution of Reciprocal Altruism," *Quarterly Review of Biology* 46 (1971):35–57.

—, "Parental Investment and Sexual Selection, 1871–1971," in B. Campbell (ed.) *Sexual Selection and the Descent of Man* (Chicago: Aldine, 1972) pp. 136–179.

Turner, John C., "Social Identification and Psychological Group Formation," in Henri Tajfel (ed.) *The Social Dimension* (Cambridge, UK: Cambridge University Press, 1984) pp. 518–538.

Tversky, Amos, and Daniel Kahneman, "Judgment under Uncertainty: Heuristics and Biases," *Science* 185 (September 1974):1124–1131.

—, "The Framing of Decisions and the Psychology of Choices," *Science* 211 (January 1981):453–458.

—, "Loss Aversion in Riskless Choice: A Reference-Dependent Model," *Quarterly Journal of Economics* 106,4 (November 1981):1039–1061.

—, "Extensional versus Intuitive Reasoning: The Conjunction Fallacy in Probability Judgment," *Psychological Review* 90 (1983):293–315.

Tversky, Amos, Paul Slovic, and Daniel Kahneman, "The Causes of Preference Reversal," *American Economic Review* 80,1 (March 1990):204–217.

van Damme, Eric, *Stability and Perfection of Nash Equilibria* (Berlin: Springer-Verlag, 1987).

Vanderschraaf, Peter, and Giacomo Sillari, "Common Knowledge," in Edward N. Zalta (ed.) *The Stanford Encyclopedia of Philosophy* (plato.stanford.edu/-archives/spr2007/entries/common-knowledge: Stanford Univerisity, 2007).

Varian, Hal R., "The Nonparametric Approach to Demand Analysis," *Econometrica* 50 (1982):945–972.

Vega-Redondo, Fernando, *Economics and the Theory of Games* (Cambridge: Cambridge University Press, 2003).

von Mises, Richard, *Probability, Statistics, and Truth* (New York: Dover, 1981).

Von Neumann, John, and Oskar Morgenstern, *Theory of Games and Economic Behavior* (Princeton, NJ: Princeton University Press, 1944).

Watabe, M., S. Terai, N. Hayashi, and T. Yamagishi, "Cooperation in the One-Shot Prisoner's Dilemma based on Expectations of Reciprocity," *Japanese Journal of Experimental Social Psychology* 36 (1996):183–196.

Weigel, Ronald M., "The Application of Evolutionary Models to the Study of Decisions Made by Children During Object Possession Conflicts," *Ethnology and Sociobiology* 5 (1984):229–238.

Wiessner, Polly, "Norm Enforcement Among the Ju/'hoansi Bushmen: A Case of Strong Reciprocity?" *Human Nature* 16,2 (June 2005):115–145.

Williams, J. H. G., A. Whiten, T. Suddendorf, and D. I Perrett, "Imitation, Mirror Neurons and Autism," *Neuroscience and Biobehavioral Reviews* 25 (2001):287–295.

Wilson, David Sloan, "Hunting, Sharing, and Multilevel Selection: The Tolerated Theft Model Revisited," *Current Anthropology* 39 (1998):73–97.

Wilson, David Sloan, and Edward O. Wilson, "Rethinking the Theoretical Foundation of Sociobiology," *The Quarterly Review of Biology* 82,4 (December 2007):327–348.

Wilson, E. O., and Bert Holldobler, "Eusociality: Origin and Consequences," *PNAS* 102,38 (2005):13367–71.

Wilson, Edward O., *Consilience: The Unity of Knowledge* (New York: Knopf, 1998).

Winter, Sidney G., "Satisficing, Selection and the Innovating Remnant," *Quarterly Journal of Economics* 85 (1971):237–261.

Woodburn, James, "Egalitarian Societies," *Man* 17,3 (1982):431–451.

Wright, Sewall, "Evolution in Mendelian Populations," *Genetics* 6 (1931):111–178.

Wrong, Dennis H., "The Oversocialized Conception of Man in Modern Sociology," *American Sociological Review* 26 (April 1961):183–193.

Yamagishi, Toshio, "The Provision of a Sanctioning System as a Public Good," *Journal of Personality and Social Psychology* 51 (1986):110–116.

—, "The Provision of a Sanctioning System in the United States and Japan," *Social Psychology Quarterly* 51,3 (1988):265–271.

—, "Seriousness of Social Dilemmas and the Provision of a Sanctioning System," *Social Psychology Quarterly* 51,1 (1988):32–42.

—, "Group Size and the Provision of a Sanctioning System in a Social Dilemma," in W. B. G. Liebrand, David M. Messick, and H. A. M. Wilke (eds.) *Social Dilemmas: Theoretical Issues and Research Findings* (Oxford: Pergamon Press, 1992) pp. 267–287.

Yamagishi, Toshio, N. Jin, and Toko Kiyonari, "Bounded Generalized Reciprocity: In-group Boasting and In-group Favoritism," *Advances in Group Processes* 16 (1999):161–197.

Young, H. Peyton, *Individual Strategy and Social Structure: An Evolutionary Theory of Institutions* (Princeton, NJ: Princeton University Press, 1998).

Zajonc, R. B., "Feeling and Thinking: Preferences Need No Inferences," *American Psychologist* 35,2 (1980):151–175.

Zajonc, Robert B., "On the Primacy of Affect," *American Psychologist* 39 (1984):117–123.

Zambrano, Eduardo, "Testable Implications of Subjective Expected Utility Theory," *Games and Economic Behavior* 53,2 (2005):262–268.

Index